ALSO BY EDWARD O. WILSON

NATURALIST

THE DIVERSITY OF LIFE

THE ANTS
(WITH BERT HÖLLDOBLER)

BIOPHILIA

ON HUMAN NATURE

SOCIOBIOLOGY:
THE NEW SYNTHESIS

THE INSECT SOCIETIES

Consilience

EDWARD O. WILSON

Consilience

THE UNITY OF KNOWLEDGE

ALFRED A. KNOPF NEW YORK 1998

THIS IS A BORZOI BOOK
PUBLISHED BY ALFRED A. KNOPF, INC.

www.randomhouse.com

Library of Congress Cataloging-in-Publication Data
Wilson, Edward Osborne.
Consilience: the unity of knowledge / Edward O. Wilson.—1st ed.
p. cm.
"A Borzoi book."
Includes index.
ISBN 0-679-45077-7
1. Philosophy. 2. Order (Philosophy). 3. Philosophy and science.
I. Title.
B72.W54 1998
121—dc21 97-2816
CIP

Manufactured in the United States of America
Published March 27, 1998
Reprinted Once
Third Printing, April 1998

Thus have I made as it were a small globe of the intellectual world, as truly and faithfully as I could discover.

FRANCIS BACON (1605)

CONTENTS

CHAPTER 1 The Ionian Enchantment 3

CHAPTER 2 The Great Branches of Learning 8

CHAPTER 3 The Enlightenment 14

CHAPTER 4 The Natural Sciences 45

CHAPTER 5 Ariadne's Thread 66

CHAPTER 6 The Mind 96

CHAPTER 7 From Genes to Culture 125

CHAPTER 8 The Fitness of Human Nature 164

CHAPTER 9 The Social Sciences 181

CHAPTER 10 The Arts and Their Interpretation 210

CHAPTER 11 Ethics and Religion 238

CHAPTER 12 To What End? 266

Notes 299

Acknowledgments 321

Index 323

Consilience

CHAPTER 1

THE IONIAN ENCHANTMENT

I REMEMBER very well the time I was captured by the dream of unified learning. It was in the early fall of 1947, when at eighteen I came up from Mobile to Tuscaloosa to enter my sophomore year at the University of Alabama. A beginning biologist, fired by adolescent enthusiasm but short on theory and vision, I had schooled myself in natural history with field guides carried in a satchel during solitary excursions into the woodlands and along the freshwater streams of my native state. I saw science, by which I meant (and in my heart I still mean) the study of ants, frogs, and snakes, as a wonderful way to stay outdoors.

My intellectual world was framed by Linnaeus, the eighteenth-century Swedish naturalist who invented modern biological classification. The Linnaean system is deceptively easy. You start by separating specimens of plants and animals into species. Then you sort species resembling one another into groups, the genera. Examples of such groups are all the crows and all the oaks. Next you label each species with a two-part Latinized name, such as *Corvus ossifragus* for the fish crow, where *Corvus* stands for the genus—all the species of crows—and *ossifragus* for the fish crow in particular. Then on to higher classification, where similar genera are grouped into families, families into orders, and so on up to phyla and finally, at the very summit, the six kingdoms—plants, animals, fungi, protists, monerans, and archaea. It is like the army: men (plus women, nowadays) into squads, squads into platoons,

platoons into companies, and in the final aggregate, the armed services headed by the joint chiefs of staff. It is, in other words, a conceptual world made for the mind of an eighteen-year-old.

I had reached the level of the Carolus Linnaeus of 1735 or, more accurately (since at that time I knew little of the Swedish master), the Roger Tory Peterson of 1934, when the great naturalist published the first edition of *A Field Guide to the Birds*. My Linnaean period was nonetheless a good start for a scientific career. The first step to wisdom, as the Chinese say, is getting things by their right names.

Then I discovered evolution. Suddenly—that is not too strong a word—I saw the world in a wholly new way. This epiphany I owed to my mentor Ralph Chermock, an intense, chain-smoking young assistant professor newly arrived in the provinces with a Ph.D. in entomology from Cornell University. After listening to me natter for a while about my lofty goal of classifying all the ants of Alabama, he handed me a copy of Ernst Mayr's 1942 *Systematics and the Origin of Species*. Read it, he said, if you want to become a real biologist.

The thin volume in the plain blue cover was one of the New Synthesis works, uniting the nineteenth-century Darwinian theory of evolution and modern genetics. By giving a theoretical structure to natural history, it vastly expanded the Linnaean enterprise. A tumbler fell somewhere in my mind, and a door opened to a new world. I was enthralled, couldn't stop thinking about the implications evolution has for classification and for the rest of biology. And for philosophy. And for just about everything. Static pattern slid into fluid process. My thoughts, embryonically those of a modern biologist, traveled along a chain of causal events, from mutations that alter genes to evolution that multiplies species, to species that assemble into faunas and floras. Scale expanded, and turned continuous. By inwardly manipulating time and space, I found I could climb the steps in biological organization from microscopic particles in cells to the forests that clothe mountain slopes. A new enthusiasm surged through me. The animals and plants I loved so dearly reentered the stage as lead players in a grand drama. Natural history was validated as a real science.

I had experienced the Ionian Enchantment. That recently coined expression I borrow from the physicist and historian Gerald Holton. It means a belief in the unity of the sciences—a conviction, far deeper than a mere working proposition, that the world is orderly and can be explained by a small number of natural laws. Its roots go back to Thales of Miletus, in Ionia, in the sixth century B.C. The legendary philosopher was considered by Aristotle two centuries later to be the founder of the physical sciences. He is of course remembered more concretely for his belief that all matter consists ultimately of water. Although the notion is often cited as an example of how far astray early Greek speculation could wander, its real significance is the metaphysics it expressed about the material basis of the world and the unity of nature.

The Enchantment, growing steadily more sophisticated, has dominated scientific thought ever since. In modern physics its focus has been the unification of all the forces of nature—electroweak, strong, and gravitation—the hoped-for consolidation of theory so tight as to turn the science into a "perfect" system of thought, which by sheer weight of evidence and logic is made resistant to revision. But the spell of the Enchantment extends to other fields of science as well, and in the minds of a few it reaches beyond into the social sciences, and still further, as I will explain later, to touch the humanities. The idea of the unity of science is not idle. It has been tested in acid baths of experiment and logic and enjoyed repeated vindication. It has suffered no decisive defeats. At least not yet, even though at its center, by the very nature of the scientific method, it must be thought always vulnerable. On this weakness I will also expand in due course.

Einstein, the architect of grand unification in physics, was Ionian to the core. That vision was perhaps his greatest strength. In an early letter to his friend Marcel Grossmann he said, "It is a wonderful feeling to recognize the unity of a complex of phenomena that to direct observation appear to be quite separate things." He was referring to his successful alignment of the microscopic physics of capillaries with the macroscopic, universe-wide physics of gravity. In later life he aimed to weld everything else into a single parsimonious system, space with time and motion, gravity with electromagnetism and cosmology. He approached but never captured that grail. All scientists, Einstein not excepted, are children of Tantalus, frustrated by the failure to grasp that which seems within reach. They are typified by those thermodynamicists who for decades have drawn ever closer to the temperature of absolute zero, when atoms cease all motion. In 1995, pushing down to within a few billionths of a degree above absolute zero, they created a Bose-Einstein condensate, a fundamental form of matter beyond the familiar gases, liquids, and solids, in which many atoms act as a single atom in one quantum state. As temperature drops and pressure is increased, a gas condenses into a liquid, then a solid; then appears the Bose-Einstein condensate. But absolute, entirely absolute zero, a temperature that exists in imagination, has still not been attained.

On a far more modest scale, I found it a wonderful feeling not just to taste the unification metaphysics but also to be released from the confinement of fundamentalist religion. I had been raised a Southern Baptist, laid backward under the water on the sturdy arm of a pastor, been born again. I knew the healing power of redemption. Faith, hope, and charity were in my bones, and with millions of others I knew that my savior Jesus Christ would grant me eternal life. More pious than the average teenager, I read the Bible cover to cover, twice. But now at college, steroid-driven into moods of adolescent rebellion, I chose to doubt. I found it hard to accept that our deepest beliefs were set in stone by agricultural societies of the eastern Mediterranean more than two thousand years ago. I suffered cognitive dissonance between the cheerfully reported genocidal wars of these people and Christian civilization in 1940s Alabama. It seemed to me that the Book of Revelation might be black magic hallucinated by an ancient primitive. And I thought, surely a loving personal God, if He is paying attention, will not abandon those who reject the literal interpretation of the biblical cosmology. It is only fair to award points for intellectual courage. Better damned with Plato and Bacon, Shelley said, than go to heaven with Paley and Malthus. But most of all, Baptist theology made no provision for *evolution*. The biblical authors had missed the most important revelation of all! Could it be that they were not really privy to the thoughts of God? Might the pastors of my childhood, good and loving men though they were, be mistaken? It was all too much, and freedom was ever so sweet. I drifted away from the church, not definitively agnostic or atheistic, just Baptist no more.

Still, I had no desire to purge religious feelings. They were bred in me; they suffused the wellsprings of my creative life. I also retained a small measure of common sense. To wit, people must belong to a tribe; they yearn to have a purpose larger than themselves. We are obliged by the deepest drives of the human spirit to make ourselves more than animated dust, and we must have a story to tell about where we came from, and why we are here. Could Holy Writ be just the first literate attempt to explain the universe and make ourselves significant within it? Perhaps science is a continuation on new and better-tested ground to attain the same end. If so, then in that sense science is religion liberated and writ large.

Such, I believe, is the source of the Ionian Enchantment: Preferring a search for objective reality over revelation is another way of satisfying religious hunger. It is an endeavor almost as old as civilization and intertwined with traditional religion, but it follows a very different course—a stoic's creed, an acquired taste, a guidebook to adventure plotted across rough terrain. It aims to save the spirit, not by surrender but by liberation of the human mind. Its central tenet, as Einstein knew, is the unification of knowledge. When we have unified enough certain knowledge, we will understand who we are and why we are here.

If those committed to the quest fail, they will be forgiven. When lost, they will find another way. The moral imperative of humanism is the endeavor alone, whether successful or not, provided the effort is honorable and failure memorable. The ancient Greeks expressed the idea in a myth of vaulting ambition. Daedalus escapes from Crete with his son Icarus on wings he has fashioned from feathers and wax. Ignoring the warnings of his father, Icarus flies toward the sun, whereupon his wings come apart and he falls into the sea. That is the end of Icarus in the myth. But we are left to wonder: Was he just a foolish boy? Did he pay the price for hubris, for pride in sight of the gods? I like to think that on the contrary his daring represents a saving human grace. And so the great astrophysicist Subrahmanyan Chandrasekhar could pay tribute to the spirit of his mentor, Sir Arthur Eddington, by saying: Let us see how high we can fly before the sun melts the wax in our wings.

CHAPTER 2

THE GREAT BRANCHES OF LEARNING

YOU WILL SEE at once why I believe that the Enlightenment thinkers of the seventeenth and eighteenth centuries got it mostly right the first time. The assumptions they made of a lawful material world, the intrinsic unity of knowledge, and the potential of indefinite human progress are the ones we still take most readily into our hearts, suffer without, and find maximally rewarding through intellectual advance. The greatest enterprise of the mind has always been and always will be the attempted linkage of the sciences and humanities. The ongoing fragmentation of knowledge and resulting chaos in philosophy are not reflections of the real world but artifacts of scholarship. The propositions of the original Enlightenment are increasingly favored by objective evidence, especially from the natural sciences.

Consilience is the key to unification. I prefer this word over "coherence" because its rarity has preserved its precision, whereas coherence has several possible meanings, only one of which is consilience. William Whewell, in his 1840 synthesis *The Philosophy of the Inductive Sciences,* was the first to speak of consilience, literally a "jumping together" of knowledge by the linking of facts and fact-based theory across disciplines to create a common groundwork of explanation. He said, "The Consilience of Inductions takes place when an Induction, obtained from one class of facts, coincides with an Induction, obtained from another different class. This Consilience is a test of the truth of the Theory in which it occurs."

The only way either to establish or to refute consilience is by methods developed in the natural sciences—not, I hasten to add, an effort led by scientists, or frozen in mathematical abstraction, but rather one allegiant to the habits of thought that have worked so well in exploring the material universe.

The belief in the possibility of consilience beyond science and across the great branches of learning is not yet science. It is a metaphysical world view, and a minority one at that, shared by only a few scientists and philosophers. It cannot be proved with logic from first principles or grounded in any definitive set of empirical tests, at least not by any yet conceived. Its best support is no more than an extrapolation of the consistent past success of the natural sciences. Its surest test will be its effectiveness in the social sciences and humanities. The strongest appeal of consilience is in the prospect of intellectual adventure and, given even modest success, the value of understanding the human condition with a higher degree of certainty.

Bear with me while I cite an example to illustrate the claim just made. Think of two intersecting lines forming a cross, and picture the four quadrants thus created. Label one quadrant environmental policy, the next ethics, the next biology, and the final one social science.

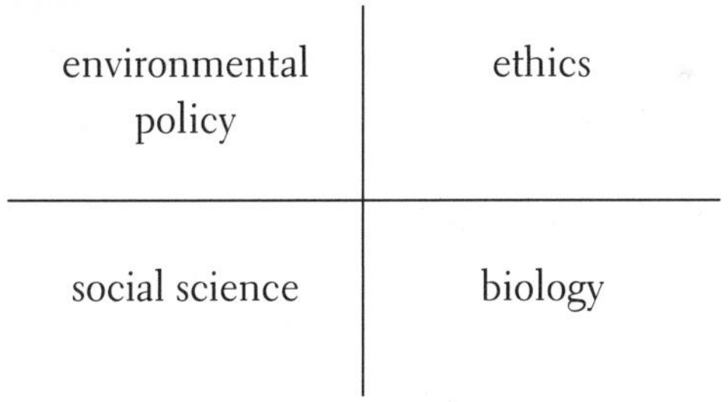

environmental policy | ethics

social science | biology

We already intuitively think of these four domains as closely connected, so that rational inquiry in one informs reasoning in the other three. Yet undeniably each stands apart in the contemporary academic mind. Each has its own practitioners, language, modes of analysis, and standards of validation. The result is confusion, and confusion was correctly identified by Francis Bacon four centuries ago as the most fatal of errors, which "occurs wherever argument or inference passes from one world of experience to another."

Next draw a series of concentric circles around the point of intersection.

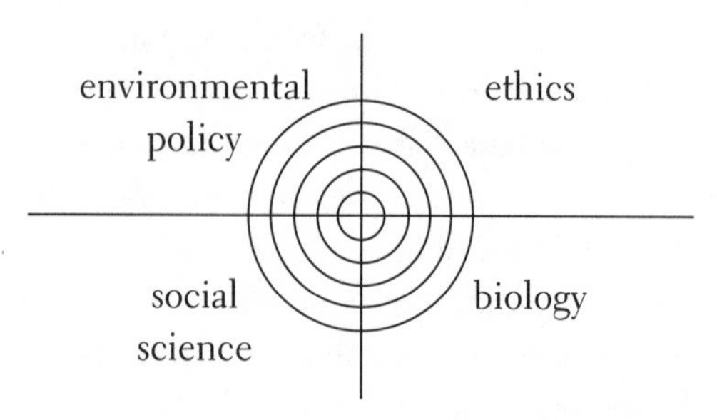

As we cross the circles inward toward the point at which the quadrants meet, we find ourselves in an increasingly unstable and disorienting region. The ring closest to the intersection, where most real-world problems exist, is the one in which fundamental analysis is most needed. Yet virtually no maps exist. Few concepts and words serve to guide us. Only in imagination can we travel clockwise from the recognition of environmental problems and the need for soundly based policy; to the selection of solutions based on moral reasoning; to the biological foundations of that reasoning; to a grasp of social institutions as the products of biology, environment, and history. And thence back to environmental policy.

Consider this example. Governments everywhere are at a loss as to the best policy for regulating the dwindling forest reserves of the world. There are few established ethical guidelines from which agreement might be reached, and those are based on an insufficient knowledge of ecology. Even if adequate scientific knowledge were available, there would still be little basis for the long-term valuation of forests. The economics of sustainable yield is still a primitive art, and the psychological benefits of natural ecosystems are almost wholly unexplored.

The time has come to achieve the tour in reality. This is not an idle exercise for the delectation of intellectuals. How wisely policy is chosen will depend on the ease with which the educated public, not just intellectuals and political leaders, can think around these and similar circuits, starting at any point and moving in any direction.

To ask if consilience can be gained in the innermost domains of the circles, such that sound judgment will flow easily from one discipline to another, is equivalent to asking whether, in the gathering of disciplines, specialists can ever reach agreement on a common body of abstract principles and evidentiary proof. I think they can. Trust in consilience is the foundation of the natural sciences. For the material world at least, the momentum is overwhelmingly toward conceptual unity. Disciplinary boundaries within the natural sciences are disappearing, to be replaced by shifting hybrid domains in which consilience is implicit. These domains reach across many levels of complexity, from chemical physics and physical chemistry to molecular genetics, chemical ecology, and ecological genetics. None of the new specialties is considered more than a focus of research. Each is an industry of fresh ideas and advancing technology.

Given that human action comprises events of physical causation, why should the social sciences and humanities be impervious to consilience with the natural sciences? And how can they fail to benefit from that alliance? It is not enough to say that human action is historical, and that history is an unfolding of unique events. Nothing fundamental separates the course of human history from the course of physical history, whether in the stars or in organic diversity. Astronomy, geology, and evolutionary biology are examples of primarily historical disciplines linked by consilience to the rest of the natural sciences. History is today a fundamental branch of learning in its own right, down to the finest detail. But if ten thousand humanoid histories could be traced on ten thousand Earthlike planets, and from a comparative study of those histories empirical tests and principles evolved, historiography—the explanation of historical trends—would already be a natural science.

The unification agenda does not sit well with a few professional philosophers. The subject I address they consider their own, to be expressed in their language, their framework of formal thought. They will draw this indictment: *conflation, simplism, ontological reductionism, scientism,* and other sins made official by the hissing suffix. To which I plead guilty, guilty, guilty. Now let us move on, thus. Philosophy plays a vital role in intellectual synthesis, and it keeps us alive to the power and continuity of thought through the centuries. It also peers into the future to give shape to the unknown—and that has always been its vocation of choice. One of its most distinguished practitioners, Alexander Rosenberg, has recently argued that philosophy in fact addresses just two issues: the questions that the sciences—physical, biological, and social—cannot answer, and the reasons for that incapacity. "Now of course," he concludes, "there may not be any questions that the sciences cannot answer eventually, in the long run, when all the facts are in, but certainly there are questions that the sciences cannot answer *yet.*" This assessment is admirably clear and honest and convincing. It neglects, however, the obvious fact that scientists are equally qualified to judge what remains to be discovered, and why. There has never been a better time for collaboration between scientists and philosophers, especially where they meet in the borderlands between biology, the social sciences, and the humanities. We are approaching a new age of synthesis, when the testing of consilience is the greatest of all intellectual challenges. Philosophy, the contemplation of the unknown, is a shrinking dominion. We have the common goal of turning as much philosophy as possible into science.

IF THE WORLD really works in a way so as to encourage the consilience of knowledge, I believe the enterprises of culture will eventually fall out into science, by which I mean the natural sciences, and the humanities, particularly the creative arts. These domains will be the two great branches of learning in the twenty-first century. The social sciences will continue to split within each of its disciplines, a process already rancorously begun, with one part folding into or becoming continuous with biology, the other fusing with the humanities. Its disciplines will continue to exist but in radically altered form. In the process the humanities, ranging from philosophy and history to moral reasoning, comparative religion, and interpretation of the arts, will draw closer to the sciences and partly fuse with them. Of these several subjects I will say more in later chapters.

I admit that the confidence of natural scientists often seems overweening. Science offers the boldest metaphysics of the age. It is a thoroughly human construct, driven by the faith that if we dream, press to discover, explain, and dream again, thereby plunging repeatedly into new terrain, the world will somehow come clearer and we will grasp the true strangeness of the universe. And the strangeness will all prove to be connected and make sense.

In his 1941 classic *Man on His Nature*, the British neurobiologist Charles Sherrington spoke of the brain as an enchanted loom, perpetually weaving a picture of the external world, tearing down and reweaving, inventing other worlds, creating a miniature universe. The communal mind of literate societies—world culture—is an immensely larger loom. Through science it has gained the power to map external reality far beyond the reach of a single mind, and through the arts the means to construct narratives, images, and rhythms immeasurably more diverse than the products of any solitary genius. The loom is the same for both enterprises, for science and for the arts, and there is a general explanation of its origin and nature and thence of the human condition, proceeding from the deep history of genetic evolution to modern culture. Consilience of causal explanation is the means by which the single mind can travel most swiftly and surely from one part of the communal mind to the other.

In education the search for consilience is the way to renew the crumbling structure of the liberal arts. During the past thirty years the ideal of the unity of learning, which the Renaissance and Enlightenment bequeathed us, has been largely abandoned. With rare exceptions American universities and colleges have dissolved their curriculum into a slurry of minor disciplines and specialized courses. While the average number of undergraduate courses per institution doubled, the percentage of mandatory courses in general education dropped by more than half. Science was sequestered in the same period; as I write, in 1997, only a third of universities and colleges require students to take at least one course in the natural sciences. The trend cannot be reversed by force-feeding students with some-of-this and some-of-that across the branches of learning. Win or lose, true reform will aim at the consilience of science with the social sciences and humanities in scholarship and teaching. Every college student should be able to answer the following question: What is the relation between science and the humanities, and how is it important for human welfare?

Every public intellectual and political leader should be able to answer that question as well. Already half the legislation coming before the United States Congress contains important scientific and technological components. Most of the issues that vex humanity daily—ethnic conflict, arms escalation, overpopulation, abortion, environment, endemic poverty, to cite several most persistently before us—cannot be solved without integrating knowledge from the natural sciences with that of the social sciences and humanities. Only fluency across the boundaries will provide a clear view of the world as it really is, not as seen through the lens of ideologies and religious dogmas or commanded by myopic response to immediate need. Yet the vast majority of our political leaders are trained exclusively in the social sciences and humanities, and have little or no knowledge of the natural sciences. The same is true of the public intellectuals, the columnists, the media interrogators, and think-tank gurus. The best of their analyses are careful and responsible, and sometimes correct, but the substantive base of their wisdom is fragmented and lopsided.

A balanced perspective cannot be acquired by studying disciplines in pieces but through pursuit of the consilience among them. Such unification will come hard. But I think it is inevitable. Intellectually it rings true, and it gratifies impulses that rise from the admirable side of human nature. To the extent that the gaps between the great branches of learning can be narrowed, diversity and depth of knowledge will increase. They will do so because of, not despite, the underlying cohesion achieved. The enterprise is important for yet another reason: It gives ultimate purpose to intellect. It promises that order, not chaos, lies beyond the horizon. I think it inevitable that we will accept the adventure, go there, and find out.

CHAPTER 3

THE ENLIGHTENMENT

THE DREAM OF INTELLECTUAL UNITY first came to full flower in the original Enlightenment, an Icarian flight of the mind that spanned the seventeenth and eighteenth centuries. A vision of secular knowledge in the service of human rights and human progress, it was the West's greatest contribution to civilization. It launched the modern era for the whole world; we are all its legatees. Then it failed.

Astonishingly—it failed. When does such a historical period come to an end? It dies when, for whatever reason, usually in the aftermath of war and revolution, its ideas no longer dominate. It is of surpassing importance, therefore, to understand the essential nature of the Enlightenment and the weaknesses that brought it down. Both can be said to be wrapped up in the life of the Marquis de Condorcet. In particular, no single event better marks the end of the Enlightenment than his death on March 29, 1794. The circumstances were exquisitely ironic. Condorcet has been called the prophet of the Laws of Progress. By virtue of his towering intellect and visionary political leadership, he seemed destined to emerge from the Revolution as the Jefferson of France. But in late 1793 and early 1794, as he was composing the ultimate Enlightenment blueprint, *Sketch for a Historical Picture of the Progress of the Human Mind*, he was instead a fugitive from the law, liable to sentence of death by representatives of the cause he had so faithfully served. His crime was political: He was perceived to be a Girondist, a member of a faction found too moderate—too reasonable—by the radical Jacobins. Worse, he had criticized the constitution drawn up by the Jacobin-dominated National Convention. He died on the floor of a cell in the jail at Bourg-la-Reine, after being mauled by villagers who had captured him on the run. They would certainly have turned him over to the Paris authorities for trial. The cause of death is unknown. Suicide was ruled out at the time. Poison, which he carried with him, is nevertheless possible; so are trauma and heart attack. At least he was spared the guillotine.

The French Revolution drew its intellectual strength from men and women like Condorcet. It was readied by the growth of educational opportunity and then fired by the idea of the universal rights of man. Yet as the Enlightenment seemed about to achieve by this means political fruition in Europe, something went terribly wrong. What seemed at first to be minor inconsistencies widened into catastrophic failures. Jean-Jacques Rousseau, in *The Social Contract* thirty years earlier, had introduced the idea that was later to inspire the rallying slogan "Liberty, Equality, Fraternity." But he had also invented the deadly abstraction of the "general will" to achieve these goals. The general will, he said, is the rule of justice agreed upon by assemblies of free people whose interest is only to serve the welfare of the society and of each person in it. When achieved, it forms a sovereign contract that is "always constant, unalterable, and pure. . . . Each of us puts his person and all his power in common under the supreme direction of the general will, and in our corporate capacity, we receive each member as an indivisible part of the whole." Those who do not conform to the general will, Rousseau continued, are deviants subject to necessary force by the assembly. There is no other way to achieve a truly egalitarian democracy and thus to break humanity out of the chains that everywhere bind it.

Robespierre, leader of the Reign of Terror that overtook the Revolution in 1793, grasped this logic all too well. He and his fellow Jacobins across France implemented Rousseau's necessary force to include summary condemnations and executions for all those who opposed the new order. Some 300,000 nobles, priests, political dissidents, and other troublemakers were imprisoned, and 17,000 died within the year. In Robespierre's universe, the goals of the Jacobins were noble and pure. They were, as he serenely wrote in February 1794 (shortly before he himself was guillotined), "the peaceful enjoyment of liberty and equality, the rule of that eternal justice whose laws have been engraved . . . upon the hearts of men, even upon the heart of the slave who knows them not and of the tyrant who denies them."

Thus took form the easy cohabitation of egalitarian ideology and savage coercion that was to plague the next two centuries. Better to exile from the tribe, the reasoning follows, those unwilling to make the commitment to the perfect society than to risk the infection of dissent. The demagogue asks only for unity of purpose on behalf of virtue: "My fellow citizens (comrades, brothers and sisters, Volk), eggs must be broken to make an omelette. To achieve that noble end, it may be necessary to wage a war." After the Revolution subsided, the principle was administered by Napoleon and the soldiers of the Revolution, who, having metamorphosed into the *grande armée*, were determined to spread the Enlightenment by conquest. Instead, they gave Europe additional cause to doubt the sovereignty of reason.

In fact, reason had never been sovereign. The decline of the Enlightenment was hastened not just by tyrants who used it for justification but by rising and often valid intellectual opposition. Its dream of a world made orderly and fulfilling by free intellect had seemed at first indestructible, the instinctive goal of all men. Its creators, among the greatest scholars since Plato and Aristotle, showed what the human mind can accomplish. Isaiah Berlin, one of their most perceptive historians, praised them justly as follows: "The intellectual power, honesty, lucidity, courage, and disinterested love of the truth of the most gifted thinkers of the eighteenth century remain to this day without parallel. Their age is one of the best and most hopeful episodes in the life of mankind." But they reached too far, and their best efforts were not enough to create the sustained effort their vision foretold.

THEIR SPIRIT WAS compressed into the life of the ill-fated Marie-Jean-Antoine-Nicolas Caritat, Marquis de Condorcet. He was the last of the French *philosophes*, the eighteenth-century public philosophers who immersed themselves in the political and social issues of their times. Voltaire, Montesquieu, d'Alembert, Diderot, Helvétius, and Condorcet's mentor, the economist and statesman Anne-Robert-Jacques Turgot, Baron de l'Aulne—all that remarkable assemblage was gone by 1789. Condorcet was the only one in their ranks who lived to see the Revolution. He embraced it totally and labored in vain to control its demonic force.

Condorcet was born in 1743 in Picardy, one of the most northerly provinces of old France, a member of an ancient noble family that originated in Dauphiné, the southeastern province from which the dauphin, eldest son of the king, took his title. The Caritats were hereditary members of the *noblesse d'épée*, order of the sword, traditionally devoted to military service, and of higher social status than the *noblesse de robe*, or high civil officials.

To the disappointment of his family, Condorcet chose not to be a soldier like his father but a mathematician. At the age of sixteen, while still a student at the Navarre College in Paris, he publicly read his first paper on the subject. But having entered the one scientific profession where talent can be confidently sorted into levels by the age of twenty, Condorcet turned out not to be a mathematician of the first rank, and certainly nowhere near the equal of his great contemporaries Leonhard Euler and Pierre Simon de Laplace. Still, he achieved enough to be elected, at the exceptionally young age of twenty-five, to the Académie des Sciences, and at thirty-two became its permanent secretary. In 1780, at age thirty-eight, he was accepted into the august Académie Française, arbiter of the literary language and pinnacle of intellectual recognition in his country.

Condorcet's principal scientific accomplishment was to pioneer the application of mathematics to the social sciences, an achievement he shared with Laplace. He was inspired by the idea, central to the Enlightenment agenda, that what had been accomplished in mathematics and physics can be extended to the collective actions of men. His 1785 *Essay on the Application of Analysis to the Probability of Majority Decisions* is a distant forerunner of present-day decision theory. As pure science, however, it is not impressive. While Laplace developed the calculus of probabilities and applied it brilliantly to physics, Condorcet made minor advances in mathematics and used the techniques he invented with little effect in the study of political behavior. Still, the concept that social action might be quantitatively analyzed and even predicted was original to Condorcet. It influenced the later development of the social sciences, especially the work of the early sociologists Auguste Comte and Adolphe Quételet in the 1800s.

Condorcet has been called the "noble philosopher," referring not just to his social rank but to his character and demeanor. Without irony his friends dubbed him "Le Bon Condorcet," Condorcet the Good. Julie de Lespinasse, who presided over his favorite salon on the rue de Belle Chasse, described him thus in a letter to a friend: "His physiognomy is sweet and calm; simplicity and negligence mark his bearing," reflecting the "absolute quality of his soul."

He was unfailingly kind and generous to others, including even the virulently jealous Jean-Paul Marat, whose own ambitions in science were unrewarded and who would gladly have seen him dead. He was passionately committed to the ideal of social justice and the welfare of others, both individually and collectively. He opposed, at considerable political risk, the colonial policies of France. With Lafayette and Mirabeau he founded the antislavery organization Society of the Friends of the Blacks. Even after he had gone into hiding during the Terror, his arguments contributed to the abolition of slavery by the National Convention.

Liberal to the bone, a follower of the English philosopher John Locke, Condorcet believed in the natural rights of men, and, like his contemporary Immanuel Kant, he sought moral imperatives that lead rather than follow the passions. He joined Tom Paine to create *Le Républicain*, a Revolutionary journal that promoted the idea of a progressive, egalitarian state. "The time will come," he later wrote, "when the sun will shine only on free men who know no other master than their reason."

Condorcet was a polymath with a near-photographic memory, for whom knowledge was a treasure to be acquired relentlessly and shared freely. Julie de Lespinasse, infatuated, praised these qualities in particular: "Converse with him, read what he has written; talk to him of philosophy, belles lettres, science, the arts, government, jurisprudence, and when you have heard him, you will tell yourself a hundred times a day that this is the most astonishing man you have ever heard; he is ignorant of nothing, not even the things most alien to his tastes and occupations; he will know . . . the genealogies of the courtiers, the details of the police and the names of the hats in fashion; in fact, nothing is beneath his attention, and his memory is so prodigious that he has never forgotten anything."

Condorcet's combination of talent and personality propelled him quickly to the highest levels of pre-Revolutionary Parisian society and established his reputation as the youngest of the *philosophes*. His taste for synthesis led him to fit into a coherent whole the principal ideas representing, if any such collection can legitimately be said to do so, the position of the late Enlightenment. On human nature he was a nurturist: He believed that the mind is molded wholly by its environment, so that humans are free to make themselves and society as they please. He was consequently a perfectibilist: The quality of human life, he insisted, can be improved indefinitely. He was politically a complete revolutionary, both anticlerical and republican, departing from Voltaire and others who would "destroy the altar but preserve the throne." In social science Condorcet was a historicist, believing that history can be read to understand the present and predict the future. As an ethicist, he was committed to the idea of the unity of the human race. And while egalitarian, he was not a multiculturalist in the present-day sense, but rather thought all societies would eventually evolve toward the high civilization of Europe. Above all, he was a humanitarian who saw politics as less a source of power than a means of implementing lofty moral principles.

With the outbreak of the Revolution in 1789, Condorcet abruptly turned from scholarship and threw himself into politics. He served two years as an elected member of the Commune of Paris, and when the Legislative Assembly was formed in 1791, he became a deputy for Paris. Immensely popular among his fellow revolutionaries, he was appointed one of the Assembly secretaries, then elected vice-president and finally president. When the Assembly was succeeded in September 1792 by the National Convention, and the Republic established, Condorcet was elected as representative for the Department of the Aisne, part of his native province of Picardy.

Throughout his brief public career, Condorcet tried to stay aloof from partisan politics. He had friends among both the moderate Girondists and the leftist Montagnards (the latter so named because their deputies sat on the higher benches, or "Mountain," of the assembly). He was identified with the Girondists nonetheless, and the more so when the Montagnards fell under the spell of the radical wing of the Jacobin Club of Paris. After the overthrow of the Girondists during the popular insurrections of 1793, the Montagnards controlled the Convention and then the Committee of Public Safety, which ruled France during the year-long Terror. It was during this spasm of official murder that Condorcet fell from hero to criminal suspect, and his arrest was ordered by the National Convention.

When he learned of the warrant, Condorcet fled to the boardinghouse of Madame Vernet, on the rue Servandoni of old Paris, where he remained in hiding for eight months. In April 1794 the refuge was discovered, and friends warned him that his arrest was imminent. He escaped once again, and for several days wandered about homeless until detected and thrown into the prison at Bourg-la-Reine.

During his stay on the rue Servandoni, Condorcet wrote his masterwork, *Sketch for a Historical Picture of the Progress of the Human Mind*. It was a remarkable achievement of both mind and will. Desperately insecure, with no books, relying only on his prodigious memory, he composed an intellectual and social history of humanity. The text, relentlessly optimistic in tone, contains little mention of the Revolution and none of his enemies in the streets of Paris. Condorcet wrote as though social progress is inevitable, and wars and revolutions were just Europe's way of sorting itself out.

His serene assurance arose from the conviction that culture is governed by laws as exact as those of physics. We need only understand them, he wrote, to keep humanity on its predestined course to a more perfect social order ruled by science and secular philosophy. These laws, he added, can be adduced from a study of past history.

Condorcet, however mistaken in details and hopelessly trusting of human nature, made a major contribution to thought through his insistence that history is an evolving material process. "The sole foundation for belief in the natural sciences," he declared, "is the idea that the general laws directing the phenomena of the universe, known or unknown, are necessary and constant. Why should this principle be any less true for the development of the intellectual and moral faculties of man than for other operations of nature?"

The idea was already in the air when those words were penned. Pascal had compared the human race to a man who never dies, always gaining knowledge, while Leibniz spoke of the Present big with the Future. Turgot, Condorcet's friend and sponsor, had written forty years before Condorcet's *Sketch* that "all epochs are fastened together by a sequence of causes and effects, linking the condition of the world to all the conditions which have gone before it." In consequence, "the human race, observed from its first beginning, seems in the eyes of the philosopher to be one vast whole, which, like each individual in it, has its own infancy and its own conditions of growth." Kant, in 1784, expressed the germ of the same concept, observing in particular that man's rational dispositions are destined to express themselves in the species as a whole, not in the individual.

Inevitable progress is an idea that has survived Condorcet and the Enlightenment. It has exerted, at different times and variously for good and evil, a powerful influence to the present day. In the final chapter of the *Sketch*, "The Tenth Stage: The Future Progress of the Human Mind," Condorcet becomes giddily optimistic about its prospect. He assures the reader that the glorious process is underway: All will be well. His vision for human progress makes little concession to the stubbornly negative qualities of human nature. When all humanity has attained a higher level of civilization, we are told, nations will be equal, and within each nation citizens will also be equal. Science will flourish and lead the way. Art will be freed to grow in power and beauty. Crime, poverty, racism, and sexual discrimination will decline. The human life span, through scientifically based medicine, will lengthen indefinitely. With the shadow of the Terror deepening without, Le Bon Condorcet concluded:

> How consoling for the philosopher who laments the errors, the crimes, the injustices which still pollute the earth and of which he is often the victim is this view of the human race, emancipated from its shackles, released from the empire of fate and from that of the enemies of its progress, advancing with a firm and sure step along the path of truth,

> virtue, and happiness! It is the contemplation of this prospect that rewards him for all his efforts to assist the progress of reason and the defense of liberty.

THE ENLIGHTENMENT GAVE RISE to the modern intellectual tradition of the West and much of its culture. Yet, while reason was supposedly the defining trait of the human species and needed only a little more cultivation to flower universally, it fell short. Humanity was not paying attention. Humanity thought otherwise. The causes of the Enlightenment's decline, which persist to the present day, illuminate the labyrinthine wellsprings of human motivation. It is worth asking, particularly in the present winter of our cultural discontent, whether the original spirit of the Enlightenment—confidence, optimism, eyes to the horizon—can be regained. And to ask in honest opposition, *should* it be regained, or did it possess in its first conception, as some have suggested, a dark-angelic flaw? Might its idealism have contributed to the Terror, which foreshadowed the horrendous dream of the totalitarian state? If knowledge can be consolidated, so might the "perfect" society be designed—one culture, one science—whether fascist, communist, or theocratic.

The Enlightenment itself, however, was never a unified movement. It was less a determined swift river than a lacework of deltaic streams working their way along twisted channels. By the time of the French Revolution it was very old. It emerged from the Scientific Revolution during the early seventeenth century and attained its greatest influence in the European academy during the eighteenth century. Its originators often clashed over fundamental issues. Most engaged from time to time in absurd digressions and speculations, such as looking for hidden codes in the Bible or for the anatomical seat of the soul. The overlap of their opinion was nevertheless extensive and clear and well reasoned enough to bear this simple characterization: They shared a passion to demystify the world and free the mind from the impersonal forces that imprison it.

They were driven by the thrill of discovery. They agreed on the power of science to reveal an orderly, understandable universe and thereby lay an enduring base for free rational discourse. They thought that the perfection of the celestial bodies discovered by astronomy and physics could serve as a model for human society. They believed in the unity of all knowledge, individual human rights, natural law, and indefinite human progress. They tried to avoid metaphysics even while the flaws and incompleteness of their explanations forced them to practice it. They resisted organized religion. They despised revelation and dogma. They endorsed, or at least tolerated, the state as a contrivance required for civil order. They believed that education and right reason would enormously benefit humanity. A few, like Condorcet, thought human beings perfectible and capable of achieving a political utopia.

We have not forgotten them. In their front rank were a disproportionate number of the tiny group of scientists and philosophers recognizable by a single name: Bacon, Hobbes, Hume, Locke, and Newton in England; Descartes and the eighteenth-century *philosophes* around Voltaire in France; Kant and Leibniz in Germany; Grotius in Holland; Galileo in Italy.

It has become fashionable to speak of the Enlightenment as an idiosyncratic construction by European males in a bygone era, one way of thinking among many different constructions generated across time by a legion of other minds in other cultures, each of which deserves careful and respectful attention. To which the only decent response is yes, of course—to a point. Creative thought is forever precious, and all knowledge has value. But what counts most in the long haul of history is seminality, not sentiment. If we ask whose ideas were the seeds of the dominant ethic and shared hopes of contemporary humanity, whose resulted in the most material advancement in history, whose were the first of their kind and today enjoy the most emulation, then in that sense the Enlightenment, despite the erosion of its original vision and despite the shakiness of some of its premises, has been the principal inspiration not just of Western high culture but, increasingly, of the entire world.

SCIENCE WAS the engine of the Enlightenment. The more scientifically disposed of the Enlightenment authors agreed that the cosmos is an orderly material existence governed by exact laws. It can be broken down into entities that can be measured and arranged in hierarchies, such as societies, which are made up of persons, whose brains consist of nerves, which in turn are composed of atoms. In principle at least, the atoms can be reassembled into nerves, the nerves into brains, and the persons into societies, with the whole understood as a system of mechanisms and forces. If you still insist on a divine intervention, continued the Enlightenment philosophers, think of the world as God's machine. The conceptual constraints that cloud our vision of the physical world can be eased for the betterment of humanity in every sphere. Thus Condorcet, in an era still unburdened by complicating fact, called for the illumination of the moral and political sciences by the "torch of analysis."

The grand architect of this dream was not Condorcet, or any of the other *philosophes* who expressed it so well, but Francis Bacon. Among the Enlightenment founders, his spirit is the one that most endures. It informs us across four centuries that we must understand nature, both around us and within ourselves, in order to set humanity on the course of self-improvement. We must do it knowing that destiny is in our hands and that denial of the dream leads back to barbarism. In his scholarship Bacon questioned the solidity of classical "delicate" learning, those medieval forms based on ancient texts and logical expatiation. He spurned reliance on ordinary scholastic philosophy, calling for a study of nature and the human condition on their own terms, without artifice. Drawing on his extraordinary insights into mental processes, he observed that because "the mind, hastily and without choice, imbibes and treasures up the first notices of things, from whence all the rest proceed, errors must forever prevail, and remain uncorrected." Thus knowledge is not well constructed but "resembles a magnificent structure that has no foundation."

> And whilst men agree to admire and magnify the false powers of the mind, and neglect or destroy those that might be rendered true, there is no other course left but with better assistance to begin the work anew, and raise or rebuild the sciences, arts, and all human knowledge from a firm and solid basis.

By reflecting on all possible methods of investigation available to his imagination, he concluded that the best among them is induction, which is the gathering of large numbers of facts and the detection of patterns. In order to obtain maximum objectivity, we must entertain only a minimum of preconceptions. Bacon proclaimed a pyramid of disciplines, with natural history forming the base, physics above and subsuming it, and metaphysics at the peak, explaining everything below—though perhaps in powers and forms beyond the grasp of man.

He was not a gifted scientist ("I can not thridd needles so well") or trained in mathematics, but a brilliant thinker who founded the philosophy of science. A Renaissance man, he took, in his own famous phrase, all knowledge to be his province. Then he stepped forward into the Enlightenment as the first taxonomist and master purveyor of the scientific method. He was *buccinator novi temporis*, the trumpeter of new times who summoned men "to make peace between themselves, and turning with united forces against the Nature of things, to storm and occupy her castles and strongholds, and extend the bounds of human empire."

Proud and reckless phrasing that, but appropriate to the age. Bacon, born in 1561, was the younger son of Sir Nicholas and Lady Ann Bacon, both of whom were well educated and extravagantly devoted to the arts. During his lifetime England, ruled successively by Elizabeth I and James I, passed tumultuously from a feudal society to a nation-state and fledgling colonial power, with its own newly acquired religion and an increasingly powerful middle class. By the year of Bacon's death, 1626, Jamestown was an established colony with the first representative government in North America, and the Pilgrims were settled at Plymouth. Bacon saw the English language come to first full flower. He ranks as one of its grand masters, even though he regarded it as a crude parochial language and preferred to write in Latin. He lived in a golden age of industry and culture, surrounded by other global overachievers, including, most famously, Drake, Raleigh, and Shakespeare.

Bacon enjoyed the privileges of rank through every step of his life. He was educated at Trinity College at Cambridge, which had been enriched some decades earlier by land grants from Henry VIII (and a century later was to serve as home to Newton). He was called to the bar in 1582 and two years afterward appointed to membership in Parliament. Virtually from infancy he was close to the throne. His father was Lord Keeper of the Seal, the highest judicial officer of the land. Elizabeth took early notice of the boy, talking with him often. Pleased by his precocious knowledge and gravity of manner, she fondly dubbed him The Young Lord Keeper.

He became a confirmed courtier for life, tying his political beliefs and fortunes to the crown. Under James I he rose, through flattery and wise counsel, to the heights commensurate to his ambition: Knighted in 1603, the year of James' accession, he was then named successively Attorney General, Lord Keeper, and, in 1618, Lord Chancellor. With the last office he was created first Baron of Verulam and soon afterward Viscount St. Alban.

Then, having flown too close too long to the royal flame, Bacon at last sustained near-fatal burns. He was targeted by a circle of determined personal enemies who found the wedge to his destruction in his tangled finances, and in 1621 successfully engineered his impeachment as Lord Chancellor. The charge, to which he pleaded guilty, was acceptance of bribes—"gifts," he said—while in high public office. He was heavily fined, escorted through the Traitor's Gate, and imprisoned in the Tower of London. Unbowed, he at once wrote the Marquis of Buckingham: "Good my Lord: Procure the warrant for my discharge this day . . . Howsoever I acknowledge the sentence just, and for reformation sake fit, [I was] the justest Chancellor that hath been in the five changes since Sir Nicholas Bacon's time."

He had been all that, and more. He was released in three days. Shorn at last of the burden of public ambition, he spent his last days totally immersed in contented scholarship. His death in the early spring of 1626 was symbolically condign, the result of an impromptu experiment to test one of his favorite ideas. "As he was taking the air in a coach with Dr. Witherborne towards High-gate," John Aubrey reported at the time, "snow lay on the ground, and it came into my Lord's thoughts, why flesh might not be preserved in snow, as in salt. They were resolved they would try the experiment presently. They alighted out of the coach and went into a poor woman's house at the bottom of High-gate hill, and bought a hen, and made the woman exenterate it, and then stuffed the body with snow, and my Lord did help to do it himself. The snow so chilled him that he immediately fell so extremely ill, that he could not return to his lodgings. . . ." He was taken instead to the Earl of Arundel's house close by. His condition remained grave, and he died on April 9, most likely of pneumonia.

The ache of disgrace had been subdued by the return to his true calling of visionary scholar. As he wrote in one of his oft-quoted adages, "He that dies in an earnest pursuit is like one that is wounded in hot blood, who for the time scarce feels the hurt." He saw his life as a contest between two great ambitions, and toward the end he regretted having invested so much effort in public service with an equivalent loss of scholarship. "My soul," he mused, "hath been a stranger in life's pilgrimage."

His genius, while of a different kind, matched that of Shakespeare. Some have believed, erroneously, that he *was* Shakespeare. He melded great literary gifts, so evident in *The Advancement of Learning*, with a passion for synthesis, two qualities most needed at the dawn of the Enlightenment. His great contribution to knowledge was that of learned futurist. He proposed a shift in scholarship away from rote learning and deductive reasoning from classical texts and toward engagement with the world. In science, he proclaimed, is civilization's future.

Bacon defined science broadly and differently from today's ordinary conception to include a foreshadowing of the social sciences and parts of the humanities. The repeated testing of knowledge by experiment, he insisted, is the cutting edge of learning. But to him experiment meant not just controlled manipulations in the manner of modern science. It was all the ways humanity brings change into the world through information, agriculture, and industry. He thought the great branches of learning to be open-ended and constantly evolving ("I do not promise you anything"), but he nonetheless focused eloquently on his belief in the underlying unity of knowledge. He rejected the sharp divisions among the disciplines prevailing since Aristotle. And fortunately, he was reticent in this enterprise when needed: He refrained from forecasting how the great branches of learning would ultimately fall out.

Bacon elaborated on but did not invent the method of induction as a counterpoint to classical and medieval deduction. Still, he deserves the title Father of Induction, on which much of his fame rested in later centuries. The procedure he favored was much more than mere factual generalizations, such as—to use a modern example—"ninety percent of plant species have flowers that are yellow, red, or white, and are visited by insects." Rather, he said, start with such an unbiased description of phenomena. Collect their common traits into an intermediate level of generality. Then proceed to higher levels of generality, such as: "Flowers have evolved colors and anatomy designed to attract certain kinds of insects, and these are the creatures that exclusively pollinate them." Bacon's reasoning was an improvement over the traditional methods of description and classification prevailing in the Renaissance, but it anticipated little of the methods of concept formation, competing hypotheses, and theory that form the core of modern science.

It was in psychology, and particularly the nature of creativity, that Bacon cast his vision farthest ahead. Although he did not use the word—it was not coined until 1653—he understood the critical importance of psychology in scientific research and all other forms of scholarship. He had a deep intuitive feel for the mental processes of discovery. He understood the means by which the processes are best systematized and most persuasively transmitted. "The human understanding," he wrote, "is no dry light, but receives an infusion from the will and affections; whence proceed sciences which may be called 'sciences as one would.' " He did not mean by this to distort perception of the real world by interposing a prism of emotion. Reality is still to be embraced directly and reported without flinching. But it is also best delivered the same way it was discovered, retaining a comparable vividness and play of the emotions. Nature and her secrets must be as stimulating to the imagination as are poetry and fables. To that end, Bacon advised us to use aphorisms, illustrations, stories, fables, analogies—anything that conveys truth from the discoverer to his readers as clearly as a picture. The mind, he argued, "is not like a wax tablet. On a tablet you cannot write the new till you rub out the old; on the mind you cannot rub out the old except by writing in the new."

Through light shed on the mental process, Bacon wished to reform reasoning across all the branches of learning. Beware, he said, of the *idols of the mind*, the fallacies into which undisciplined thinkers most easily fall. They are the real distorting prisms of human nature. Among them, idols of the *tribe* assume more order than exists in chaotic nature; those of the imprisoning *cave*, the idiosyncrasies of individual belief and passion; of the *marketplace*, the power of mere words to induce belief in nonexistent things; and of the *theater*, unquestioning acceptance of philosophical beliefs and misleading demonstrations. Stay clear of these idols, he urged, observe the world around you as it truly is, and reflect on the best means of transmitting reality as you have experienced it; put into it every fiber of your being.

I do not wish by ranking Francis Bacon so highly in this respect to portray him as a thoroughly modern man. He was far from that. His younger friend William Harvey, a physician and a real scientist who made a fundamental discovery, the circulation of the blood, noted drily that Bacon wrote philosophy like a Lord Chancellor. His phrases make splendid marble inscriptions and commencement flourishes. The unity of knowledge he conceived was remote from the present-day concept of consilience, far from the deliberate, systematic linkage of cause and effect across the disciplines. His stress lay instead upon the common means of inductive inquiry that might optimally serve all the branches of learning. He searched for the techniques that best convey the knowledge gained, and to that end he argued for the full employment of the humanities, including art and fiction, as the best means for developing and expressing science. Science, as he broadly defined it, should be poetry, and poetry science. That, at least, has a pleasingly modern ring.

Bacon envisioned a disciplined and unified learning as the key to improvement of the human condition. Much of the veritable library that accumulated beneath his pen still makes interesting reading, from his often quoted essays and maxims to *Advancement of Learning* (1605), *Novum Organum* (The New Logic, 1620), and *New Atlantis* (1627), the latter a utopian fable about a science-based society. Most of his philosophical and fictional writing was planned to implement the scheme of the unification of know-ledge, which he called *Instauratio Magna*, literally the Great Instauration, or the New Beginning.

His philosophy raised the sights of a small but influential public. It helped to prime the scientific revolution that was to blossom spectacularly in the decades ahead. To this day his vision remains the heart of the scientific-technological ethic. He was a magnificent figure standing alone by necessity of circumstance, who achieved that affecting combination of humility and innocent arrogance present only in the greatest scholars. Beneath the title of *Novum Organum* he had the publisher print these lines:

FRANCIS OF VERULAM
REASONED THUS WITH HIMSELF
and judged it to be for the interest of the present and future
generations that they should be made acquainted
with his thoughts.

ALL HISTORIES THAT live in our hearts are peopled by archetypes in mythic narratives, and such I believe is part of Francis Bacon's appeal and why his fame endures. In the tableau of the Enlightenment, Bacon is the herald of adventure. There is a new world waiting, he announced; let us begin the long and difficult march into its unmapped terrain. René Descartes, the founder of algebraic geometry and modern philosophy and France's preeminent scholar of all time, is the mentor in the narrative. Like Bacon before him, he summoned scholars to the scientific enterprise, among whom was soon to follow the young Isaac Newton. Descartes showed how to do science with the aid of precise deduction, cutting to the quick of each phenomenon and skeletonizing it. The world is three-dimensional, he explained, so let our perception of it be framed in three coordinates—Cartesian coordinates they are called today. With them the length, breadth, and height of any object can be exactly specified and subjected to mathematical operations to explore its essential qualities. He accomplished this step in elementary form by reformulating algebraic notation so that it could be used to solve complex problems of geometry and, further, to explore realms of mathematics beyond the visual realm of three-dimensional space.

Descartes' overarching vision was one of knowledge as a system of interconnected truths that can be ultimately abstracted into mathematics. It all came to him, he said, through a series of dreams on a November night in 1619, when somehow in a flurry of symbols (thunderclaps, books, an evil spirit, a delicious melon) he perceived that the universe is both rational and united throughout by cause and effect. He believed that this conception could be applied from physics to medicine—hence biology—and even to moral reasoning. In this respect, he laid the groundwork for the belief in the unity of learning that was to influence Enlightenment thought profoundly in the eighteenth century.

Descartes insisted upon systematic doubt as the first principle of learning. By his light all knowledge was to be laid out and tested upon the iron frame of logic. He allowed himself only one undeniable premise, captured in the celebrated phrase *Cogito ergo sum,* I think therefore I am. The system of Cartesian doubt, which still thrives in modern science, is one in which all assumptions possible are systematically eliminated so as to leave only one set of axioms upon which rational thought can be logically based, and experiments can be rigorously designed.

Descartes nonetheless made a fundamental concession to metaphysics. A lifelong Catholic, he believed in God as an absolutely perfect being, manifested by the power of the idea of such a being in his own mind. That given, he went on to argue for the complete separation of mind and matter. The stratagem freed him to put spirit aside to concentrate on matter as pure mechanism. In works published over the years 1637–49, Descartes introduced reductionism, the study of the world as an assemblage of physical parts that can be broken apart and analyzed separately. Reductionism and analytic mathematical modeling were destined to become the most powerful intellectual instruments of modern science. (The year 1642 was a signal one in the history of ideas: With Descartes' *Meditationes de Prima Philosophia* just published and *Principia Philosophiae* soon to follow, Galileo died and Newton was born.)

As Enlightenment history unfolded, Isaac Newton came to rank with Galileo as the most influential of the heroes who answered Bacon's call. A restless seeker of horizons, stunningly resourceful, he invented calculus before Gottfried Leibniz, whose notation was nevertheless clearer and is the one used today. Calculus proved to be, in company with analytic geometry, one of the two crucial mathematical techniques in physics and, later, in chemistry, biology, and economics. Newton was also an inventive experimentalist, one of the first to recognize that the general laws of science might be discovered by manipulating physical processes. While investigating prisms, he demonstrated the relation of the refrangibility of light to color and from that the compound nature of sunlight and the origin of rainbows. As in many great experiments of science, this one is simple; anyone can quickly repeat it. With a prism bend a beam of sunlight so that its different wavelengths fall out into the colors of the visible spectrum. Now bend the colors back together again to create the beam of sunlight. Newton applied his findings in the construction of the first reflecting telescope, a superior instrument perfected a century later by the British astronomer William Herschel.

In 1684 Newton formulated the mass and distance laws of gravity, and in 1687 the three laws of motion. With these mathematical formulations he achieved the first great breakthrough in modern science. He showed that the planetary orbits postulated by Copernicus and proved elliptical by Kepler can be predicted from the first principles of mechanics. His laws were exact and equally applicable to all inanimate matter, from the solar system down to grains of sand, and of course to the falling apple that had triggered his thinking on the subject twenty years previously—apparently a true story. The universe, he said, is not just orderly but also intelligible. At least part of God's grand design could be written with a few lines on a piece of paper. His triumph enshrined Cartesian reductionism in the conduct of science.

Because Newton established order where magic and chaos had reigned before, his impact on the Enlightenment was enormous. Alexander Pope celebrated him with a famous couplet:

> Nature and Nature's laws lay hid in night:
> God said, "*Let Newton be!*" and all was light.

Well—not all, not yet. But the laws of gravity and motion were a powerful beginning. And they started Enlightenment scholars thinking: Why not a Newtonian solution to the affairs of men? The idea grew into one of the mainstays of the Enlightenment agenda. As late as 1835, Adolphe Quételet was proposing "social physics" as the basis of the discipline soon to be named sociology. Auguste Comte, his contemporary, believed a true social science to be inevitable. "Men," he said, echoing Condorcet, "are not allowed to think freely about chemistry and biology, so why should they be allowed to think freely about political philosophy?" People, after all, are just extremely complicated machines. Why shouldn't their behavior and social institutions conform to certain still-undefined natural laws?

Reductionism, given its unbroken string of successes during the next three centuries, may seem today the obvious best way to have constructed knowledge of the physical world, but it was not so easy to grasp at the dawn of science. Chinese scholars never achieved it. They possessed the same intellectual ability as Western scientists, as evidenced by the fact that, even though far more isolated, they acquired scientific information as rapidly as did the Arabs, who had all of Greek knowledge as a launching ramp. Between the first and thirteenth centuries they led Europe by a wide margin. But according to Joseph Needham, the principal Western chronicler of Chinese scientific endeavors, their focus stayed on holistic properties and on the harmonious, hierarchical relationships of entities, from stars down to mountains and flowers and sand. In this world view the entities of Nature are inseparable and perpetually changing, not discrete and constant as perceived by the Enlightenment thinkers. As a result the Chinese never hit upon the entry point of abstraction and break-apart analytic research attained by European science in the seventeenth century.

Why no Descartes or Newton under the Heavenly Mandate? The reasons were historical and religious. The Chinese had a distaste for abstract codified law, stemming from their unhappy experience with the Legalists, rigid quantifiers of the law who ruled during the transition from feudalism to bureaucracy in the Ch'in dynasty (221–206 B.C.). Legalism was based on the belief that people are fundamentally antisocial and must be bent to laws that place the security of the state above their personal desires. Of probably even greater importance, Chinese scholars abandoned the idea of a supreme being with personal and creative properties. No rational Author of Nature existed in their universe; consequently the objects they meticulously described did not follow universal principles, but instead operated within particular rules followed by those entities in the cosmic order. In the absence of a compelling need for the notion of general laws—thoughts in the mind of God, so to speak—little or no search was made for them.

Western science took the lead largely because it cultivated reductionism and physical law to expand the understanding of space and time beyond that attainable by the unaided senses. The advance, however, carried humanity's self-image ever further from its perception of the remainder of the universe, and as a consequence the full reality of the universe seemed to grow progressively more alien. The ruling talismans of twentieth-century science, relativity and quantum mechanics, have become the ultimate in strangeness to the human mind. They were conceived by Albert Einstein, Max Planck, and other pioneers of theoretical physics during a search for quantifiable truths that would be known to extraterrestrials as well as to our species, and hence certifiably independent of the human mind. The physicists succeeded magnificently, but in so doing they revealed the limitations of intuition unaided by mathematics; an understanding of Nature, they discovered, comes very hard. Theoretical physics and molecular biology are acquired tastes. The cost of scientific advance is the humbling recognition that reality was not constructed to be easily grasped by the human mind. This is the cardinal tenet of scientific understanding: Our species and its ways of thinking are a product of evolution, not the purpose of evolution.

WE NOW PASS to the final archetype of the epic tableau, the keepers of the innermost room. The more radical Enlightenment writers, alert to the implications of scientific materialism, moved to reassess God Himself. They invented a Creator obedient to His own natural laws, the belief known as deism. They disputed the theism of Judaeo-Christianity, whose divinity is both omnipotent and personally interested in human beings, and they rejected the nonmaterial world of heaven and hell. At the same time, few dared go the whole route and embrace atheism, which seemed to imply cosmic meaninglessness and risked outraging the pious. So by and large they took a middle position. God the Creator exists, they conceded, but He is allowed only the entities and processes manifest in His own handiwork.

Deistic belief, by persisting in attenuated form to the present day, has given scientists a license to search for God. More precisely, it has prompted a small number to make a partial sketch of Him (Her? It? Them?) from their professional meditations. He is material in another plane but not personal. He is, perhaps, the manager of alternative universes popping out of black holes, Who adjusts physical laws and parameters in order to observe the outcome. Maybe we see a faint trace of Him in the pattern of ripples in cosmic background radiation, dating back to the first moments of our own universe. Alternatively, we may be predestined to reach Him billions of years in the future at an omega point of evolution—total unity, total knowledge—toward which the human species and extraterrestrial life forms are converging. I must say that I have read many such schemes, and even though they are composed by scientists, I find them depressingly non-Enlightenment. That the Creator lives outside this universe and will somehow be revealed at its end is what the theologians have been telling us all along.

Few scientists and philosophers, however, let alone religious thinkers, take the playful maunderings of scientific theology very seriously. A more coherent and interesting approach, possibly within the reach of theoretical physics, is to try to answer the following question: Is a universe of discrete material particles possible only with one specific set of natural laws and parameter values? In other words, does human imagination, which can conceive of other laws and values, thereby exceed possible existence? Any act of Creation may be only a subset of the universes we can imagine. To this point Einstein is reported to have remarked to his assistant Ernst Straus, in a moment of neo-deistic reflection, "What really interests me is whether God had any choice in the creation of the world." That line of reasoning can be extended rather mystically to formulate the "anthropic principle," which notes that the laws of nature, in *our* universe at least, had to be set a certain precise way so as to allow the creation of beings able to ask about the laws of nature. Did Someone decide to do it that way?

The dispute between Enlightenment deism and theology can be summarized as follows. The traditional theism of Christianity is rooted in both reason and revelation, the two conceivable sources of knowledge. According to this view, reason and revelation cannot be in conflict, because in areas of opposition, revelation is given the higher role—as the Inquisition reminded Galileo in Rome when they offered him a choice between orthodoxy and pain. In contrast, deism grants reason the edge, and insists that theists justify revelation with the use of reason.

Traditional theologians of the eighteenth century, faced with the Enlightenment challenge, refused to yield an inch of ground. Christian faith, they argued back, cannot submit itself to the debasing test of rationality. Deep truths exist that are beyond the grasp of the unaided human mind, and God will reveal them to our understanding when and by whatever means He chooses.

Given the centrality of religion in everyday life, the stand of the theists against reason seemed . . . well, it seemed reasonable. Believers in the eighteenth century saw no difficulty in conducting their lives by both ratiocination and revelation. The theologians won the argument simply because there was no compelling reason to adopt a new metaphysics. For the first time, the Enlightenment visibly stumbled.

The fatal flaw in deism is thus not rational at all, but emotional. Pure reason is unappealing because it is bloodless. Ceremonies stripped of sacred mystery lose their emotional force, because celebrants need to defer to a higher power in order to consummate their instinct for tribal loyalty. In times of danger and tragedy especially, unreasoning ceremony is everything. There is no substitute for surrender to an infallible and benevolent being, the commitment called salvation. And no substitute for formal recognition of an immortal life force, the leap of faith called transcendence. It follows that most people would very much like science to prove the existence of God but not to take the measure of His capacity.

Deism and science also failed to colonize ethics. The sparkling Enlightenment promise of an objective basis for moral reasoning could not be met. If an immutable secular field of ethical premises exists, the human intellect during the Enlightenment seemed too weak and shifting to locate it. So theologians and philosophers stuck to their original positions, either by deferring to religious authority or by articulating subjectively perceived natural rights. There was no logical alternative open to them. The millennium-old rules sacralized by religion seemed to work, more or less, and in any case there was no time to figure it all out. You can defer reflection on the celestial spheres indefinitely but not on daily matters of life and death.

THERE WAS and remains another, more purely rationalist objection to the Enlightenment program. Grant for argument's sake that the most extravagant claims of the Enlightenment supporters proved true, so that it became possible for scientists to look into the future and to see what course of action is best for humanity. Wouldn't that trap us in a cage of logic and revealed fate? The thrust of the Enlightenment, like the Greek humanism that prefigured it, was Promethean: The knowledge it generated was to liberate mankind by lifting it above the savage world. But the opposite might occur. If scientific inquiry diminishes the conception of divinity while prescribing immutable natural laws, then humanity can lose what freedom it already possesses. Perhaps there is only one "perfect" social order, and scientists will find it—or worse, falsely claim to have found it. Religious authority, the Hadrian's Wall of civilization, will be breached and the barbarians of totalitarian ideology will pour in. Such is the dark side of Enlightenment secular thought, unveiled in the French Revolution and expressed more recently by theories of "scientific" socialism and racialist fascism.

And there is another concern: that a science-driven society risks upsetting the natural order of the world set in place by God or, if you prefer, by billions of years of evolution. Science given too much authority risks conversion into a self-destroying impiety. The godless creations of science and technology are in fact powerful and arresting images of modern culture. Frankenstein's monster and Hollywood's Terminator, the latter an all-metal and microchip-guided Frankenstein's monster, wreak destruction on their creators, including the naive geniuses in lab coats who arrogantly forecast a new age ruled by science. Storms rage, hostile mutants spread, life dies. Nations menace one another with world-destroying technology. Even Winston Churchill, whose country was saved by radar, worried after the atom bombing of Japan that the stone age might return "on the gleaming wings of Science."

FOR THOSE WHO for so long thus feared science as Faustian rather than Promethean, the Enlightenment program posed a grave threat to spiritual freedom, even to life itself. What is the answer to such a threat? Revolt! Return to natural man, reassert the primacy of individual imagination and confidence in immortality. Find an escape to a higher realm through art, promote a Romantic Revolution. In 1807 William Wordsworth, in words typical of the movement then spreading over Europe, evoked the aura of a more primal and serene existence beyond Reason's grasp:

> Our Souls have sight of that immortal sea
> Which brought us hither,
> Can in a moment travel thither,
> And see the Children sport upon the shore,
> And hear the mighty waters rolling evermore.

With Wordsworth's "breathings for incommunicable powers," the eyes close, the mind soars, the inverse square distance law of gravity falls away. The spirit enters another reality beyond the reach of weight and measure. If the constraining universe of matter and energy cannot be denied, at least it can be ignored with splendid contempt. There is no question that Wordsworth and his fellow English Romantic poets of the first half of the nineteenth century conjured works of great beauty. They spoke truths in another tongue, and guided the arts still further from the sciences.

Romanticism also flowered in philosophy, where it placed a premium on rebellion, spontaneity, intense emotion, and heroic vision. Searching for aspirations available only to the heart, its practitioners dreamed of man as part of boundless nature. Rousseau, while often listed as an Enlightenment *philosophe*, was really instead the founder and most extreme visionary of the Romantic philosophical movement. For him learning and social order are the enemies of humanity. In works from 1749 (*Discourse on the Sciences and the Arts*) to 1762 (*Émile*), he extolled the "sleep of reason." His utopia is a minimalist state in which people abandon books and other accouterments of intellect in order to cultivate enjoyment of the senses and good health. Humanity, Rousseau claimed, was originally a race of noble savages in a peaceful state of nature, who were later corrupted by civilization—and by scholarship. Religion, marriage, law, and government are deceptions created by the powerful for their own selfish ends. The price paid by the common man for this high-level chicanery is vice and unhappiness.

Where Rousseau invented a stunningly inaccurate form of anthropology, the German Romantics, led by Goethe, Hegel, Herder, and Schelling, set out to reinsert metaphysics into science and philosophy. The product, *Naturphilosophie*, was a hybrid of sentiment, mysticism, and quasi-scientific hypothesis. Johann Wolfgang von Goethe, preeminent among its expositors, wanted most of all to be a great scientist. He placed that ambition above literature, where in fact he became an immortal contributor. His respect for science as an idea, an approach to tangible reality, was unreserved, and he understood its basic tenets. Analysis and synthesis, he liked to say, should be alternated as naturally as breathing in and breathing out. At the same time he was critical of the mathematical abstractions of Newtonian science, thinking physics far too ambitious in its goal of explaining the universe. He was also often contemptuous of the "technical tricks" employed by experimental scientists. In fact, he tried to repeat Newton's optical experiments but with poor results.

Goethe can be easily forgiven. After all, he had a noble purpose, no less than the coupling of the soul of the humanities to the engine of science. He would have grieved had he foreseen history's verdict: great poet, poor scientist. He failed in his synthesis through lack of what is today called the scientist's instinct. Not to mention the necessary technical skills. Calculus baffled him, and it is said he could not tell a lark from a sparrow. But he loved Nature in a profoundly spiritual sense. One must cultivate a close, deep feeling for her, he proclaimed. "She loves illusion. She shrouds man in mist, and she spurs him toward the light. Those who will not partake of her illusions she punishes as a tyrant would punish. Those who accept her illusions she presses to her heart. To love her is the only way to approach her." In the philosophers' empyrean I imagine Bacon has long since lectured Goethe on the idols of the mind. Newton will have lost patience immediately.

Friedrich Schelling, leading philosopher of the German Romantics, attempted to bind the scientific Prometheus to immobility not with poetry but with reason. He proposed a cosmic unity of all things, beyond the understanding of man. Facts by themselves can never be more than partial truths. Those we perceive are only fragments of the universal flux. Nature is alive, Schelling concluded; she is a creative spirit that unites knower and known, progressing through greater and greater understanding and feeling toward an eventual state of complete self-realization.

In America, German philosophical Romanticism was mirrored in New England transcendentalism, whose most celebrated proponents were Ralph Waldo Emerson and Henry David Thoreau. The transcendentalists were radical individualists who rejected the overwhelming commercialism that came to prevail in American society during the Jacksonian era. They envisioned a spiritual universe built entirely within their personal ethos. They nevertheless found science more congenial than did their European counterparts—witness the many accurate natural history observations in *Faith in a Seed* and other writings by Thoreau. Their ranks even included one full-fledged scientist: Louis Agassiz, director of the Museum of Comparative Zoology at Harvard University, founding member of the National Academy of Science, geologist, zoologist, and supremely gifted lecturer. This great man, in a metaphysical excursion paralleling that of Schelling, conceived the universe as a vision in the mind of God. The deities of science in his universe were essentially the same as those of theology. In 1859, at the height of his career, Agassiz was scandalized by the appearance of Darwin's *Origin of Species*, which advanced the theory of evolution by natural selection and saw the diversity of life as self-assembling. Surely, he argued before rapt audiences in cities along the Atlantic seaboard, God would not create the living world by random variation and survival of the fittest. Our view of life must not be allowed to descend from cosmic grandeur to the grubby details of ponds and woodlots. Even to think of the human condition in such a manner, he argued, is intolerable.

NATURAL SCIENTISTS, chastened by such robust objections to the Enlightenment agenda, mostly abandoned the examination of human mental life, yielding to philosophers and poets another century of free play. In fact, the concession turned out to be a healthy decision for the profession of science, because it steered researchers away from the pitfalls of metaphysics. Throughout the nineteenth century, knowledge in the physical and biological sciences grew at an exponential rate. At the same time the social sciences—sociology, anthropology, economics, and political theory—newly risen like upstart duchies and earldoms, vied for territory in the space created between the hard sciences and the humanities. The great branches of learning emerged in their present form—natural sciences, social sciences, and the humanities—out of the unified Enlightenment vision generated during the seventeenth and eighteenth centuries.

The Enlightenment, defiantly secular in orientation while indebted and attentive to theology, had brought the Western mind to the threshold of a new freedom. It waved aside everything, every form of religious and civil authority, every imaginable fear, to give precedence to the ethic of free inquiry. It pictured a universe in which humanity plays the role of perpetual adventurer. For two centuries God seemed to speak in a new voice to humankind. That voice had been foreshadowed in 1486 by Giovanni Pico della Mirandola, Renaissance forerunner of the Enlightenment thinkers, in this benediction:

> We have made thee neither of heaven nor of earth, neither mortal nor immortal, so that with freedom of choice and with honor, as though the maker and molder of thyself, thou mayest fashion thyself in whatever shape thou shalt prefer.

BY THE EARLY 1800S, however, the splendid image was fading. Reason fractured, intellectuals lost faith in the leadership of science, and the prospect of the unity of knowledge sharply declined. It is true that the spirit of the Enlightenment lived on in political idealism and the hopes of individual thinkers. In the ensuing decades new schools sprang up like shoots from the base of a shattered tree: the utilitarian ethics of Bentham and Mill, the historical materialism of Marx and Engels, the pragmatism of Charles Peirce, William James, and John Dewey. But the core agenda seemed irretrievably abandoned. The grand conception that had riveted thinkers during the previous two centuries lost most of its credibility.

Science traveled its own way. It continued to double every fifteen years in practitioners, discoveries, and technical journals, as it had since the early 1700s, finally beginning to level off only around 1970. Its continuously escalating success began to give credence again to the idea of an ordered, intelligible universe. This essential Enlightenment premise grew stronger in the disciplines of mathematics, physics, and biology, where it had first been conceived by Bacon and Descartes. Yet the enormous success of reductionism, its key method, worked perversely against any recovery of the Enlightenment program as a whole. Precisely because scientific information was growing at a geometric pace, most individual researchers were not concerned with unification, and even less with philosophy. They thought, what works, works, and so what need is there to reflect more deeply on the matter? They were even slower to address the taboo-laden physical basis of mind, a concept hailed in the late 1700s as the gateway from biology to the social sciences.

There was another, humbler reason for the lack of interest in the big picture: Scientists simply didn't have the requisite intellectual energy. The vast majority of scientists have never been more than journeymen prospectors.

That is even more the case today. They are professionally focused; their education does not orient them to the wide contours of the world. They acquire the training they need to travel to the frontier and make discoveries of their own, and as fast as possible, because life at the growing edge is expensive and chancy. The most productive scientists, installed in million-dollar laboratories, have no time to think about the big picture and see little profit in it. The rosette of the United States National Academy of Sciences, which the two thousand elected members wear on their lapels as a mark of achievement, contains a center of scientific gold surrounded by the purple of natural philosophy. The eyes of most leading scientists, alas, are fixed on the gold.

It is therefore not surprising to find physicists who do not know what a gene is, and biologists who guess that string theory has something to do with violins. Grants and honors are given in science for discoveries, not for scholarship and wisdom. And so has it ever been. Francis Bacon, using the political skills that lofted him to the Lord Chancellorship, personally importuned the English monarchs for funds to carry forth his great scheme of unifying knowledge. He never got a penny. At the height of his fame Descartes was ceremoniously awarded a stipend by the French royal court. But the account remained unfunded, helping to drive him to the more generous Swedish court in the "land of bears between rock and ice," where he soon died of pneumonia.

The same professional atomization afflicts the social sciences and humanities. The faculties of higher education around the world are a congeries of experts. To be an original scholar is to be a highly specialized world authority in a polyglot Calcutta of similarly focused world authorities. In 1797, when Jefferson took the president's chair at the American Philosophical Society, all American scientists of professional caliber and their colleagues in the humanities could be seated comfortably in the lecture room of Philosophical Hall. Most could discourse reasonably well on the entire world of learning, which was still small enough to be seen whole. Their successors today, including 450,000 holders of the doctorate in science and engineering alone, would overcrowd Philadelphia. Professional scholars in general have little choice but to dice up research expertise and research agendas among themselves. To be a successful scholar means spending a career on membrane biophysics, the Romantic poets, early American history, or some other such constricted area of formal study.

Fragmentation of expertise was further mirrored in the twentieth century by modernism in the arts, including architecture. The work of the masters—Braque, Picasso, Stravinsky, Eliot, Joyce, Martha Graham, Gropius, Frank

Lloyd Wright, and their peers—was so novel and discursive as to thwart generic classification, except perhaps for this: The modernists tried to achieve the new and provocative at any cost. They identified the constraining bonds of tradition and self-consciously broke them. Many rejected realism in expression in order to explore the unconscious. Freud, as much a literary stylist as a scientist, inspired them and can be justifiably included in their ranks. Psychoanalysis was a force that shifted the attention of modernist intellectuals and artists from the social and political to the private and psychological. Subjecting every topic within their domain to the "ruthless centrifuge of change," in Carl Schorske's phrase, they meant to proudly assert the independence of twentieth-century high culture from the past. They were not nihilists; rather, they sought to create a new level of order and meaning. They were complete experimentalists who wished to participate in a century of radical technological and political change and to fashion part of it entirely on their own terms.

Thus the free flight bequeathed by the Enlightenment, which disengaged the humanities during the Romantic era, had by the middle of the twentieth century all but erased hope for the unification of knowledge with the aid of science. The two cultures described by C. P. Snow in his 1959 Rede Lecture, the literary and the scientific, were no longer on speaking terms.

ALL MOVEMENTS TEND to extremes, which is approximately where we are today. The exuberant self-realization that ran from romanticism to modernism has given rise now to philosophical postmodernism (often called poststructuralism, especially in its more political and sociological expressions). Postmodernism is the ultimate polar antithesis of the Enlightenment. The difference between the two extremes can be expressed roughly as follows: Enlightenment thinkers believe we can know everything, and radical postmodernists believe we can know nothing.

The philosophical postmodernists, a rebel crew milling beneath the black flag of anarchy, challenge the very foundations of science and traditional philosophy. Reality, they propose, is a state constructed by the mind, not perceived by it. In the most extravagant version of this constructivism, there is no "real" reality, no objective truths external to mental activity, only prevailing versions disseminated by ruling social groups. Nor can ethics be firmly grounded, given that each society creates its own codes for the benefit of the same oppressive forces.

If these premises are correct, it follows that one culture is as good as any other in the expression of truth and morality, each in its own special way. Political multiculturalism is justified; each ethnic group and sexual preference in the community has equal validity. And, more than mere tolerance, it deserves communal support and mandated representation in educational agendas, not because it has general importance to the society but because it exists. That is—again—if the premises are correct. And they must be correct, say their promoters, because to suggest otherwise is bigotry, which is a cardinal sin. Cardinal, that is, if we agree to waive in this one instance the postmodernist prohibition against universal truth, and all agree to agree for the common good. Thus, Rousseau redivivus.

Postmodernism is expressed more explicitly still in deconstruction, a technique of literary criticism. Each author's meaning is unique to himself, goes the underlying premise; nothing of his true intention or anything else connected to objective reality can be reliably assigned to it. His text is therefore open to fresh analysis and commentary issuing from the equally solipsistic world in the head of the reviewer. But then the reviewer is in turn subject to deconstruction, as well as the reviewer of the reviewer, and so on in infinite regress. That is what Jacques Derrida, the creator of deconstruction, meant when he stated the formula *Il n'y a pas de hors-texte* (There is nothing outside the text). At least, that is what I think he meant, after reading him, his defenders, and his critics with some care. If the radical postmodernist premise is correct, we can never be sure that is what he meant. Conversely, if that *is* what he meant, it is not certain we are obliged to consider his arguments further. This puzzle, which I am inclined to set aside as the "Derrida paradox," is similar to the Cretan paradox (a Cretan says "all Cretans are liars"). It awaits solution, though one need not feel any great sense of urgency in the matter.

Nor is it certain from Derrida's ornately obscurantist prose that he himself knows what he means. Some observers think his writing is meant as a *jeu d'esprit*, a kind of joke. His new "science" of grammatology is the opposite of science, rendered in fragments with the incoherence of a dream, at once banal and fantastical. It is innocent of the science of mind and language developed elsewhere in the civilized world, rather like the pronouncements of a faith healer unaware of the location of the pancreas. He seems, in the end, to be conscious of this omission, but contents himself with the stance of Rousseau, self-professed enemy of books and writing, whose work *Émile* he quotes: ". . . the dreams of a bad night are given to us as philosophy. You will say I too am a dreamer; I admit it, but I do what others fail to do, I give my dreams as dreams, and leave the reader to discover whether there is anything in them which may prove useful to those who are awake."

Scientists, awake and held responsible for what they say while awake, have not found postmodernism useful. The postmodernist posture toward science in return is one of subversion. There appears to be a provisional acceptance of gravity, the periodic table, astrophysics, and similar stanchions of the external world, but in general the scientific culture is viewed as just another way of knowing, and, moreover, contrived mostly by European and American white males.

It is tempting to relegate postmodernism to history's curiosity cabinet alongside theosophy and transcendental idealism, but it has seeped by now into the mainstream of the social sciences and humanities. It is viewed there as a technique of metatheory (theory about theories), by which scholars analyze not so much the subject matter of the scientific discipline as the cultural and psychological reasons particular scientists think the way they do. The analyst places emphasis on "root metaphors," those ruling images in the thinker's mind by which he designs theory and experiments. Here, for example, is Kenneth Gergen explaining how modern psychology is dominated by the metaphor of human beings as machines:

> Regardless of the character of the person's behavior, the mechanist theorist is virtually obliged to segment him from the environment, to view the environment in terms of stimulus or input elements, to view the person as reactive to and dependent on these input elements, to view the domain of the mental as structured (constituted of interacting elements), to segment behavior into units that can be coordinated to the stimulus inputs, and so on.

Put briefly, and to face the issue squarely, psychology is at risk of becoming a natural science. As a possible remedy for those who wish to keep it otherwise, and there are many scholars who do, Gergen cites other, perhaps less pernicious root metaphors of mental life that might be considered, such as the marketplace, dramaturgy, and rule-following. Psychology, if not allowed to be contaminated with too much biology, can accommodate endless numbers of theoreticians in the future.

As the diversity of metaphors has been added to ethnic diversity and gender dualism to create new workstations in the postmodernist academic industry, and then politicized, schools and ideologies have multiplied explosively. Usually leftist in orientation, the more familiar modes of general postmodernist thought include Afrocentrism, constructivist social anthropology, "critical" (i.e., socialist) science, deep ecology, ecofeminism, Lacanian psychoanalysis, Latourian sociology of science, and neo-Marxism. To which add all the bewildering varieties of deconstruction techniques and New Age holism swirling round about and through them.

Their adherents fret upon the field of play, sometimes brilliantly, usually not, jargon-prone and elusive. Each in his own way seems to be drifting toward that *mysterium tremendum* abandoned in the seventeenth century by the Enlightenment. And not without the expression of considerable personal anguish. Of the late Michel Foucault, the great interpreter of political power in the history of ideas, poised "at the summit of Western intellectual life," George Scialabba has perceptively written,

> Foucault was grappling with the deepest, most intractable dilemmas of modern identity. . . . For those who believe that neither God nor natural law nor transcendent Reason exists, and who recognize the varied and subtle ways in which material interest—power—has corrupted, even constituted, every previous morality, how is one to live, to what values can one hold fast?

How and what indeed? To solve these disturbing problems, let us begin by simply walking away from Foucault, and existentialist despair. Consider this rule of thumb: To the extent that philosophical positions both confuse and close doors to further inquiry, they are likely to be wrong.

To Foucault I would say, if I could (and without meaning to sound patronizing), it's not so bad. Once we get over the shock of discovering that the universe was not made with us in mind, all the meaning the brain can master, and all the emotions it can bear, and all the shared adventure we might wish to enjoy, can be found by deciphering the hereditary orderliness that has borne our species through geological time and stamped it with the residues of deep history. Reason will be advanced to new levels, and emotions played in potentially infinite patterns. The true will be sorted from the false, and we will understand one another very well, the more quickly because we are all of the same species and possess biologically similar brains.

And to others concerned about the growing dissolution and irrelevance of the intelligentsia, which is indeed alarming, I suggest there have always been two kinds of original thinkers, those who upon viewing disorder try to create order, and those who upon encountering order try to protest it by creating disorder. The tension between the two is what drives learning forward. It lifts us upward through a zigzagging trajectory of progress. And in the Darwinian contest of ideas, order always wins, because—simply—that is the way the real world works.

Nevertheless, here is a salute to the postmodernists. As today's celebrants of corybantic Romanticism, they enrich culture. They say to the rest of us: Maybe, just maybe, you are wrong. Their ideas are like sparks from firework explosions that travel away in all directions, devoid of following energy, soon to wink out in the dimensionless dark. Yet a few will endure long enough to cast light on unexpected subjects. That is one reason to think well of postmodernism, even as it menaces rational thought. Another is the relief it affords those who have chosen not to encumber themselves with a scientific education. Another is the small industry it has created within philosophy and literary studies. Still another, the one that counts the most, is the unyielding critique of traditional scholarship it provides. We will always need postmodernists or their rebellious equivalents. For what better way to strengthen organized knowledge than continually to defend it from hostile forces? John Stuart Mill correctly noted that teacher and learner alike fall asleep at their posts when there is no enemy in the field. And if somehow, against all the evidence, against all reason, the linchpin falls out and everything is reduced to epistemological confusion, we will find the courage to admit that the postmodernists were right, and in the best spirit of the Enlightenment, we will start over again. Because, as the great mathematician David Hilbert once said, capturing so well that part of the human spirit expressed through the Enlightenment, *Wir müssen wissen. Wir werden wissen.* We must know, we will know.

CHAPTER 4

The Natural Sciences

By any reasonable measure of achievement, the faith of the Enlightenment thinkers in science was justified. Today the greatest divide within humanity is not between races, or religions, or even, as widely believed, between the literate and illiterate. It is the chasm that separates scientific from prescientific cultures. Without the instruments and accumulated knowledge of the natural sciences—physics, chemistry, and biology—humans are trapped in a cognitive prison. They are like intelligent fish born in a deep, shadowed pool. Wondering and restless, longing to reach out, they think about the world outside. They invent ingenious speculations and myths about the origin of the confining waters, of the sun and the sky and the stars above, and the meaning of their own existence. But they are wrong, always wrong, because the world is too remote from ordinary experience to be merely imagined.

Science is neither a philosophy nor a belief system. It is a combination of mental operations that has become increasingly the habit of educated peoples, a culture of illuminations hit upon by a fortunate turn of history that yielded the most effective way of learning about the real world ever conceived.

With instrumental science humanity has escaped confinement and prodigiously extended its grasp of physical reality. Once we were nearly blind;

now we can see—literally. Visible light, we have learned, is not the sole illuminating energy of the universe, as prescientific common sense decreed. It is instead an infinitesimal sliver of electromagnetic radiation, comprising wavelengths of 400 to 700 nanometers (billionths of a meter), within a spectrum that ranges from gamma waves trillions of times shorter to radio waves trillions of times longer. Radiation over most of this span, in wildly varying amounts, continually rains down on our bodies. But without instruments we were oblivious to its existence. Because the human retina is rigged to report only 400–700 nanometers, the unaided brain concludes that only visible light exists.

Many kinds of animals know better. They live in a different visual world, oblivious to part of the human visible spectrum, sensitive to some wavelengths outside it. Below 400 nanometers, butterflies find flowers and pinpoint pollen and nectar sources by the pattern of ultraviolet rays reflected off the petals. Where we see a plain yellow or white blossom, they see spots and concentric circles in light and dark. The patterns have evolved in plants to guide insect pollinators to the anthers and nectar pools.

With the aid of appropriate instruments we can now view the world with butterfly eyes.

Scientists have entered the visual world of animals and beyond because they understand the electromagnetic spectrum. They can translate any wavelength into visible light and audible sound, and generate most of the spectrum from diverse energy sources. By manipulating selected segments of the electromagnetic spectrum they peer downward to the trajectories of subatomic particles and outward to star birth in distant galaxies whose incoming light dates back to near the beginning of the universe. They (more accurately we, since scientific knowledge is universally available) can visualize matter across thirty-seven orders of magnitude. The largest galactic cluster is larger than the smallest known particle by a factor of the number one with about thirty-seven zeroes following it.

I mean no disrespect when I say that prescientific people, regardless of their innate genius, could never guess the nature of physical reality beyond the tiny sphere attainable by unaided common sense. Nothing else ever worked, no exercise from myth, revelation, art, trance, or any other conceivable means; and notwithstanding the emotional satisfaction it gives, mysticism, the strongest prescientific probe into the unknown, has yielded zero. No shaman's spell or fast upon a sacred mountain can summon the electromagnetic spectrum. Prophets of the great religions were kept unaware of its existence, not because of a secretive god but because they lacked the hard-won knowledge of physics.

Is this a paean to the god of science? No—to human ingenuity, to the capacity in all of us, freed at last in the modern era. And to the fortunate comprehensibility of the universe. The signature achievement of humanity has been to find its way without assistance through a world that proved surprisingly well ordered.

All our other senses have been expanded by science. Once we were deaf; now we can hear everything. The human auditory range is 20 to 20,000 Hz, or cycles of air compression per second. Above that range, flying bats broadcast ultrasonic pulses into the night air and listen for echoes to locate moths and other insects on the wing. Many of their potential prey listen with ears tuned to the same frequencies as the bats. When they hear the telltale pulses, they dip and wheel in evasive maneuvers or else power-dive to the ground. Before the 1950s, zoologists were unaware of this nocturnal contest. Now, with receivers, transformers, and night-time photography they can follow every squeak and aerial roll-out.

We have even uncovered basic senses entirely outside the human repertory. Where humans detect electricity only indirectly by a tingling of skin or flash of light, the electric fishes of Africa and South America, a medley of freshwater eels, catfish, and elephant-nosed fishes, live in a galvanic world. They generate charged fields around their bodies with trunk muscle tissue that has been modified by evolution into organic batteries. The power is controlled by a neural switch. Each time the switch turns on the field, individual fish sense the resulting power with electroreceptors distributed over their bodies. Perturbations caused by nearby objects, which cast electric shadows over the receptors, allow them to judge size, shape, and movement. Thus continuously informed, the fish glide smoothly past obstacles in dark water, escape from enemies, and target prey. They also communicate with one another by means of coded electrical bursts. Zoologists, using generators and detectors, can join the conversation. They are able to talk as through a fish's skin.

FROM THESE AND countless other examples can be drawn an informal rule of biological evolution important to the understanding of the human condition: If an organic sensor can be imagined that picks up any signal from the environment, there exists a species somewhere that possesses it. The bountiful powers of life expressed in such diversity raise a question about the incapacity of the unaided human senses: Why can't our species, the supposed *summum bonum* of Creation, do as much as all the animals combined, and more? Why were we brought into the world physically handicapped?

Evolutionary biology offers a simple answer. Natural selection, defined as the differential survival and reproduction of different genetic forms, prepares organisms only for necessities. Biological capacity evolves until it maximizes the fitness of organisms for the niches they fill, and not a squiggle more. Every species, every kind of butterfly, bat, fish, and primate, including *Homo sapiens*, occupies a distinctive niche. It follows that each species lives in its own sensory world. In shaping that world, natural selection is guided solely by the conditions of past history and by events occurring moment to moment then and now. Because moths are too small and indigestible to be energetically efficient food for large primates, *Homo sapiens* never evolved echolocation to catch them. And since we do not live in dark water, an electrical sense was never an option for our species.

Natural selection, in short, does not anticipate future needs. But this principle, while explaining so much so well, presents a difficulty. If the principle is universally true, how did natural selection prepare the mind for civilization before civilization existed? That is the great mystery of human evolution: how to account for calculus and Mozart.

Later I will attempt an answer by expanding the evolutionary explanation to embrace culture and technological innovation. For the moment, let me soften the problem somewhat by addressing the peculiar nature of the natural sciences as a product of history. Three preconditions, three strokes of luck in the evolutionary arena, led to the scientific revolution. The first was the boundless curiosity and creative drive of the best minds. The second was the inborn power to abstract the essential qualities of the universe. This ability was possessed by our Neolithic ancestors, but (again, here the primary puzzle) seemingly developed beyond their survival needs. In just three centuries, from 1600 to 1900, too short a time for improvement of the human brain by genetic evolution, humankind launched the technoscientific age.

The third enabling precondition is what the physicist Eugene Wigner once called the unreasonable effectiveness of mathematics in the natural sciences. For reasons that remain elusive to scientists and philosophers alike, the correspondence of mathematical theory and experimental data in physics in particular is uncannily close. It is so close as to compel the belief that mathematics is in some deep sense the natural language of science. "The enormous usefulness of mathematics in the natural sciences," Wigner wrote, "is something bordering on the mysterious and there is no rational explanation for it. It is not at all natural that 'laws of nature' exist, much less that man is able to discover them. The miracle of the appropriateness of the language of mathematics for the formulation of the laws of physics is a wonderful gift which we neither understand nor deserve."

The laws of physics are in fact so accurate as to transcend cultural differences. They boil down to mathematical formulae that cannot be given Chinese or Ethiopian or Mayan nuances. Nor do they cut any slack for masculinist or feminist variations. We may even reasonably suppose that any advanced extraterrestrial civilizations, if they possess nuclear power and can launch spacecraft, have discovered the same laws, such that their physics could be translated isomorphically, point to point, set to point, and point to set, into human notation.

The greatest exactitude of all has been obtained in measurements of the electron. A single electron is almost unimaginably small. Abstracted into a probabilistic packet of wave energy, it is also nearly impossible to visualize (as is the case generally for phenomena in quantum physics) within the conventional cognitive framework of objects moving in three-dimensional space. Yet we know with confidence that it has a negative charge of 0.16 billion-billionth (-1.6×10^{-19}) coulomb and a rest mass of 0.91 billion-billion-billionth (9.1×10^{-28}) gram. From these and other verifiable quantities have been accurately deduced the properties of electric currents, the electromagnetic spectrum, the photoelectric effect, and chemical bonding.

The theory that unites such basic phenomena is an interlocking set of graphical representations and equations called quantum electrodynamics (Q.E.D.). Q.E.D. treats the position and momentum of each electron as both a wave function and a discrete particle in space. The electron is further envisioned in Q.E.D. as randomly emitting and reabsorbing photons, the unique massless particles that carry the electromagnetic force.

In one property of the electron, its magnetic moment, theory and experiment have been matched to the most extreme degree ever achieved in the physical sciences. The magnetic moment is a measure of the interaction between an electron and a magnetic field. More precisely, it is the maximum torque experienced by the electron divided by the magnetic induction acting on it. The quantity of interest is the gyromagnetic ratio, the magnetic moment divided in turn by the angular momentum. Theoretical physicists predicted the value of the gyromagnetic ratio with calculations incorporating both special relativity and perturbations from photon emission and resorption, the two phenomena expected from Q.E.D. to cause small deviations from the ratio previously predicted by classical atomic physics.

For their part, and independently, atomic scientists directly measured the gyromagnetic ratio. In a technical tour de force, they trapped single electrons inside a magnetic-electric bottle and studied them for long periods of time. Their data matched the theoretical prediction to one part in a hundred billion. Together the theoretical and experimental physicists accomplished the equivalent of launching a needle due east from San Francisco and correctly calling in advance where it would strike (near Washington, D.C.) to within the width of a human hair.

THE DESCENT TO minutissima, the search for ultimate smallness in entities such as electrons, is a driving impulse of Western natural science. It is a kind of instinct. Human beings are obsessed with building blocks, forever pulling them apart and putting them back together again. The impulse goes as far back as 400 B.C. to the first protoscience, when Leucippus and his student Democritus speculated, correctly as it turned out, that matter is made of atoms. Reduction to microscopic units has been richly consummated in modern science.

The search for the ultimate has been aided through direct visual observation by steady advances in the resolving power of microscopes. This technological enterprise satisfies a second elemental craving: to see all the world with our own eyes. The most powerful of modern instruments, invented during the 1980s, are the scanning-tunneling microscope and atomic force microscope, which provide an almost literal view of atoms bonded into molecules. A DNA double helix can now be viewed exactly as it is, including every twist and turn into which a particular molecule fell as the technician prepared it for study. Had such visual techniques existed fifty years ago, the infant science of molecular biology would have escalated even more sharply than it has. In science, as in whist and bridge, one peek is worth a hundred finesses.

Atomic-level imaging is the end product of three centuries of technological innovation in search of the final peek. Microscopy began with the primitive optical instruments of Anton van Leeuwenhoek, which in the late 1600s revealed bacteria and other objects a hundred times smaller than the resolution of the human eye. It has arrived at methods for showing objects a million times smaller.

The passion for dissecting and reassembling has resulted in the invention of nanotechnology, the manufacture of devices composed of a relatively small number of molecules. Among the more impressive recent achievements are:

• Etching stainless steel pins with ion beams, Bruce Lamartine and Roger Stutz of the Los Alamos National Laboratory have created high-density ROMs ("read-only memories"), whose lines are cut so fine, down to 150 billionths of a meter, as to allow the storage of two gigabytes of data on a pin 25 millimeters long and 1 millimeter wide. Since the materials are nonmagnetic, the information thus stored is nearly indestructible. Yet there is still a long way to go. In theory at least, atoms can be ordered to store knowledge.

• A fundamental question in chemistry since the work of Lavoisier in the eighteenth century has been the following: How long does it take a pair of molecules to meet and bond when different reagents are mixed together? By confining solutions to extremely small spaces, Mark Wightman and his fellow researchers at the University of North Carolina observed flashes of light that mark the contact of oppositely charged reagent molecules, enabling the chemists to time the reactions with unprecedented accuracy.

• Molecule-sized machines that assemble themselves under the direction of technicians have for many years been considered a theoretical possibility. Now the ensembles are being realized in practice. One of the most promising techniques, engineered by George M. Whitesides of Harvard University and other organic chemists, consists in self-assembled monolayers. The SAMs (for short) consist of sausage-shaped molecules such as long hydrocarbon chains called alkanethiols. After synthesis in the laboratory the substances are painted onto a gold surface. One end of each molecule has properties that cause it to adhere to the gold; the other end, built of atoms with different properties, projects outward into space. Thus lined up like soldiers on parade, molecules of the same kind create a single layer only one to two nanometers thick. Molecules of a different construction are next laid down to create a second layer on top of the first, and so on, compound by compound, to produce a stratified film of desired thickness and chemical properties. SAMs share some of the basic properties of membranes of living cells. Their construction suggests one possible step in the eventual assembly of simple artificial organisms. Although far from being alive, SAMs are simulacra of elemental pieces of life. Given enough such components assembled the right way, chemists may someday produce a passable living cell.

THE INTELLECTUAL THRUST of modern science and its significance for the consilient world view can be summarized as follows. In the ultimate sense our brain and sensory system evolved as a biological apparatus to preserve and multiply human genes. But they enable us to navigate only through the tiny segment of the physical world whose mastery serves that primal need. Instrumental science has removed the handicap. Still, science in its fullness is much more than just the haphazard expansion of sensory capacity by instruments. The other elements in its creative mix are classification of data and their interpretation by theory. Together they compose the rational processing of sensory experience enhanced by instrumentation.

Nothing in science—nothing in life, for that matter—makes sense without theory. It is our nature to put all knowledge into context in order to tell a story, and to re-create the world by this means. So let us visit the topic of theory for a moment. We are enchanted by the beauty of the natural world. Our eye is caught by the dazzling visual patterns of polar star trails, for example, and the choreography of chromosomes in dividing root tip cells of a plant. Both disclose processes that are also vital to our lives. In unprocessed form, however, without the theoretical frameworks of heliocentric astronomy and Mendelian heredity, they are no more than beautiful patterns of light.

Theory: a word hobbled by multiple meanings. Taken alone without *a* or *the*, it resonates with erudition. Taken in everyday context, it is shot through with corrupting ambiguity. We often hear that such and such an assertion is only a theory. Anyone can have a theory; pay your money and take your choice among the theories that compete for your attention. Voodoo priests sacrificing chickens to please spirits of the dead are working with a theory. So are millenarian cultists watching the Idaho skies for signs of the Second Coming. Because scientific theories contain speculation, they too may seem just more guesswork, and therefore built on sand. That, I suspect, is the usual postmodernist conception: Everyone's theory has validity and is interesting. Scientific theories, however, are fundamentally different. They are constructed specifically to be blown apart if proved wrong, and if so destined, the sooner the better. "Make your mistakes quickly" is a rule in the practice of science. I grant that scientists often fall in love with their own constructions. I know; I have. They may spend a lifetime vainly trying to shore them up. A few squander their prestige and academic political capital in the effort. In that case—as the economist Paul Samuelson once quipped—funeral by funeral, theory advances.

Quantum electrodynamics and evolution by natural selection are examples of successful big theories, addressing important phenomena. The entities they posit, such as photons, electrons, and genes, can be measured. Their statements are designed to be tested in the acid washes of skepticism, experiments, and the claims of rival theories. Without this vulnerability, they will not be accorded the status of scientific theories. The best theories are rendered lean by Occam's razor, first expressed in the 1320s by William of Occam. He said, "What can be done with fewer assumptions is done in vain with more." Parsimony is a criterion of good theory. With lean, tested theory we no longer need Phoebus in a chariot to guide the sun across the sky, or dryads to populate the boreal forests. The practice grants less license for New Age dreaming, I admit, but it gets the world straight.

Still, scientific theories are a product of imagination—*informed* imagination. They reach beyond their grasp to predict the existence of previously unsuspected phenomena. They generate hypotheses, disciplined guesses about unexplored topics whose parameters the theories help to define. The best theories generate the most fruitful hypotheses, which translate cleanly into questions that can be answered by observation and experiment. Theories and their progeny hypotheses compete for the available data, which comprise the limiting resource in the ecology of scientific knowledge. The survivors in this tumultuous environment are the Darwinian victors, welcomed into the canon, settling in our minds, guiding us to further exploration of physical reality, more surprises. And yes, more poetry.

Science, to put its warrant as concisely as possible, is the *organized, systematic enterprise that gathers knowledge about the world and condenses the knowledge into testable laws and principles.* The diagnostic features of science that distinguish it from pseudoscience are first, repeatability: The same phenomenon is sought again, preferably by independent investigation, and the interpretation given to it is confirmed or discarded by means of novel analysis and experimentation. Second, economy: Scientists attempt to abstract the information into the form that is both simplest and aesthetically most pleasing—the combination called elegance—while yielding the largest amount of information with the least amount of effort. Third, mensuration: If something can be properly measured, using universally accepted scales, generalizations about it are rendered unambiguous. Fourth, heuristics: The best science stimulates further discovery, often in unpredictable new directions; and the new knowledge provides an additional test of the original principles that led to its discovery. Fifth and finally, consilience: The explanations of different phenomena most likely to survive are those that can be connected and proved consistent with one another.

Astronomy, biomedicine, and physiological psychology possess all these criteria. Astrology, ufology, creation science, and Christian Science, sadly, possess none. And it should not go unnoticed that the true natural sciences lock together in theory and evidence to form the ineradicable technical base of modern civilization. The pseudosciences satisfy personal psychological needs, for reasons I will explain later, but lack the ideas or the means to contribute to the technical base.

THE CUTTING EDGE of science is reductionism, the breaking apart of nature into its natural constituents. The very word, it is true, has a sterile and invasive ring, like scalpel or catheter. Critics of science sometimes portray reductionism as an obsessional disorder, declining toward a terminal stage one writer recently dubbed "reductive megalomania." That characterization is an actionable misdiagnosis. Practicing scientists, whose business is to make verifiable discoveries, view reductionism in an entirely different way: It is the search strategy employed to find points of entry into otherwise impenetrably complex systems. Complexity is what interests scientists in the end, not simplicity. Reductionism is the way to understand it. The love of complexity without reductionism makes art; the love of complexity with reductionism makes science.

Here is how reductionism works most of the time, as it might appear in a user's manual. *Let your mind travel around the system. Pose an interesting question about it. Break the question down and visualize the elements and questions it implies. Think out alternative conceivable answers. Phrase them so that a reasonable amount of evidence makes a clear-cut choice possible. If too many conceptual difficulties are encountered, back off. Search for another question. When you finally hit a soft spot, search for the model system—say a controlled emission in particle physics or a fast-breeding organism in genetics—on which decisive experiments can be most easily conducted. Become thoroughly familiar—no, better, become obsessed—with the system. Love the details, the feel of all of them, for their own sake. Design the experiment so that no matter what the result, the answer to the question will be convincing. Use the result to press on to new questions, new systems. Depending on how far others have already gone in this sequence (and always keep in mind, you must give them complete credit), you may enter it at any point along the way.*

Followed more or less along these lines, reductionism is the primary and essential activity of science. But dissection and analysis are not all that scientists do. Also crucial are synthesis and integration, tempered by philosophical reflection on significance and value. Even the most narrowly focused researchers, including those devoted to the search for elemental units, still think all the time about complexity. To make any progress they must meditate on the networks of cause and effect across adjacent levels of organization—from subatomic particles to atoms, say, or organisms to species—and they must think on the hidden design and forces of the networks of causation. Quantum physics thus blends into chemical physics, which explains atomic bonding and chemical reactions, which form the foundation of molecular biology, which demystifies cell biology.

Behind the mere smashing of aggregates into smaller pieces lies a deeper agenda that also takes the name of reductionism: to fold the laws and principles of each level of organization into those at more general, hence more fundamental levels. Its strong form is total consilience, which holds that nature is organized by simple universal laws of physics to which all other laws and principles can eventually be reduced. This transcendental world view is the light and way for many scientific materialists (I admit to being among them), but it could be wrong. At the least, it is surely an oversimplification. At each level of organization, especially at the living cell and above, phenomena exist that require new laws and principles, which still cannot be predicted from those at more general levels. Perhaps some of them will remain forever beyond our grasp. Perhaps prediction of the most complex systems from more general levels is impossible. That would not be all bad. I will confess with pleasure: The challenge and the crackling of thin ice are what give science its metaphysical excitement.

SCIENCE, its imperfections notwithstanding, is the sword in the stone that humanity finally pulled. The question it poses, of universal and orderly materialism, is the most important that can be asked in philosophy and religion. Its procedures are not easy to master, even to conceptualize; that is why it took so long to get started, and then mostly in one place, which happened to be western Europe. The work is also hard and for long intervals frustrating. You have to be a bit compulsive to be a productive scientist. Keep in mind that new ideas are commonplace, and almost always wrong. Most flashes of insight lead nowhere; statistically, they have a half-life of hours or maybe days. Most experiments to follow up the surviving insights are tedious and consume large amounts of time, only to yield negative or (worse!) ambiguous results. Over the years I have been presumptuous enough to counsel new Ph.D.'s in biology as follows: If you choose an academic career you will need forty hours a week to perform teaching and administrative duties, another twenty hours on top of that to conduct respectable research, and still another twenty hours to accomplish really important research. This formula is not boot-camp rhetoric. More than half the Ph.D.'s in science are stillborn, dropping out of original research after at most one or two publications. Percy Bridgman, the founder of high-pressure physics—no pun intended—put the guideline another way: "The scientific method is doing your damnedest, no holds barred."

Original discovery is everything. Scientists as a rule do not discover in order to know but rather, as the philosopher Alfred North Whitehead observed, they know in order to discover. They learn what they need to know, often remaining poorly informed about the rest of the world, including most of science for that matter, in order to move speedily to some part of the frontier of science where discoveries are made. There they spread out like foragers on a picket line, each alone or in small groups probing a carefully chosen, narrow sector. When two scientists meet for the first time the usual conversation entry is, "What do you work on?" They already know what generally bonds them. They are fellow prospectors pressing deeper into an abstracted world, content most of the time to pick up an occasional nugget but dreaming of the mother lode. They come to work each day thinking subconsciously, *It's there, I'm close, this could be the day.*

They know the first rule of the professional game book: Make an important discovery, and you are a successful scientist in the true, elitist sense in a profession where elitism is practiced without shame. You go into the textbooks. Nothing can take that away; you may rest on your laurels for the rest of your life. But of course you won't; almost no one driven enough to make an important discovery ever rests. And any discovery at all is thrilling. There is no feeling more pleasant, no drug more addictive, than setting foot on virgin soil.

Fail to discover, and you are little or nothing in the culture of science, no matter how much you learn and write about science. Scholars in the humanities also make discoveries, of course, but their most original and valuable scholarship is usually the interpretation and explanation of already existing knowledge. When a scientist begins to sort out knowledge in order to sift for meaning, and especially when he carries that knowledge outside the circle of discoverers, he is classified as a scholar in the humanities. Without scientific discoveries of his own, he may be a veritable archangel among intellectuals, his broad wings spread above science, and still not be in the circle. The true and final test of a scientific career is how well the following declarative sentence can be completed: *He (or she) discovered that* . . . A fundamental distinction thus exists in the natural sciences between process and product. The difference explains why so many accomplished scientists are narrow, foolish people, and why so many wise scholars in the field are considered weak scientists.

Yet, oddly, there is very little science *culture*, at least in the strict tribal sense. Few rites are performed to speak of. There is at most only a scattering of icons. One does, however, hear a great deal of bickering over territory and status. The social organization of science most resembles a loose confederation of petty fiefdoms. In religious belief, individual scientists vary from born-again Christians, admittedly rare, to hard-core atheists, very common. Few are philosophers. Most are intellectual journeymen, exploring locally, hoping for a strike, living for the present. They are content to work at discovery, often teaching science at the college level, pleased to be relatively well-paid members of one of the more contentious but overall least conspiratorial of professions.

In character they are as variable as the population at large. Take any random sample of a thousand and you will find the near-full human range on every axis of measurement—generous to predatory, well adjusted to psychopathic, casual to driven, grave to frivolous, gregarious to reclusive. Some are as stolid as tax accountants in April, while a few are clinically certifiable as manic-depressives (or bipolars, to use the ambiguous new term).

In motivation they run from venal to noble. Einstein classified scientists very well during the celebration of Max Planck's sixtieth birthday in 1918. In the temple of science, he said, are three kinds of people. Many take to science out of a joyful sense of their superior intellectual power; for them, research is a kind of sport that satisfies personal ambition. A second class of researchers engages in science to achieve purely utilitarian ends. But of the third: If "the angel of the Lord were to come and drive all the people belonging to these two categories out of the temple, a few people would be left, including Planck, and that is why we love him."

Scientific research is an art form in this sense: It does not matter how you make a discovery, only that your claim is true and convincingly validated. The ideal scientist thinks like a poet and works like a bookkeeper, and I suppose that if gifted with a full quiver, he also writes like a journalist. As a painter stands before bare canvas or a novelist recycles past emotion with eyes closed, he searches his imagination for subjects as much as for conclusions, for questions as much as for answers. Even if his highest achievement is only to perceive the need for a new instrument or theory, that may be enough to open the door to a new industry of research.

This level of creativity in science, as in art, depends as much on self-image as on talent. To be highly successful the scientist must be confident enough to steer for blue water, abandoning sight of land for a while. He values risk for its own sake. He keeps in mind that the footnotes of forgotten treatises are strewn with the names of the gifted but timid. If on the other hand he chooses, like the vast majority of his colleagues, to hug the coast, he must be fortunate enough to possess what I like to define as optimum intelligence for normal science: bright enough to see what needs to be done but not so bright as to suffer boredom doing it.

The scientist's style of investigation is the product of the discipline he chooses, further narrowed by aptitude and taste. If a naturalist at heart, he saunters at random, sometimes through real woods thick with trees, or, more commonly nowadays, cells thick with molecules, in search of objects and happenings still unimagined. His instinct is that of the hunter. If on the other hand the scientist is a mathematical theorist, he creates a mental picture of a known but still poorly understood process, skeletonizes it into what intuition suggests are its essential elements, and recasts it in diagrams and equations. He looks for vindication, by saying to the experimentalists: If this is the way the process works, even if we cannot see it directly, then here are the parameters for an indirect probe, and the language by which we might come to explain the results.

Differences in validation criteria across the disciplines are accordingly vast. Systematic biologists need only stumble upon an unusual new species, and recognize its novelty, to make an important discovery. In 1995 two Danish zoologists erected an entirely new phylum of animals, the thirty-fifth known, from a species of tiny rotiferlike creatures found living on the mouthparts of lobsters. In a wholly different domain, and style, biochemists regularly trace the natural syntheses of hormones and other biologically important molecules by duplicating the steps with enzymatically mediated reactions in the laboratory. Experimental physicists, even further removed than chemists from direct perception, and hence the most esoteric among the scientific multitude, deduce (to take a properly esoteric example) the spatial distribution of quarks from high-energy collisions of electrons with protons of atomic nuclei.

Advice to the novice scientist: There is no fixed way to make and establish a scientific discovery. Throw everything you can at the subject, as long as the procedures can be duplicated by others. Consider repeated observations of a physical event under varying circumstances, experiments in different modes and styles, correlation of supposed causes and effects, statistical analyses to reject null hypotheses (those deliberately raised to threaten the conclusion), logical argument, and attention to detail and consistency with the results published by others. All these actions, singly and in combination, are part of the tested and true armamentarium of science. As the work comes together, also think about the audience to whom it will be reported. Plan to publish in a reputable, peer-reviewed journal. One of the strictures of the scientific ethos is that a discovery does not exist until it is safely reviewed and in print.

SCIENTIFIC EVIDENCE IS accretionary, built from blocks of evidence joined artfully by the blueprints and mortar of theory. Only very rarely, as in the theories of natural selection and relativity, does an idea change our conception of the world in one quantal leap. Even the revolution of molecular biology was accretionary, building upon but not fundamentally altering physics and chemistry.

Few claims in science, and particularly those entailing concepts, are accepted as final. But as evidence piles upon evidence and theories interlock more firmly, certain bodies of knowledge do gain universal acceptance. In seminar patois they ascend a scale of credibility from "interesting" to "suggestive" to "persuasive" and finally "compelling." And given enough time thereafter, "obvious."

No objective yardstick exists on which to mark these degrees of acceptance; there is no body of external objective truth by which they can be calibrated. There is only warranted assertibility, to use William James' phrase, within which particular descriptions of reality grow ever more congenial to scientists until objections cease. A proof, as the mathematician Mark Kac once put it, is that which convinces a reasonable man; a rigorous proof is that which convinces an unreasonable man.

It is occasionally possible to encapsulate a method of science as a recipe. The most satisfying is that based on multiple competing hypotheses, also known as strong inference. It works only on relatively simple processes under restricted circumstances and particularly in physics and chemistry, where context and history are unlikely to affect the outcome. The phenomenon under scrutiny is known to occur but cannot be seen directly, with the result that its exact nature can only be guessed. Investigators think out every possible way the process might occur—the multiple competing hypotheses—and devise tests that will eliminate all but one.

In a celebrated 1958 example, Matthew Meselson and Franklin Stahl, then at the California Institute of Technology, used the method to demonstrate the steps by which DNA molecules duplicate themselves. I will first give their conclusion: The double helix splits lengthwise to create two single helices; each single helix then assembles a new partner to create another double helix. Alternative hypotheses, that the double helix duplicates itself in its entirety or that the single helices are broken and dispersed by the duplication process, must be discarded.

Now the proof, which despite its technical content is elegantly simple. Having phrased what in retrospect turns out to have been the right question, Meselson and Stahl devised the right experiment to make a choice among the competing alternatives. They first let bacteria that had manufactured DNA molecules in a heavy-nitrogen medium continue their multiplication in a normal-nitrogen medium. The researchers then extracted the molecules and centrifuged them in a cesium chloride solution that formed a gradient of density. DNA molecules built by the bacteria with heavy nitrogen settled deeper into the cesium chloride density gradient than did otherwise identical DNA molecules built by the same bacteria with normal nitrogen. When equilibrium was reached, the DNA had separated out into sharply defined bands in a pattern that exactly fit the hypothesis of single-helix separation and double-helix regeneration. The pattern eliminated the two competing hypotheses of whole-molecule duplication and fragmentation followed by dispersion of the fragments.

Science, even in the relatively tidy world of molecular genetics, is a patchwork of such arguments and proofs. But perhaps there are common elements in its methods. Can we devise a universal litmus test for scientific statements and with it eventually attain the grail of objective truth? Current opinion holds that we cannot and never will. Scientists and philosophers have largely abandoned the search for absolute objectivity and are content to ply their trade elsewhere.

I think otherwise and will risk heresy: The answer could well be yes. Criteria of objective truth might be attainable through empirical investigation. The key lies in clarifying the still poorly understood operations composing the mind and in improving the piecemeal approach science has taken to its material properties.

Here is the argument. Outside our heads there is freestanding reality. Only madmen and a scattering of constructivist philosophers doubt its existence. Inside our heads is a reconstitution of reality based on sensory input and the self-assembly of concepts. Input and self-assembly, rather than an independent entity in the brain—the "ghost in the machine," in the philosopher Gilbert Ryle's famous derogation—constitute the mind. The alignment of outer existence with its inner representation has been distorted by the idiosyncrasies of human evolution, as I noted earlier. That is, natural selection built the brain to survive in the world and only incidentally to understand it at a depth greater than is needed to survive. *The proper task of scientists is to diagnose and correct the misalignment.* The effort to do so has only begun. No one should suppose that objective truth is impossible to attain, even when the most committed philosophers urge us to acknowledge that incapacity. In particular it is too early for scientists, the foot soldiers of epistemology, to yield ground so vital to their mission.

Although seemingly chimerical at times, no intellectual vision is more important and daunting than that of objective truth based on scientific understanding. Or more venerable. Argued at length in Greek philosophy, it took modern form in the eighteenth-century Enlightenment hope that science would find the laws governing all physical existence. Thus empowered, the savants believed, we could clear away the debris of millennia, including all the myths and false cosmologies that encumber humanity's self-image. The Enlightenment dream faded before the allure of Romanticism; but, even more important, science could not deliver in the domain most crucial to its promise, the physical basis of mind. The two failings worked together in a devastating combination: People are innate romantics, they desperately need myth and dogma, and scientists could not explain why people have this need.

As the nineteenth century closed, the dream of objective truth was rekindled by two philosophies. The first, European in origin, was positivism, the conviction that the only certain knowledge is the exact description of what we perceive with our senses. The second, American in origin, was pragmatism, the belief that truth is what consistently works in human action. From the outset both positions were symbiotic with science. They drew major strength from the spectacular advances in the physical sciences then underway, which vindicated them by the varied actions—electromagnetic motors, X-rays, reagent chemistry—that exact, practical knowledge made possible.

The dream of objective truth peaked soon afterward with the formulation of logical positivism, a variation on general positivism that attempted to define the essence of scientific statements by means of logic and the analysis of language. Although many thinkers contributed to the movement, its driving force was the Vienna Circle, a group of mostly Austrian intellectuals founded by the philosopher Moritz Schlick in 1924. Regular meetings of the

Circle continued until Schlick's death in 1936 and the subsequent dispersion of its members and correspondents, some of whom emigrated to America as exiles from the Nazi regime.

On September 3–9, 1939, many of the scholars sympathetic to logical positivism met at Harvard University to attend the fifth International Congress for the Unity of Science. It was a scintillating assemblage of names now enshrined in the history of ideas: Rudolf Carnap, Phillip Frank, Susanne Langer, Richard von Mises, Ernest Nagel, Otto Neurath, Talcott Parsons, Willard van Quine, and George Sarton. The conferees must have been badly distracted by the invasion of Poland, which began two days before the meeting started. Where the Napoleonic campaigns weakened the plausibility of the original Enlightenment, now a savage war of territorial conquest fired by a pseudoscientific theory of racial superiority threatened to make a still greater mockery of the power of reason. The scholars persisted, however, in exploring the idea that rationally acquired knowledge is the best hope of humanity.

How then, they asked, to distill the scientific ethos? The movement created by the Vienna Circle had worked at two levels over the years. First was the reaffirmation of the core Enlightenment ideal that the cause of the human species is best served by unblinking realism. Having "no protectors or enemies," in Carnap's expression, humanity must find its way to transcendent existence solely by its own intelligence and will. Science is simply the best instrument at our disposal. As the Vienna Circle had declared a decade earlier, "the scientific world conception serves life and in turn is taken up by life."

The second level, requisite to the first, was the search for pure standards against which scientific knowledge can be judged. Every symbol, the logical positivists concluded, should denote something real. It should be consistent with the total structure of established facts and theories, with no revelations or free-flight generalizing allowed. Theory must follow in lockstep with facts. Finally, the informational content of language is to be carefully distinguished from its emotional content. To these various ends verification is all important—indeed, the very meaning of a statement is its method of verification. If the guidelines are progressively refined and followed, we will in time close in on objective truth. While this happens, ignorance-based metaphysics will back away step by step, like a vampire before the lifted cross.

The logical positivists who met in Cambridge knew that pure mathematics was on the road to the grail but not the prize itself. Mathematics, for all its unchallengeable power in framing theory, is tautological. That is, every conclusion follows completely from its own premises, which may or may not have anything to do with the real world. Mathematicians invent and prove lemmas and theorems that lead to other lemmas and theorems, and onward with no end in sight. Some fit data from the material world, some do not. The greatest mathematicians are intellectual athletes of dazzling skill. Sometimes they hit upon concepts that open new domains of abstract thought. Complex numbers, linear transformations, and harmonic functions are among those that have proved most interesting mathematically as well as useful to science.

Pure mathematics is the science of all conceivable worlds, a logically closed system yet infinite in all directions allowed by starting premises. With it we might, if given unlimited time and computational capacity, describe every imaginable universe. But mathematics alone cannot inform us of the very special world in which we live. Only observation can disclose the periodic table, the Hubble constant, and all the other certainties of our existence, which may be different or nonexistent in other universes. Because physics, chemistry, and biology are constrained by the parameters of this universe, the one we see from inside the Milky Way, they compose the science of all possible phenomena tangible to us.

Still, because of its effectiveness in the natural sciences, mathematics seems to point arrowlike toward the ultimate goal of objective truth. The logical positivists were especially impressed by the tight meshing of observation with abstract mathematical theory in quantum and relativistic physics. This greatest of twentieth-century triumphs inspired new confidence in the inborn power of the human brain. Think of it. Here is *Homo sapiens*, a primate species barely out of its stone-age villages, correctly divining phenomena almost unimaginably beyond ordinary experience. Surely, the theorists reasoned, we are close to a general formula for objective truth.

Yet the grail eluded them. Logical positivism stumbled and halted. Today its analyses, while favored by a few, are more commonly studied in philosophy, as dinosaur fossils are studied in paleontology laboratories, to understand the causes of extinction. Its last stand may have been a seldom-read 1956 monograph by Carnap in *Minnesota Studies in the Philosophy of Science*. The fatal flaw was in the semantic linchpin of the whole system: The founders and their followers could not agree on the basic distinctions between fact and concept, between empirical generalization and mathematical truth, between theory and speculation, and from a collation of all these fog-shrouded dichotomies, the differences between scientific and nonscientific statements.

Logical positivism was the most valiant concerted effort ever mounted by modern philosophers. Its failure, or put more generously, its shortcoming, was caused by ignorance of how the brain works. That in my opinion is the whole story. No one, philosopher or scientist, could explain the physical acts of observation and reasoning in other than highly subjective terms. Not much has improved in the past fifty years. The mindscape is now under active exploration but still largely unmapped. Scientific discourse, the focus of logical positivism, comprises the most complex of mental operations, and the brain is a messy place at best even when handling the most elementary of ideas. Scientists themselves do not think in straight lines. They contrive concepts, evidence, relevance, connections, and analysis as they go along, parsing it all into fragments and in no particular order. Herbert Simon, a Nobelist who has devoted part of his career to the subject, says of the complexity of concept formation: "What chiefly characterizes creative thinking from more mundane forms are (i) willingness to accept vaguely defined problem statements and gradually structure them, (ii) continuing preoccupation with problems over a considerable period of time, and (iii) extensive background knowledge in relevant and potentially relevant areas."

To put that in a nutshell: knowledge, obsession, daring. The creative process is an opaque mix. Perhaps only openly confessional memoirs, still rare to nonexistent, might disclose how scientists actually find their way to a publishable conclusion. In one sense scientific articles are deliberately misleading. Just as a novel is better than the novelist, a scientific report is better than the scientist, having been stripped of all the confusions and ignoble thought that led to its composition. Yet such voluminous and incomprehensible chaff, soon to be forgotten, contains most of the secrets of scientific success.

The canonical definition of objective scientific knowledge avidly sought by the logical positivists is not a philosophical problem nor can it be attained, as they hoped, by logical and semantic analysis. It is an empirical question that can be answered only by a continuing probe of the physical basis of the thought process itself. The most fruitful procedures will almost certainly include the use of artificial intelligence, aided in time by the still embryonic field of artificial emotion, to simulate complex mental operations. This modeling system will be joined to an already swiftly maturing neurobiology of the brain, including the high-resolution scanning of computational networks active in various forms of thought. Important advances will also come from the molecular biology of the learning process.

If the exact biological processes of concept formation can be defined, we might devise superior methods of inquiry into both the brain and the world outside it. As a consequence we could expect to tighten the connectedness between the events and laws of nature and the physical basis of human thought processes. Might it be possible then to take the final step and devise an unassailable definition of objective truth? Perhaps not. The very idea is risky. It smells of absolutism, the dangerous Medusa of science and the humanities alike. Its premature acceptance is likely to be more paralyzing than its denial. But should we then be prepared to give up? Never! Better to steer by a lodestar than to drift across a meaningless sea. I think we will know if we come close to the goal of our predecessors, even if unattainable. Its glow will be caught in the elegance and beauty and power of our shared ideas and, in the best spirit of philosophical pragmatism, the wisdom of our conduct.

CHAPTER 5

ARIADNE'S THREAD

WITH THE AID of the scientific method, we have gained an encompassing view of the physical world far beyond the dreams of earlier generations. The great adventure is now beginning to turn inward, toward ourselves. In the last several decades the natural sciences have expanded to reach the borders of the social sciences and humanities. There the principle of consilient explanation guiding the advance must undergo its severest test. The physical sciences have been relatively easy; the social sciences and humanities will be the ultimate challenge. This uncertain conjunction of the disciplines has mythic elements that would have pleased the ancient Greeks: treacherous road, heroic journey, secret instructions that lead us home. The elements have been assembled into many narratives over the centuries. Among them is the story of the Cretan labyrinth, which can also serve as a metaphor of consilience.

Into the heart of the Cretan labyrinth walks Theseus, Heracles-like champion of Athens. Through each corridor, past uncounted twists and turns, he unravels a ball of thread given him by Ariadne, lovestruck daughter of Crete's King Minos. Somewhere in the hidden passages he meets the Minotaur, the cannibal half man, half bull to whom seven youths and maidens are sacrificed each year as Athens' tribute to Crete. Theseus kills the Minotaur with his bare hands. Then, following Ariadne's thread, he retraces his steps through and out of the labyrinth.

The labyrinth, its likely origin a prehistoric conflict between Crete and Attica, is a fitting mythic image of the uncharted material world in which humanity was born and which it forever struggles to understand. Consilience among the branches of learning is the Ariadne's thread needed to traverse it. Theseus is humanity, the Minotaur our own dangerous irrationality. Near the entrance of the labyrinth of empirical knowledge is physics, comprising one gallery, then a few branching galleries that all searchers undertaking the journey must follow. In the deep interior is a nebula of pathways through the social sciences, humanities, art, and religion. If the thread of connecting causal explanations has been well laid, it is nonetheless possible to follow any pathway quickly in reverse, back through the behavioral sciences to biology, chemistry, and finally physics.

With time, we discover that the labyrinth has a troubling peculiarity that makes its complete mastery impossible. While there is an entrance, more or less, there is no center, only an immense number of end points deep within the maze. In tracking the thread backward, from effect to cause, assuming we have enough knowledge to do so, we can begin with only one end point. The labyrinth of the real world is thus a Borgesian maze of almost infinite possibility. We can never map it all, never discover and explain everything. But we can hope to travel through the known parts swiftly, from the specific back to the general, and—in resonance with the human spirit—we can go on tracing pathways forever. We can connect threads into broadening webs of explanation, because we have been given the torch and the ball of thread.

There is another defining character of consilience: It is far easier to go backward through the branching corridors than to go forward. After segments of explanation have been laid one at a time, one level of organization to the next, to many end points (say, geological formations or species of butterflies) we can choose any thread and reasonably expect to follow it through the branching points of causation all the way back to the laws of physics. But the opposite journey, from physics to end points, is extremely problematic. As the distance away from physics increases, the options allowed by the antecedent disciplines increase exponentially. Each branching point of causal explanation multiplies the forward-bound threads. Biology is almost unimaginably more complex than physics, and the arts equivalently more complex than biology. To stay on course all the way seems impossible. And worse, we cannot know before departure whether the complete journey we have imagined even exists.

The accelerating growth of forward-bound complexity, from entrance to end points, is illustrated with textbook clarity by cell biology. Researchers have used the reductionist principles of physics and chemistry to explain cellular structure and activity in admirably brilliant detail, with no discernible room left for rival approaches. They expect in time to explain everything about any particular kind of cell chosen for study, reducing it organelle by organelle and finally reassembling it holistically, thus traveling toward the labyrinth entrance and simplicity. But they nourish faint hope of *predicting*—as opposed to explaining and reconstructing retrodictively—the character of any complete cell from physics and chemistry, hence traveling away from the labyrinth entrance toward rising complexity. To recite one of the mantras of science, the explanations of the physical sciences are necessary but not sufficient. There is too much idiosyncrasy in the arrangement of a particular cell's nucleus and other organelles as well as the molecules composing them, and too much complexity in the cell's constantly shifting chemical exchanges with the environment, to accomplish such a conceptual traverse. And beyond these particularities awaits the still-hidden history of the prescriptive DNA, stretched across countless generations.

Put briefly, the questions of interest are how the cell is put together and what was the evolutionary history that led to its prescription. In order to proceed, biologists are compelled first to describe complexity in the cell, then break it down. To go the other way is conceivable, but the biologists all agree it will be forbiddingly difficult.

To dissect a phenomenon into its elements, in this case cell into organelles and molecules, is consilience by reduction. To reconstitute it, and especially to predict with knowledge gained by reduction how nature assembled it in the first place, is consilience by synthesis. That is the two-step procedure by which natural scientists generally work: top down across two or three levels of organization at a time by analysis, then bottom up across the same levels by synthesis.

The procedure can be simply illustrated with a modest example from my own research. Ants alert one another to danger at a distance. When a worker ant is jostled, pinned to the ground, or otherwise threatened, nestmates up to several inches away somehow sense her distress and rush to her aid. ("Her," I say, because all workers are female.) Alarm can be communicated by sight but only rarely, since confrontations usually occur in the dark and in any case many kinds of ants are blind. The signal can also be transmitted by sound. Agitated workers make squeaking noises by rubbing their waists against a rear segment of their bodies, or else repeatedly pump their bodies up and down to strike the ground. But again, sound is used only by some species, and then only on special occasions.

Knowing these facts in the 1950s, as a beginning entomologist, I speculated that the key alarm signals are chemical. The substances are what researchers in those days called chemical releasers and today are known as pheromones. To test my idea, I collected colonies of red harvester ants and a few other species whose natural history I knew well. Then I installed them in artificial nests not much different from a child's ant farm. With the aid of a dissecting microscope and watchmaker's forceps I dissected freshly killed workers to obtain organs that might contain alarm pheromones. I crushed each one of these barely visible white gobbets of tissue onto the sharpened tips of applicator sticks and presented them in turn to resting groups of workers. In that way I learned that at least two of the glands are active. One opens at the base of the mandibles and the other next to the anus. The ants were galvanized by the substances released from the glands. They raced back and forth in whirligig loops around the applicator sticks, pausing only occasionally to examine and snap at the crushed tissue.

I had pinpointed the origin of the pheromones. But what were they? I enlisted the help of Fred Regnier, a chemist of like age just starting his own career. He was expert in the skills most needed at that time to advance the study of ant communication, the analysis of extremely small organic samples. Using the latest techniques of the day, gas chromatography and mass spectrometry, Regnier identified the active substances as a medley of simple compounds called alkanes and terpenoids. He then obtained samples of identical compounds that had been synthesized in the laboratory, guaranteeing their purity. Presenting minute quantities to the ant colonies, we obtained the same responses I had observed in my first experiments, and confirmed that the glandular components Regnier had identified were the alarm pheromones.

This information was the first step to the understanding of broader and more basic phenomena. I next enlisted the help of William Bossert, a young mathematician. (We were all young in those days; young scientists have the best ideas and, more important, the most time.) Intrigued by the novelty of the problem, as well as by the small stipend I offered him, he agreed to construct physical models of the diffusion of the pheromones. We knew that chemicals evaporate from the gland openings. The molecules closest to the openings are dense enough to be smelled by the ants. The three-dimensional domain within which this occurs we called the active space. The geometrical form of the active space can be predicted from knowledge of the physical properties of the molecules and confirmed by the time required for the expanding cloud of molecules to alert the ants. We used both the models and experiments to measure the rate of spread of the molecules and the sensitivity of the ants to them, and established with reasonable certainty that workers release evaporated pheromones in order to communicate.

The steps in reasoning we followed are universal in scientific research. They follow from the consilience of the disciplines established by generations of earlier scientists. To solve the problem of alarm communication in ants, we employed reduction, working our way down from one level of specific organization, namely the organism, to a more general level, the molecule. We tried to explain a phenomenon in biology with physics and chemistry. Luckily, our ideas succeeded, this time.

The same approach to pheromone research continued to be rewarded in the decades to follow. Scores of biologists working independently established that ants organize their colonies with many chemical systems like those used to transmit alarm. Their bodies, we discovered, are walking batteries of glands filled with semiotic compounds. When ants dispense their pheromones, singly or in combination and in varying amounts, they say to other ants, in effect: *danger, come quickly*; or *danger, disperse*; or *food, follow me*; or *there is a better nest site, follow me*; or *I am a nestmate, not an alien*; or *I am a larva*; and on through a repertoire of ten to twenty messages, with the number differing according to caste (such as soldier or minor worker) and species. So pervasive and powerful are these codes of taste and smell that all together they bind ant colonies into a single operational unit. As a result each colony can be viewed as a superorganism, a congeries of conventional organisms acting like a single and much larger organism. The colony is a primitive semiotic web that crudely resembles a nerve net, a hundred-mouthed hydra writ large. Touch one ant, one strand of the net, and the displacement spreads out to engage the communal intelligence.

We had crossed four levels—superorganism, to organism, to glands and sense organs, to molecules. Was it possible then to turn around and travel in the opposite direction, predicting the outcome without advance knowledge of the biology of the ants? Yes, at least in the form of a few broad principles. From the theory of natural selection, molecules serving as pheromones can be expected to possess certain properties that allow efficient manufacture and transmission. Adding in principles of organic chemistry, we concluded that the molecules will likely contain 5 to 20 carbon atoms and have molecular weights between 80 and 300. Molecules acting as alarm pheromones in particular will usually be on the light side. They will be produced in comparatively large quantities, for example millionths rather than billionths of a gram in each ant, and the responding workers will be less sensitive to them than to most other kinds of pheromones. This combination of traits allows quick transmission followed by a rapid fade-out of the signal after the danger passes. In contrast, trail substances, which are followed by the ants from nest to food and back, can be predicted to consist of molecules with the opposite qualities. Their traits allow long duration of the signal, as well as insuring privacy of transmission. This privacy prevents predators from locking onto the signals and hunting down the senders. In war—and Nature is a battlefield, make no mistake—one needs secret codes.

These predictions, or educated guesses if you prefer, qualify as consilience by synthesis. With some puzzling exceptions, they have been confirmed. But biologists cannot predict from physics and chemistry alone the exact structure of the pheromone molecules or the identity of the glands that manufacture them. For that matter, in advance of experiments, they cannot stipulate whether a given signal is used or not used by a particular species of ant. To attain that level of accuracy, to travel all the way from physics and chemistry near the entrance of the labyrinth to an end point in the social life of ants, we need detailed collateral knowledge of the evolutionary history of the species and of the environment in which it lives.

PREDICTIVE SYNTHESIS, in short, is formidably difficult. On the other hand, I believe that explanation in the opposite direction, by reduction, can in some instances be achieved across all levels of organization and hence all branches of learning. As a demonstration, I will now attempt to trace a magician's dream all the way down to an atom.

Serpents are in the magician's dream, transfigured from real-life snakes. I have not placed them there capriciously. They belong as the wild creatures most frequently conjured around the world in dreams and drug-induced hallucinations. Coming with ease to Zulu and Manhattanite alike, serpents are powerful images of human fantasy, flesh-and-blood snakes transformed into flickering images of the subconscious mind. There, depending on the culture and experience of the individual dreamer, they are conjured variously as predators, menacing demons, guardians of a hidden world, oracles, spirits of the dead, and gods. The slithering bodies and lethal strikes of real snakes make them ideal for magic. Their images evoke blends of emotion that fall on a triangular gradient defined by the three points of fear, revulsion, and reverential awe. Where the real snake frightens, the dream serpent transfixes. In the dreamer's paralytic state of sleep the serpent cannot be escaped.

Snakes are abundant and diverse in the rain forests of western Amazonia.

Serpents, their dream equivalents, figure prominently in the cultures of the Amerindian and mestizo inhabitants. Shamans preside over the taking of hallucinogenic drugs and interpret the meaning of the serpents and other apparitions that subsequently emerge. The Jívaro of Ecuador use *maikua*, the juice from the green bark of a member of the nightshade family, *Datura arborea*. Warriors drink it to summon *arutams*, ancestors living in the spirit world. If the seeker is fortunate, a spirit emerges from the depths of the forest, often in the form of two giant anacondas, which in real life is the species *Eunectes murinus*, heaviest of the world's snakes, big enough to kill a human being. The dream serpents roll toward him, entwined in combat. When they come within twenty or thirty feet the Jívaro must run forward and touch them. Otherwise they will explode "like dynamite," and disappear.

After receiving his vision the Jívaro must tell no one, or else the spell will end. That night he sleeps on the bank of the nearest river, and as he dreams the *arutam* returns to him as an old man. It says, "I am your ancestor. Just as I have lived a long time, so will you. Just as I have killed many times, so will you." The apparition then disappears, and as it does its soul enters the body of the dreamer. The Jívaro rises at dawn with an enhanced feeling of bravery and grace in bearing. His new demeanor is noted by others in the scattered households of the local Jívaro community. If he wishes, he can don the bird-bone shoulder ornament that symbolizes *arutam* soul power. In the old days he would have been considered fit to serve as a warrior on headhunting expeditions.

Five hundred miles southeast in Amazonian Peru lives Pablo Amaringo, mestizo shaman and artist. Drawing on the traditions of his Amerindian forebears, the Cocama and Quechua speakers of Amazonas and Cajamarca, Amaringo conjures visions and depicts them in paintings. His drug of choice, widely used in communities of the Río Ucayali region, is *ayahuasca*, extracted from the jungle vine *Banisteriopsis*. His dreams are populated with serpents in most of their Amazonian cultural roles: mounts of gods, forest spirits, ambush predators of animals and people, impregnators of women, landlords of lakes and forests, and sometimes the sinuous *ayahuasca* vine itself transmuted into animal form.

In the rich local Shipibo tradition followed by Amaringo in his paintings, the serpents, as well as other real and supernatural beings, are decorated with intricate geometric designs in primary colors. The paintings also share the Shipibo *horror vacui*: Every available space is crowded with detail. The style fits the Amazon region, which teems with life of stupendous variety.

Amaringo's subjects are loosely eclectic. Spirits and conjurers and fantastical animals from ancient Amerindian myths are thrown together with contemporary Peruvians and industrial artifacts. Ships and airplanes pass by; even flying saucers hover above the rain forest canopy. The images, surreal and disturbing, freed from normal sensory input, are incarnate emotions in search of theater and narrative. Their craziness illustrates the principle that during trances and dreaming, any metaphor serves and any fragment of memory able to slip into the unguarded mind becomes part of the story.

The sacred plants, which have been analyzed by chemists, are no longer mysterious. Their juices are laced with neuromodulators that in large oral doses produce a state of excitation, delirium, and vision. The primary effects are often followed by narcosis and dreaming of similar kind. In the Jívaros' *Datura* they are the structurally similar alkaloids atropine and scopolamine. In *Banisteriopsis* of the mestizos they include beta-carbolines, to which the shamans usually add dimethyltryptamine from another plant species. The substances are psychotropic, stimulating a flurry of images intense enough to break through the controlled processes of ordinary conscious thought. They alter the brain in the same manner as the natural neuromodulator molecules that regulate normal dreaming. The difference is that under their influence people enter a semicomatose trance in which dreaming, uncontrolled and often vivid and urgent, is no longer confined to sleep.

It is tempting to patronize the spiritual searches of the Amazonian *vegetalistas*, just as it is easy to dismiss the counterculture's innocent faith in drug-soaked gurus and sorcerers during the 1960s and 1970s. Outside of a few cults, few people today believe in the late drug guru Timothy Leary, or even remember Carlos Castañeda and his once-famous *The Teachings of Don Juan*. Yet it would be a mistake to underestimate the importance of such visions. They tell us something important about biology and human nature. For millennia the use of hallucinogens to enhance inner awareness has been widespread through the cultures of the world. Natural sleep and drug-induced dreams have long been viewed in Western civilization as a portal to the divine. They appear at pivotal moments in both the Old and New Testaments. We learn from Matthew 1:20, for example, that as Joseph pondered Mary's pregnancy, the conception of Jesus, "behold, an angel of the Lord appeared to him in a dream" to reveal the Holy Spirit as progenitor. Joseph's witness established one of the two essential pillars of Christian belief, the other being the disciples' account of the Resurrection, also dreamlike.

Emanuel Swedenborg, the eighteenth-century scientist and theologian whose followers founded the Church of the New Jerusalem, believed that dreams contain secrets of the divine. God does not restrict his word to Holy

Scripture. If the sacred code cannot be found under the microscope (as the Swedish savant discovered to his disappointment), it might yet be forthcoming in the scenarios of the dreamworld. Swedenborg recommended irregular hours and sleep deprivation as a means of inducing sharper and more frequent images. At least he had his physiology right; I suspect that he would have enjoyed a stiff dose of *ayahuasca*.

CONSIDER THEN the dreams of a magician, a sorcerer, a shaman. They are more than just unique productions of a single mind; they exhibit qualities general to the human species. The art of Pablo Amaringo is worthy of analysis in the manner of the natural sciences. His paintings are a test case of consilience, an arresting fragment of culture that might be explained and thereby given added meaning at the next, biological level down in complexity from artistic inspiration.

It is the habit of scientists to look for elements available as entry points for such analysis. To this end I have chosen two elements from Amaringo's paintings that present themselves for convenient explanation: the dreamscape as a whole, and the serpents that conspicuously populate it.

Mysticism and science meet in dreams. Freud, aware of the conjunction, composed a hypothesis to explain their meaning. He said that our dreams are disguises for unconscious wishes. When we sleep, the ego releases its grip on the id, which is the embodiment of instinct, and our most primitive fears and desires then escape into the conscious mind. They are not, however, experienced in raw form. Like characters in a bad Victorian novel, they are altered by the mind's censor into symbols so as not to disrupt sleep. The average person cannot expect to read their meaning accurately upon awakening. He must turn, Freud argued, to a psychoanalyst, who will guide him through free association in order to decipher the codes. As the translations are made, the connections of the symbols to childhood experience become clear. If the revelation unfolds correctly, the patient enjoys an easing of neuroses and other psychological disturbances that stem from his repressed memories.

Freud's conception of the unconscious, by focusing attention on hidden irrational processes of the brain, was a fundamental contribution to culture. It became a wellspring of ideas flowing from psychology into the humanities. But it is mostly wrong. Freud's fatal error was his abiding reluctance to test his own theories—to stand them up against competing explanations—then revise them to accommodate controverting facts. He also suffered from the luck of the draw. The actors of his drama—id, ego, and superego—and the roles they played in suppression and transference might have evolved smoothly into the elements of a modern scientific theory had he guessed their basic nature correctly. Darwin's theory of natural selection prospered that way, even though the great naturalist had no idea of particulate heredity carried by genes. Only later did modern genetics verify his insight concerning the evolutionary process. In dreams Freud was faced with a far more complex and intractable set of elements than genes, and—to put it as kindly as possible—he guessed wrong.

The competing and more modern hypothesis of the basic nature of dreaming is the activation-synthesis model of biology. As created during the past two decades by J. Allan Hobson of Harvard Medical School and other researchers, it pieces together our deepening knowledge of the actual cellular and molecular events that occur in the brain during dreaming.

In brief, dreaming is a kind of insanity, a rush of visions, largely unconnected to reality, emotion-charged and symbol-drenched, arbitrary in content, and potentially infinite in variety. Dreaming is very likely a side effect of the reorganization and editing of information in the memory banks of the brain. It is not, as Freud envisioned, the result of savage emotions and hidden memories that slip past the brain's censor.

The facts behind the activation-synthesis hypothesis can be interpreted as follows. During sleep, when almost all sensory input ceases, the conscious brain is activated internally by impulses originating in the brain stem. It scrambles to perform its usual function, which is to create images that move through coherent narratives. But lacking moment-by-moment input of sensory information, including stimuli generated by body motion, it remains unconnected to external reality. Therefore, it does the best it can: It creates fantasy. The conscious brain, regaining control upon awakening, and with all its sensory and motor inputs restored, reviews the fantasy and tries to give it a rational explanation. The explanation fails, and as a result dream interpretation itself becomes a kind of fantasy. That is the reason psychoanalytic theories relating to dreaming, as well as parallel supernatural interpretations arising in myth and religion, are at one and the same time emotionally convincing and factually incorrect.

The molecular basis of dreaming is understood in part. Sleep descends upon the brain when chemical nerve cell transmitters of a certain kind, amines such as norepinephrine and serotonin, decline in amount. Simultaneously a transmitter of a second kind, acetylcholine, rises in amount. Both wash the junctions of nerve cells specialized to be sensitive to them. The two kinds of neurotransmitters exist in a dynamic balance. The amines waken the brain and mediate its control of the sensory systems and voluntary muscles. Acetylcholine shuts these organs down. As acetylcholine gains ascendancy, the activities of the conscious brain are reduced. So are other functions of the body except for circulation, respiration, digestion, and—remarkably—movement of the eyeballs. The voluntary muscles of the body are paralyzed during sleep. Temperature regulation is also diminished. (That is why it can be dangerous to fall asleep while the body is cold.)

In a normal nocturnal cycle, sleep is at first deep and dreamless. Then at intervals, consuming overall about 25 percent of the total sleep period, it turns shallow. During the shallow periods the sleeper is more easily awakened. His eyes move erratically in their sockets, the condition called rapid eye movement, or REM. The conscious brain stirs and dreams but remains sealed off from external stimuli. Dreaming is triggered when acetylcholine nerve cells in the brain stem begin to fire wildly, initiating what are called PGO waves. The electrical membrane activity, still mediated by acetylcholine at the nerve junctions, moves from the pons (the P of PGO), a bulbous mass of nerve centers located at the top of the brain stem, upward to the lower center of the brain mass, where it enters the geniculate nuclei (G) of the thalamus, which are major switching centers in the visual neuronal pathways. The PGO waves then pass on to the occipital cortex (O), at the rear of the brain, where integration of visual information takes place.

Because the pons is also a principal control station for motor activity when the brain is awake, the signals it passes through the PGO system falsely report to the cortex that the body is in motion. But of course the body is immobile—in fact it is paralyzed. What the visual brain does then is to hallucinate. It pulls images and stories out of the memory banks and integrates them in response to the waves arriving from the pons. Unconstrained by information from the outside world, deprived of context and continuity in real space and time, the brain hastily constructs images that are often phantasmagoric and engaged in events that are impossible. We fly through the air, swim in the deep sea, walk on a distant planet, converse with a long-dead parent. People, wild animals, and nameless apparitions come and go. Some constitute the materialization of our emotions triggered by the PGO surges, so that from dream to dream our mood is variously calm, fearstruck, angry, erotic, maudlin, humorous, lyrical, but most of the time just anxious. There seems to be no limit to the combinatorial power of the dreaming brain. And whatever we see we believe, at least while sleeping; it rarely occurs to us to doubt even the most bizarre events into which we have been involuntarily thrust. Someone has defined insanity as an inability to choose among false alternatives. In dreams we are insane. We wander across our limitless dreamscapes as madmen.

Strong stimuli can break through the sensory barrier. If they do not wake us, they are fitted into the dream story. Let real thunder roll into our bedroom from lightning a mile away. To take one of endless possible responses, our dream switches to a bank robbery, a gun is fired, we are shot. No, another person is shot, has fallen, but no again—we realize it is us, displaced to someone else's body. Oddly, we feel no pain. Then the scene changes. We are walking down a long corridor, lost, anxious to get home, another shot is fired. This time we come awake, tense, to lie still in the real world and listen to real thunder roll in from the approaching storm outside.

In dreams we seldom experience the physical discomforts of pain, nausea, thirst, or hunger. A few people suffer apnea, a temporary halt in breathing, which may be turned into visions of suffocation or drowning. There is no smell or taste in dreams; the channels of these sensory circuits are shut down by the acetylcholine wash of the sleeping brain. Unless we wake soon afterward, we remember no details of any kind. Ninety-five to 99 percent of dreams are forgotten completely. A small minority of persons believe, erroneously, that they do not dream at all. This amazing amnesia is apparently due to the low concentration of amine transmitters, which are needed to convert short-term memories into longer ones.

WHAT IS the function of dreaming? Biologists have tentatively concluded from detailed studies of animals and humans that the information learned while the brain is awake is sorted and consolidated while it is asleep. There is further evidence that at least some of this processing, particularly the sharpening of cognitive skills by repetition, is limited to periods of REM sleeping, and therefore to dreamtime. The flow of acetylcholine itself may be a crucial part of the process. The fact that dreaming activates such intense inward motor and emotional activity has led some researchers to suggest that REM sleep has an even more profound, Darwinian function. When we dream, we deepen moods and improve responses basic to survival and sexual activity.

The findings from neurobiology and experimental psychology nevertheless say nothing about the content of the dreams. Are the fantasies *all* temporary insanity, the sum of quickly forgotten epiphenomena during the consolidation of learning? Or can we search in some neo-Freudian manner for deep meaning in the symbols from which dreams are composed? Because dreams are not entirely random, the truth must lie somewhere in between.

The composition may be irrational, but the details comprise fragments of information appropriate to the emotions activated by the PGO waves. It is quite possible that the brain is genetically predisposed to fabricate certain images and episodes more than others. These fragments may correspond in a loose way to Freud's instinctual drives and to the archetypes of Jungian psychoanalysis. Both theories can perhaps be made more concrete and verifiable by neurobiology.

Genetic predisposition and evolution lead to the second element I have chosen from the Amaringo paintings: serpents. The form of our understanding of these creatures of the night is the exact opposite of that concerning the nature of dreams in general. As I have just explained, biologists now understand in very general terms how dreams happen—they have puzzled through many of the key cellular and molecular events of dreaming. They are much less sure of the good that dreams do mind and body. In the case of the prevalence of serpents, the situation is reversed. Biologists have a sound working hypothesis on the function of the images but as yet no idea of their molecular and cellular basis beyond the general control of dreaming. The mystery as to the exact mechanism is due to our ignorance of the cellular processes by which specific memories such as those of serpents are assembled and colored by emotion.

What we know about serpents as dream images can be expressed by the two key modes of analysis used in biology. The first mode exposes proximate causes, the entities and physiological processes that create the phenomenon. Proximate explanations answer the question of *how* biological phenomena work, usually at the cellular and molecular levels. The second mode of explanation addresses *why* they work—their ultimate causes, which are the advantages the organism enjoys as a result of evolution that created the mechanisms in the first place. Biologists aim for both proximate and ultimate explanations. To put the study of dreaming in a nutshell, we understand a good deal about the proximate causes of dreaming in general, but very little about its ultimate causes, while the reverse is true for the presence of serpents in dreams.

The account I will now give of the ultimate cause of the bond between snake and man has been pieced together from accounts of animal and human behavior by many researchers, and most fully by the American anthropologist and art historian Balaji Mundkur. Fear of snakes is deep and primordial among the Old World primates, the phylogenetic group to which *Homo sapiens* belongs. When vervets and other guenons, common long-tailed arboreal monkeys of Africa, encounter certain kinds of snakes, they emit a unique chuttering call. They are evidently good instinctive herpetologists, because the response, which appears to be inborn, is limited to the poisonous cobras, mambas, and puff adders. The response is not made to harmless snakes. Others of the monkey group come to the side of the caller, and together they watch the intruder until it leaves the neighborhood. They are also ready with an inborn eagle call, causing all the troop members to scramble down from the trees and out of danger, and an inborn leopard call, triggering a rush in the opposite direction to parts of the canopy that big cats cannot reach.

Common chimpanzees, a species believed to share a common ancestor with prehumans as recently as five million years ago, are unusually apprehensive in the presence of snakes, even if they have had no previous experience. They back off to a safe distance and follow the intruder with a fixed stare while alerting companions with a *Wah!* warning call. The response gradually intensifies during adolescence.

Human beings also possess an innate aversion to snakes, and, as in the chimpanzee, it grows stronger during adolescence. The reaction is not a hard-wired instinct. It is a bias in development of the kind psychologists call prepared learning. Children simply learn fear of snakes more easily than they remain indifferent or learn affection for snakes. Before the age of five they feel no special anxiety. Later they grow increasingly wary. Then just one or two bad experiences—a snake writhing nearby through the grass or a frightening story—can make them deeply and permanently afraid. The propensity is deep-set. Other common fears—of the dark, strangers, loud noises—start to wane after seven years of age. In contrast, the tendency to avoid snakes grows stronger with time. It is possible to turn in the opposite direction, learning to handle snakes without fear or even to like them in some special way. I did, as a boy, and once thought seriously of becoming a professional herpetologist. But the adaptation was for me forced and self-conscious. People's special sensitivity can just as easily turn into full-blown ophidiophobia, the pathological extreme in which the close proximity of a snake brings on panic, cold sweat, and waves of nausea.

The neural pathways of snake aversion have not been explored. We do not know the proximate cause of the phenomenon except to classify it as "prepared learning." In contrast, the probable ultimate cause, the survival value of the aversion, is well understood. Throughout human history a few kinds of snakes have been a major cause of sickness and death. Every continent except Antarctica has poisonous snakes. Over most of Africa and Asia the known death rate from snakebite is 5 persons per 100,000 each year. The local record is held by a province in Burma (lately called Myanmar), with 36.8 deaths per 100,000 in a year. Australia has an exceptional abundance of deadly snakes, most of whose species are evolutionary relatives of the cobras. Unless you are an expert, it is wise to stay clear of every snake in Australia, just as it is wise to avoid wild mushrooms anywhere in the world. In South and Central America live deadly snakes well known to the Jívaro and *vegetalista* shamans, including the bushmaster, fer-de-lance, and jaracara, which are among the largest and most aggressive of the pit vipers. Possessing skins patterned and colored like fallen leaves, and fangs long enough to pass through a human hand, they wait in ambush on the floor of the tropical forest for small birds and mammals and are quick to deliver defensive strikes at passing humans.

Snakes and dream serpents provide an example of how agents of nature can be translated into the symbols of culture. For hundreds of thousands of years, time enough for genetic changes in the brain to program the algorithms of prepared learning, poisonous snakes have been a significant source of injury and death to human beings. The response to the threat is not simply to avoid it, in the way that certain berries are recognized as poisonous through painful trial and error, but to feel the kind of apprehension and morbid fascination displayed in the presence of snakes by the nonhuman primates. The snake image also attracts many extraneous details that are purely learned, and as a result the intense emotion it evokes enriches cultures around the world. The tendency of the serpent to appear suddenly in trances and dreams, its sinuous form, and its power and mystery are logical ingredients of myth and religion.

Amaringoan images stretch back through the millennia. Prior to the pharaonic dynasties the kings of Lower Egypt were crowned at Buto by the cobra goddess Wadjet. In Greece there was Ouroboros, the serpent that continuously devoured itself tail-first while regenerating from the inside. For gnostics and alchemists of later centuries this self-cannibal came to symbolize the eternal cycle of destruction and re-creation of the world. One day in 1865, while dozing by a fire, the German chemist Friedrich August Kekule von Stradonitz dreamed of Ouroboros and thereby conceived of the benzene molecule as a circle of six carbon atoms, each bonded to a hydrogen atom. Because of that inspiration some of the most puzzling data of nineteenth-century organic chemistry fell into place. In the Aztec pantheon, Quetzalcoatl, the plumed serpent with a human head, ruled as the god of the morning and evening star, and thus of death and resurrection. He was the inventor of the calendar and patron of learning and the priesthood. Tlaloc, god of rain and lightning, was another serpentine chimera, with humanoid upper lips formed from two rattlesnake heads. Such apparitions could have been born only in dreams and trances.

IN MIND AND CULTURE the serpent transcends the snake. An understanding of its transformation from an earthly reptile can be viewed as one of many pathways through the borderlands that separate science from the humanities. Having followed the serpent a considerable distance in our journey from magician to atom, we next enter the interior of the biological sciences. Here better maps are available, and progress considerably easier. Scores of Nobel prizes, the fruit of millions of hours of labor and billions of dollars allocated to biomedical research, point the way on down through the sciences from body and organ through cell to molecule and atom. The general structure of the human nerve cell has now been charted in considerable detail. Its electric discharge and synaptic chemistry are partly understood and can be expressed in formulas obedient to the principles of physics and chemistry. The stage has been set to attack the master unsolved problem of biology: how the hundred billion nerve cells of the brain work together to create consciousness.

I say the master problem, because the most complex systems known to exist in the universe are biological, and by far the most complex of all biological phenomena is the human mind. If brain and mind are at base biological phenomena, it follows that the biological sciences are essential to achieving coherence among all the branches of learning, from the humanities on down to the physical sciences. The task is made somewhat easier by the fact that disciplines within biology itself are now generally consilient and growing more so each year. I would like to explain how this has been accomplished.

Consilience among the biological sciences is based on a thorough understanding of scale in time and space. Passing from one level to the next, say molecule to cell or organ to organism, requires the correct orchestration of changes in time and space. To make the point, I will return for a last time to Pablo Amaringo, magician, artist, and fellow organism. Imagine that we can speed or slow the time we spend with him, while expanding or shrinking the space we see in and around his person. So we enter his house, we shake his hand, and Amaringo shows us a painting. The actions consume seconds or minutes. An obvious fact, so why mention it? The question makes more sense when put in another form: Why did these familiar actions not consume millionths of seconds, or months, instead? The answer is that human beings are constructed of billions of cells that communicate across membranes by chemical surges and electrical impulses. To see and speak with Amaringo entails a sequence of these units covering seconds to minutes, not microseconds or months. We think of that span of time as normal and somehow standard for the world in which we live. It is not. Because it involves Amaringo and us, all of whom are organic machines, it is only organismic time. And because the full apparatus of our communication takes up from millimeters to meters of surface and volume, not nanometers or kilometers, our unaided minds dwell entirely in organismic space.

Imagine now that with the best of our instruments (and his permission!) we can look into the brain of Pablo Amaringo. By magnifying the image, his smallest nerves come into view. Then we see the constituent cells, and finally the molecules and atoms. We watch as a nerve cell discharges: Along the length of its membrane, the voltage drops as sodium ions flow inward. At each point on the shaft of the nerve cell, the events consume only several thousandths of a second, while the electrical signal they create—the drop in voltage—speeds along the shaft at ten meters a second, as fast as an Olympic sprinter. With our field of vision now brought to a space only one ten-thousandth that of our original field, the events occur too swiftly to be seen. An electric discharge of the cell membrane crosses the field of vision faster than a rifle bullet. To see it—remember, as human observers we are still in organismic time—we must record and slow the action down enough to witness events that originally occurred in a few thousandths of a second or less. We are now in biochemical time, a necessity if we are to observe events in biochemical space.

In the midst of this magic Amaringo keeps talking, but is scarcely aware of the changes that occurred as our own actions accelerated a thousandfold. Only enough of his time has elapsed for one or two words to be spoken. We turn the dials in the opposite direction, shift time and space until his full image reappears and his words flow through our minds at an audible pace. We turn the dials further. Amaringo shrinks in proportionate size and speedwalks jerkily out of the room, like an actor in an early silent film. Perhaps he does so in frustration because we are now frozen in position like marble statues. Our vision continues to expand. Let us rise in the air to gather more space. Our view grows to encompass the town of Pucallpa and then a large stretch of the Río Ucayali valley. Houses disappear, new ones pop up. Day blends with night into continuous twilight as the flicker-fusion frequency of our organismic-time vision is exceeded. Amaringo grows old, he dies. His children grow old, they die. Nearby the rain forest is changing. Clearings appear as great trees fall, saplings spring up, the gaps close. We are now in ecological time. We turn the dials again, and space-time expands still more. Individual persons and other organisms are no longer distinguishable, only blurred populations—of anacondas, *ayahuasca* vines, the people of central Peru—and these can be seen across the passage of generations. A century of their time collapses into a minute of ours. Some of their genes are changing, in both kind and relative frequency. Detached from other human beings and shorn of their emotions, godlike at last, we witness the world in evolutionary time and space.

This conception of scale is the means by which the biological sciences have become consilient during the past fifty years. According to the magnitude of time and space adopted for analysis, the basic divisions of biology are from top to bottom as follows: evolutionary biology, ecology, organismic biology, cellular biology, molecular biology, and biochemistry. That sequence is also the basis of the organization of professional societies and of the curricula of colleges and universities. The degree of consilience can be measured by the degree to which the principles of each division can be telescoped into those of the others.

THE INTERLOCKING of the biological disciplines, a tidy concept, is still compromised in execution by the dilemma of the labyrinth with which I began this chapter. Consilience among the disciplines grows more smoothly from the top down as more links are laid in place, from the most specific of the entities, such as the brain of Amaringo, all the way to the most general, his atoms and molecules. But to establish consilience the other way, from general to more specific, is vastly more difficult. In short, it is far easier to analyze Amaringo than to synthesize him.

The greatest obstacle to consilience by synthesis, the approach often loosely called holism, is the exponential increase in complexity encountered during the upward progress through levels of organization. I have already described how an entire cell cannot yet be predicted from a knowledge of its scrambled molecules and organelles alone. Let me now indicate how bad the problem really is. It is not even possible to predict the three-dimensional structure of a protein from a complete knowledge of its constituent atoms. The composition of amino acids can be determined, and the exact position of each atom can be mapped precisely with the aid of X-ray crystallography. We know, to choose one of the simplest proteins, that the insulin molecule is a sphere containing fifty-one amino acids. Such reconstruction is one of the many triumphs of reductionist biology. But this knowledge of the sequence of all the amino acids and of the atoms composing them is not enough to predict the shape of the sphere or its internal structure as revealed by X-ray crystallography.

In principle the prediction of the form of proteins is possible. Synthesis at the level of macromolecules is a technical, not a conceptual, problem. The effort to solve it is in fact an important industry in biochemistry. To have such knowledge would be a major breakthrough in medicine. Synthetic proteins, some perhaps more effective than the natural molecules, could be created upon demand to combat disease-causing organisms and remedy enzyme deficiencies. In practice, however, the difficulties seem almost insurmountable. Making the prediction requires a summation of all the energy relationships among nearby atoms. That alone is daunting. But then the interactions of more distantly separated atoms in the molecule must be added. The forces shaping the molecule comprise an immensely complex web of thousands of energy contributions, all of which must be integrated simultaneously in order to form the whole. Some biochemists believe that to achieve that final step, each energy contribution in turn must be calculated with an accuracy still beyond the grasp of the physical sciences.

Even greater difficulties exist in the environmental sciences. The paramount challenge to ecology for the foreseeable future is the cracking apart and resynthesis of the assemblages of organisms that occupy ecosystems, particularly the most complex ecosystems such as estuaries and rain forests. Most studies in ecology focus on only one or two species of organisms at a time, out of the thousands occupying a typical habitat. The researchers, forced into reductionism by practical necessity, start with small fragments of the whole ecosystem. Yet they are aware that the fate of each species is determined by the diverse actions of scores or hundreds of other species that variously photosynthesize, browse, graze, decompose, hunt, fall prey, and turn soil around the target species. The ecologists know this principle very well, but they still can do little about predicting its precise manifestation in any particular case. Even more than biochemists manipulating atoms in a large molecule, the ecologists face immensurable dynamic relationships among still largely unknown combinations of species.

Consider this example of the complexity they face. When Gatun Lake was created during the construction of the Panama Canal in 1912, the rising waters cut off a piece of elevated land. The new isolate, covered by tropical evergreen forest, was named Barro Colorado Island and made into a biological research station. In the following decades it became the most closely studied ecosystem of its kind in the world. The size of the island, seventeen square kilometers, was too small to sustain jaguars and pumas. The prey of the great cats had consisted partly of agoutis and pacas, outsized rodents that vaguely resemble jackrabbits and small deer. These animals, freed from a major cause of mortality, multiplied to ten times their original numbers. They over-exploited their own food, which consists mostly of large seeds that fall from the forest canopy, which caused a reduction in the reproduction and abundance of the tree species that produce the seeds. The effect rippled outward. Other tree species whose seeds are too small to be of interest to the agoutis and pacas benefited from the reduced competition. Their seeds set more abundantly and their seedlings flourished, and a larger number of the young trees reached full height and reproductive age. It was inevitable then that the animal species specialized to feed on small-seed trees also prospered, that the predators depending on these animals increased, that the fungi and bacteria parasitizing the small-seed trees and associated animals spread, that the microscopic animals feeding on the fungi and bacteria grew denser, that the predators of these creatures increased in turn, and so on across the food web and back again as the ecosystem reverberated from the restriction of its area and consequent loss of its top carnivores.

THE GREATEST CHALLENGE today, not just in cell biology and ecology but in all of science, is the accurate and complete description of complex systems. Scientists have broken down many kinds of systems. They think they know most of the elements and forces. The next task is to reassemble them, at least in mathematical models that capture the key properties of the entire ensembles. Success in this enterprise will be measured by the power researchers acquire to predict emergent phenomena when passing from general to more specific levels of organization. That in simplest terms is the great challenge of scientific holism.

Physicists, whose subject matter is the simplest in science, have already succeeded in part. By treating individual particles such as nitrogen atoms as random agents, they have deduced the patterns that emerge when the particles act together in large assemblages. Statistical mechanics, originated in the nineteenth century by James Clerk Maxwell (who also pioneered the theory of electromagnetic radiation) and Ludwig Boltzmann, accurately predicted the behavior of gases at different temperatures by the application of classical mechanics to the large numbers of freely moving molecules that compose the gases. Other researchers, by moving back and forth between the same two levels of organization, in other words between molecules and gases, were further able to define viscosity, heat conduction, phase transition, and other macroscopic properties as expressions of the forces between the molecules. At the next level down, quantum theorists in the early 1900s connected the collective behavior of electrons and other subatomic particles to the classical physics of atoms and molecules. Through many such advances during the past century, physics has been welded into the most exact of the sciences.

At higher, more specific levels of organization, beyond the traditional realm of physics, the difficulties of synthesis are almost inconceivably more difficult. Entities such as organisms and species, unlike electrons and atoms, are indefinitely variable. Worse, each particular one changes during development and evolution. Consider this example: Among the vast array of molecules that an organism can manufacture to serve its needs are simple hydrocarbons of the methane series, composed entirely of carbon and hydrogen atoms. With one carbon atom, only a single kind of molecule is possible. With 10 carbon atoms the number is 75, with 20 it is 366,319, and with 40 it is 62 trillion. Add oxygen atoms here and there on the hydrocarbon chains to produce alcohols, aldehydes, and ketones, and the number rises even more rapidly with molecular size. Now select various subsets and imagine multiple ways they can be derived by enzyme-mediated manufacture, and you have potential complexity beyond the powers of present-day imagination.

Biologists, it has been said, suffer from physics envy. They build physics-like models that lead from the microscopic to the macroscopic, but find it difficult to match them with the messy systems they experience in the real world. Theoretical biologists are nevertheless easily seduced. (I confess to being one, and having been responsible for more than my share of failures.) Armed with sophisticated mathematical concepts and high-speed computers, they can generate unlimited numbers of predictions about proteins, rain forests, and other complex systems. With the passage to each higher level of organization, they need to contrive new algorithms, which are sets of exactly defined mathematical operations pointed to the solution of given problems. And so with artfully chosen procedures they can create virtual worlds that evolve into more highly organized systems. Wandering through the Cretan labyrinth of cyberspace they inevitably encounter emergence, the appearance of complex phenomena not predictable from the basic elements and processes alone, and not initially conceivable from the algorithms. And behold! Some of the productions actually look like emergent phenomena found in the real world.

Their hopes soar. They report the results at conferences of like-minded theoreticians. After a bit of questioning and probing, heads nod in approval: "Yes, original, exciting, and important—if true." If true . . . if true. *Folie de grandeur* is their foible, the big picture their illusion. They are on the edge of a breakthrough! But how do they know that nature's algorithms are the same as their own, or even close? Many procedures may be false and yet produce an approximately correct answer. The biologists are at special risk of committing the fallacy of affirming the consequent: It is wrong to assume that because a correct result was obtained by means of theory, the steps used to obtain it are necessarily the same as those that exist in the real world.

To see this point clearly, think of a blossom in a painting rendered photographic in detail and as beautiful as life. In our minds the macroscopic entity has truth because it matches real flowers sprung from the soil. From a distance we might easily confuse the image with the real thing. But the algorithms that created it are radically different. Its microscopic elements are flakes of paint instead of chromosomes and cells. Its developmental pathways exist in the brain of the artist, not in prescription by DNA of the unfolding of tissues. How do theoreticians know that their computer simulations are not just the paintings of flowers?

These and other difficulties endemic to higher systems have not gone unnoticed. Researchers from several scientific disciplines have joined to take the measure of the problems, forming a loose enterprise variously designated as complexity, complexity studies, or complexity theory. Complexity theory (the best expression, in my opinion) can be defined as the search for algorithms used in nature that display common features across many levels of organization. At the very least, according to the proponents of complexity theory, the commonalities can be expected to provide an explorer's guide for quicker movement when passing from simple to more complex systems through the real-world labyrinth. The commonalities will assist in pruning all the algorithms that can be conceived down to the ones that nature has chosen. At their best, they might lead to deep new laws that account for the emergence of such phenomena as cells, ecosystems, and minds.

By and large, the theoreticians have focused their attention on biology, and that makes sense. Organisms and their assemblages are the most complex systems known. They are also self-assembling and adaptive. Living systems in general, by constructing themselves from molecule to cell to organism to ecosystem, surely display whatever deep laws of complexity and emergence lie within our reach.

Complexity theory was born in the 1970s, gathered momentum in the early 1980s, and was enveloped in controversy by the mid-1990s. The issues of contention are almost as tangled as the systems the theorists hoped to unravel. I think it possible to cut to the heart of the matter, as follows. The great majority of scientists, their minds focused narrowly on well-defined phenomena, do not care about complexity theory. Many have not yet heard of it. They, the uninvolved, can be ignored, lest all of contemporary science be thought of as a boiling cauldron of argument. Those who care can be divided into three camps. The first comprises a heterogeneous scattering of skeptics. They believe that brains and rain forests are too complicated ever to be reduced to elementary processes, let alone reconstituted in a manner that predicts the whole. Some of the skeptics doubt the existence of deep laws of complexity, at least any that can be grasped by the human mind.

In the second camp are the fervent advocates, a band of audacious complexity theorists, exemplified by Stuart Kauffman (author of *The Origins of Order*) and Christopher Langton, who work at the Sante Fe Institute in New Mexico, unofficial headquarters of the complexity movement. They believe not only that deep laws exist but that their discovery is on the near horizon. Some of the essential elements of the laws, they say, are already emerging from mathematical theories that use exotic conceptions such as chaos, self-criticality, and adaptive landscapes. These abstractions bring into vivid focus the way complex systems might build themselves up, persist for a while, and then disintegrate. Their architects—computer-oriented, abstraction-absorbed, light on natural history, heavy on nonlinear transformations—think they smell success. They believe that massive computer-aided simulations, exploring many possible worlds, will reveal the methods and principles needed to leapfrog conventional science, including most of contemporary biology, to achieve a comprehensive understanding of the higher productions of the material world. Their grail is a set of hoped-for master algorithms that will speed passage from atom to brain and ecosystem, consistent with reality but requiring far less factual knowledge than would be needed without the algorithms.

The third group of scientists, of which I am a reluctant member, has settled along positions strung between the two extremes of rejection and unbridled support. I say reluctant, because I would like to be a true believer: I really am impressed by the sophistication and élan of the complexity theorists, and my heart is with them. But my mind is not, at least not yet. I believe with many other centrists that they are on the right track—but only more or less, maybe, and still far short of success. Doubt and dissension on important issues have broken out even within their own ranks. The basic difficulty, to put the matter plainly, is an insufficiency of facts. The complexity theorists do not yet have enough information to carry with them into cyberspace. The postulates they start with clearly need more detail. Their conclusions thus far are too vague and general to be more than rallying metaphors, and their abstract conclusions tell us very little that is really new.

Take the "edge of chaos," one of the most frequently cited paradigms of complexity theory. It starts with the observation that in a system containing perfect internal order, such as a crystal, there can be no further change. At the opposite extreme, in a chaotic system such as boiling liquid, there is very little order to change. The system that will evolve the most rapidly must fall between, and more precisely on the edge of chaos, possessing order but with the parts connected loosely enough to be easily altered either singly or in small groups.

Kauffman applied the concept to the evolution of life in his NK model. N is the number of parts in an organism, such as the number of genes or of amino acids, that contribute to its survival and reproduction, hence its representation in future generations. K is the number of parts of such kind (genes or amino acids) in the same organism that affect the contribution of any one of the parts. A gene, for example, does not act alone to guide the development of a cell. It acts with other genes, typically in a complicated fashion. Kauffman pointed out that if genes were completely interconnected in their effects, with K equaling N, there could be little or no evolution in a population of organisms, because one thing in the heredity of the organisms cannot be changed without changing everything. In the extreme opposite case, where there are no connections among the genes, so that K equals zero, the population is in evolutionary chaos. If each gene is on its own, the population of organisms evolves randomly across a near-infinity of possible gene combinations, never stable in evolutionary time, never settling on one adaptive type. When connections exist but are very few—the edge of chaos—evolving populations can settle on adaptive peaks but are still capable of evolving with relative ease to other, nearby adaptive peaks. A species of bird, for example, might shift from eating seeds to eating insects; a savanna plant might acquire the ability to grow in the desert. Being on the edge of chaos, Kauffman reasoned, provides the greatest evolvability. Perhaps species adjust the number of connections so as to remain in this most fluid of adaptive zones.

Kauffman has spread the NK models to touch on a wide array of topics in molecular and evolutionary biology. His arguments, like those of other leading complexity theorists, are original and directed at important problems. First time around, they sound good. But as an evolutionary biologist familiar with genetics, I have learned little from them. While wading through Kauffman's equations and peculiarly fustian prose, I realized that I already knew most of the results in a different context. They are essentially a reinvention of the wheel, a re-creation in a difficult new language of principles already outlined in the mainstream literature of biology. Unlike the important theories of physics, the NK formulations do not shift the foundations of thought or offer predictions in measurable quantities. So far they contain nothing to take into the field or laboratory.

This personal and possibly unfair reaction to a single example is not to belittle the ultimate prospects of the complexity theorists. Some of the elementary concepts they have advanced, most notably chaos and fractal geometry, have assisted in understanding broad sectors of the physical world. In ecology, for example, the British biologist and mathematician Robert May has used realistic difference equations to derive patterns of population fluctuation of the kinds actually observed in plants and animals. As the rate of population growth increases, or as the environment relaxes its control of population growth, the number of individuals passes from a nearly steady state to a smooth up-and-down cycle. Then, as growth rate and environmental control change further, the number of individuals shifts to complex cycles with multiple peaks in time. Finally, the number slides into a chaotic regime, zigzagging up and down in no discernible pattern. The most interesting feature of chaos in populations is that it can be produced by exactly defined properties of real organisms. Contrary to previous belief, chaotic patterns are not necessarily the product of randomly acting forces of the environment that rock the population up and down. In this case and in many other complex physical phenomena, chaos theory provides an authentically deep principle of nature. It says that extremely complicated, outwardly indecipherable patterns can be determined by small, measurable changes within the system.

But, again, which systems, which changes? That is the nub of the problem. None of the elements of complexity theory has anything like the generality and the fidelity to factual detail we wish from theory. None has triggered an equivalent cascade of theoretical innovations and practical applications. What does complexity theory need to be successful in biology?

COMPLEXITY THEORY NEEDS more empirical information. Biology can supply it. Three hundred years in the making, having recently been wedded to physics and chemistry, biology is now a mature science. But its researchers may not require a body of special theory to master complexity.

They have refined reductionism into a high art and begun to achieve partial syntheses at the level of the molecule and organelle. Even if complete cells and organisms are still beyond them, they know they can reconstruct some of the elements one at a time. They foresee no need for overarching grand explanations as a prerequisite for creating artificial life. An organism is a machine, and the laws of physics and chemistry, most believe, are enough to do the job, given sufficient time and research funding.

Putting a living cell together will be a moon shot, not an Einsteinian revolution of space and time. Complexity in real organisms is being taken apart swiftly enough to enliven the pages of *Nature* and *Science* each week and drain away the need for conceptual revolution. A great vaulting revolution may occur, and suddenly, but the busy and well-fed masses of researchers are not awaiting it in desperate suspense.

The machine the biologists have opened up is a creation of riveting beauty. At its heart are the nucleic acid codes, which in a typical vertebrate animal may comprise about 50,000 to 100,000 genes. Each gene is a string of 2,000 to 3,000 base pairs (genetic letters). Among the base pairs composing active genes, each triplet (set of three) translates into an amino acid. The final molecular products of the genes, as transcribed outward through the cell by scores of perfectly orchestrated chemical reactions, are sequences of amino acids folded into giant protein molecules. There are about 100,000 kinds of protein in a vertebrate animal. Where the nucleic acids are the codes, the proteins are the substance of life, making up half the animal's dry weight. They give form to the body, hold it together by collagen sinews, move it by muscle, catalyze all its animating chemical reactions, transport oxygen to all its parts, arm the immune system, and carry the signals by which the brain scans the environment and mediates behavior.

The role a protein molecule plays is determined not just by its primary structure, not just by the sequence of amino acids within it, but also by its shape. The amino acid string of each kind is folded upon itself in a precise manner, coiled about like twine and crumpled together like a piece of wadded paper. The total molecule bears resemblance to forms as variable as clouds in the sky. Looking at these forms, we readily imagine lumpy spheres, donuts, dumbbells, rams' heads, angels with wings spread, and corkscrews.

The resulting contours of the surface are particularly critical for the function of enzymes, the proteins that catalyze the body's chemistry. Somewhere on the surface is the active site, a pocket or groove consisting of a few of the amino acids, held in place by the architecture of the remaining amino acids.

Only substrate molecules of a very specific form can fit the active site and submit to catalysis. As soon as one docks in the correct alignments, its active site alters shape slightly. The two molecules bind more closely, like hands clasped in greeting. Within an instant the substrate molecule is changed chemically and released. In the embrace of the enzyme sucrase, for example, sucrose is cleaved into fructose and glucose. Just as swiftly the active site of the enzyme molecule returns to its original shape, with its chemical structure unchanged. The productivity of most types of enzyme molecules, snapping in and out of the active state, is prodigious. A single one can process a thousand substrate molecules every second.

How to put all these nanometer components and millisecond reactions together into a coherent picture? Biologists are determined to do it from the ground up, molecule by molecule, and metabolic pathway by metabolic pathway. They have begun to assemble the data and mathematical tools needed to model an entire cell. When they succeed, they will also have reached the level of entire simple organisms, the single-cell bacteria, archaea, and protists.

Most biologists favor middle-level models in their theory of cell integration—neither primarily mathematical nor purely descriptive but instead front-loaded with large amounts of empirical information and conceived as genetic networks. The flavor of this state-of-the-art approach has been nicely captured by two of the researchers, William Loomis and Paul Sternberg, as follows:

> The nodes of such networks are genes or their RNA and protein products. The connections are the regulatory and physical interactions among the RNAs, proteins, and cis-regulatory DNA sequences of each gene. Modern molecular genetic techniques have greatly increased the rate at which genes are being recognized and their primary sequences determined. The challenge is to link the genes and their products into functional pathways, circuits, and networks. Analyses of regulatory networks (such as those involving signal transduction and transcriptional regulation cascades) illustrate combinatorial action that implements, for example, digital logic, analog-digital conversions, cross-talk and insulation, and signal integration. Although the existence of sophisticated network elements has been suggested by decades of physiological studies, what is new is the scale and detail becoming available for the components. Much of current molecular biology focuses on identifying new components, defining the regulatory inputs and outputs of each node, and delineating the physiologically relevant pathways.

The complexity conceived in this single paragraph exceeds that in supercomputers, million-part space vehicles, and all the other artifacts of human technology. Can scientists manage to explain it, and in a microscopic system to boot? The answer is undoubtedly *yes*. Yes, for social reasons if no other. Scientists have been charged with conquering cancer, genetic disease, and viral infection, all of which are cellular disorders, and they are massively funded to accomplish these tasks. They know roughly the way to reach the goals demanded by the public, and they will not fail. Science, like art, and as always through history, follows patronage.

Rapidly improving instrumentation already allows biologists to probe the interior of living cells and inspect the molecular architecture directly. They are discovering some of the simplicities by which adaptive systems organize themselves. Among the most notable simplicities are the rules used to fold the flexuous strings of amino acids into serviceable shapes of protein molecules, and the powered filtering devices by which membranes admit selected substances in and out of the cell and organelles. Scientists are also acquiring the computational capacity needed to simulate these and even more complex processes. In 1995 an American team using two linked Intel paragon computers set a world speed record of 281 billion calculations per second. The U.S. federal high-performance program has upped the goal to a trillion calculations per second by the end of the century. By the year 2020, petacrunchers, capable of reaching a thousand trillion calculations per second, may be possible, although new technologies and programming methods will be needed to reach that level. At this point brute-force simulation of cell mechanics, tracking every active molecule and its web of interactions, should be attainable—even without the simplifying principles envisioned in complexity theory.

Scientists also foresee early solutions to the self-assembly of finished cells into tissues and whole multicellular organisms. In 1994 the editors of *Science*, celebrating the inauguration of developmental biology by Wilhelm Roux a century earlier, asked one hundred contemporary researchers in the field to identify what they considered the crucial unanswered questions in the discipline. Their responses, in rank order of attributed importance, were:

1. The molecular mechanisms of tissue and organ development.
2. The connection between development and genetic evolution.
3. The steps by which cells become committed to a particular fate.
4. The role of cell-to-cell signaling in tissue development.

5. The self-assembly of tissue patterns in the early embryo.
6. The manner in which nerve cells establish their specific connections to create the nerve cord and brain.
7. The means by which cells choose to divide and to die in the sculpting of tissues and organs.
8. The steps by which the processes controlling transcription (the transmission of DNA information within the cell) affect the differentiation of tissues and organs.

Remarkably, the biologists considered research on all of these topics to be in a state of rapid advance, with partial success in at least some of them close at hand.

LET US SUPPOSE that early in the next century the hopes of the molecular and cellular biologists are fully realized. Suppose further that the researchers succeed in breaking a human cell down into all its component parts, track the processes, and accurately model the whole system from the molecules up. And suppose finally that the developmental biologists, whose focus is on tissues and organs, enjoy similar success. The stage will then be set for the final assault on the still more complex systems of mind and behavior. They are, after all, products of the selfsame kinds of molecules, tissues, and organs.

Let us see how such explanatory power might be acquired. With a close approximation of organic processes in a few species completed, it will be possible to infer how life is reproduced and maintained in an indefinite number of other species. With such an expansion of comparative holistic biology, a picture could be drawn of life as it is today, as it also was in the earliest stages of its evolution, and as it might be on other planets with different but habitable environments. In visualizing habitable environments, we will need to be liberal, keeping in mind that algae grow within rocks in Antarctica and microorganisms thrive in the boiling water of deep-sea thermal vents.

At some point deep and powerful principles of complexity may well emerge from the large ensemble of simulations. They will reveal the algorithms conserved across many levels of organization up to the most complex systems conceivable. These systems will be self-assembled, sustainable, and constantly changing yet perfectly reproducing. In other words, they will be living organisms.

At this time, if it comes, and I believe it will come, we will have a true theory of biology, as opposed to thick descriptions of particular living processes that now constitute the science. Its principles will accelerate inquiry into mind, behavior, and ecosystems, which are products of organisms and, by virtue of their extreme complexity, the ultimate challenge.

So the important questions are, first, do general organizing principles exist that allow a living organism to be reconstituted in full without recourse to brute force simulation of all its molecules and atoms? Second, will the same principles apply to mind, behavior, and ecosystems? Third, is there a body of mathematics that will serve as a natural language for biology, parallel to the one that works so well for physics? Fourth, even if the correct principles are discovered, how detailed must factual information be in order to use those principles in the desired models? In all these matters we see today as through a glass, darkly. In time, to complete the biblical allusion, we will come face to face with it all—and perhaps see it clearly. In any case, the search for answers will test the full powers of the human intellect.

CHAPTER 6

The Mind

Belief in the intrinsic unity of knowledge—the reality of the labyrinth—rides ultimately on the hypothesis that every mental process has a physical grounding and is consistent with the natural sciences. The mind is supremely important to the consilience program for a reason both elementary and disturbingly profound: Everything that we know and can ever know about existence is created there.

The loftier forms of such reflection and belief may seem at first to be the proper domain of philosophy, not science. But history shows that logic launched from introspection alone lacks thrust, can travel only so far, and usually heads in the wrong direction. Much of the history of modern philosophy, from Descartes and Kant forward, consists of failed models of the brain. The shortcoming is not the fault of the philosophers, who have doggedly pushed their methods to the limit, but a straightforward consequence of the biological evolution of the brain. All that has been learned empirically about evolution in general and mental process in particular suggests that the brain is a machine assembled not to understand itself, but to survive. Because these two ends are basically different, the mind unaided by factual knowledge from science sees the world only in little pieces. It throws a spotlight on those portions of the world it must know in order to live to the next day, and surrenders the rest to darkness. For thousands of generations people lived and reproduced with no need to know how the machinery of the brain works. Myth and self-deception, tribal identity and ritual, more than objective truth, gave them the adaptive edge.

That is why even today people know more about their automobiles than they do about their own minds—and why the fundamental explanation of mind is an empirical rather than a philosophical or religious quest. It requires a journey into the brain's interior darkness with preconceptions left behind. The ships that brought us here are to be left scuttled and burning at the shore.

THE BRAIN IS a helmet-shaped mass of gray and white tissue about the size of a grapefruit, one to two quarts in volume, and on average weighing three pounds (Einstein's brain, for example, was 2.75 pounds). Its surface is wrinkled like that of a cleaning sponge, and its consistency is custardlike, firm enough to keep from puddling on the floor of the brain case, soft enough to be scooped out with a spoon.

The brain's true meaning is hidden in its microscopic detail. Its fluffy mass is an intricately wired system of about a hundred billion nerve cells, each a few millionths of a meter wide and connected to other nerve cells by hundreds or thousands of endings. If we could shrink ourselves to the size of a bacterium and explore the brain's interior on foot, as philosophers since Leibniz in 1713 have imagined doing, we might eventually succeed in mapping all the nerve cells and tracking all the electrical circuits. But we could never thereby understand the whole. Far more information is needed. We need to know what the electric patterns mean, as well as how the circuits were put together and, most puzzling of all, for what purpose.

What we know of the heredity and development of the brain shows them to be almost unimaginably complicated. The human genome database accumulated to 1995 reveals that the brain's structure is prescribed by at least 3,195 distinctive genes, 50 percent more than for any other organ or tissue (the total number of genes in the entire human genome is estimated to be 50,000 to 100,000). The molecular processes that guide the growth of neurons to their assigned places have only begun to be deciphered. Overall, the human brain is the most complex object known in the universe—known, that is, to itself.

It rose by evolution to its present form swiftly, even by the standards of the generally hurried pace of mammalian phylogeny evident in the fossil record. Across three million years, from the ancestral man-apes of Africa to the earliest anatomically modern *Homo sapiens*, who lived about 200,000 years ago, the brain increased in volume four times over. Much of the growth occurred in the neocortex, the seat of the higher functions of mind, including, especially, language and its symbol-based product, culture.

The result was the capacity to take possession of the planet. Advanced humans, their big spherical skulls teetering precariously on fragile stems of compacted cervical vertebrae, walked, paddled, and sailed out of Africa through Europe and Asia and thence to all the remaining continents and great archipelagoes except uninhabitable Antarctica. By 1000 A.D. they reached the outermost islands of the Pacific and Indian Oceans. Only a handful of remote mid-Atlantic islands, including St. Helena and the Azores, remained pristine for a few centuries longer.

It is, I must acknowledge, unfashionable in academic circles nowadays to speak of evolutionary progress. *All the more reason to do so.* In fact, the dilemma that has consumed so much ink can be evaporated with a simple semantic distinction. If we mean by progress the advance toward a preset goal, such as that composed by intention in the human mind, then evolution by natural selection, which has no preset goals, is not progress. But if we mean the production through time of increasingly complex and controlling organisms and societies, in at least some lines of descent, with regression always a possibility, then evolutionary progress is an obvious reality. In this second sense, the human attainment of high intelligence and culture ranks as the last of the four great steps in the overall history of life. They followed one upon the other at roughly one-billion-year intervals. The first was the beginning of life itself, in the form of simple bacteriumlike organisms. Then came the origin of the complex eukaryotic cell through the assembly of the nucleus and other membrane-enclosed organelles into a tightly organized unit. With the eukaryotic building block available, the next advance was the origin of large, multicellular animals such as crustaceans and mollusks, whose movements were guided by sense organs and central nervous systems. Finally, to the grief of most preexisting life forms, came humanity.

VIRTUALLY ALL contemporary scientists and philosophers expert on the subject agree that the mind, which comprises consciousness and rational process, is the brain at work. They have rejected the mind-brain dualism of René Descartes, who in *Meditationes* (1642) concluded that "by the divine power the mind can exist without the body, and the body without the mind." According to the great philosopher, the noncorporeal mind and hence the immortal soul repose somewhere in the corporeal and mortal body. Its location, he suggested, might be the pineal gland, a tiny organ located at the base of the brain. In this early neurobiological model, the brain receives information from all over the body and feeds it into the pineal headquarters, where it is translated somehow into conscious thought. Dualism was congenial to the philosophy and science of Descartes' time, appealing as it did to the materialistic explanation of the universe while remaining safely pious. In one form or other, it has persisted into the late twentieth century.

The brain and its satellite glands have now been probed to the point where no particular site remains that can reasonably be supposed to harbor a nonphysical mind. The pineal gland, for example, is known to secrete the hormone melatonin and to assist in regulating the body's biological clock and daily rhythms. But even as mind-body dualism is being completely abandoned at long last, in the 1990s, scientists remain unsure about the precise material basis of mind. Some are convinced that conscious experience has unique physical and biological properties that remain to be discovered. A few among them, archly called the mysterians by their colleagues, believe that conscious experience is too alien, too complex, or both, ever to be comprehended.

No doubt, the transcendent difficulty of the subject inspires this kind of denial. As late as 1970 most scientists thought the concept of mind a topic best left to philosophers. Now the issue has been joined where it belongs, at the juncture of biology and psychology. With the aid of powerful new techniques, researchers have shifted the frame of discourse to a new way of thinking, expressed in the language of nerve cells, neurotransmitters, hormone surges, and recurrent neural networks.

The cutting edge of the endeavor is cognitive neuroscience, also and more popularly known as the brain sciences, an alliance formed by neurobiologists, cognitive psychologists, and a new school of empirically minded philosophers sometimes referred to as neurophilosophers. Their research reports are dispatched weekly to premier scientific journals, and their theories and impassioned disagreements fill the pages of such open-commentary periodicals as *Behavioral and Brain Sciences*. Many of the popular books and articles they write rank among the best in contemporary science exposition.

Such traits are the hallmark of the heroic period, or romantic period as it is often called, experienced by every successful scientific discipline during its youth. For a relatively brief interval, usually a decade or two, rarely more than half a century, researchers are intoxicated with a mix of the newly discovered and the imaginable unknown. For the first time the really important questions are asked in a form that can be answered, thus: *What are the cellular events that compose the mind?* Not create the mind—too vague, that expression—but compose the mind. The pioneers are paradigm hunters. They are risk takers, who compete with rival theorists for big stakes and are willing to endure painful shake-outs. They bear comparison with explorers of the sixteenth century, who, having discovered a new coastline, worked rivers up to the fall line, drew crude maps, and commuted home to beg for more expeditionary funds. And governmental and private patrons of the brain scientists, like royal geographic commissions of past centuries, are generous. They know that history can be made by a single sighting of coastland, where inland lies virgin land and the future lineaments of empire.

Call the impulse Western if you wish, call it androcentric, and by all means dismiss it as colonialist if you feel you must. I think it instead basic to human nature. Whatever its source, the impulse drives major scientific advance. During my career I have been privileged to witness close at hand the heroic periods of molecular biology, plate tectonics in geology, and the modern synthesis of evolutionary biology. Now it is the turn of the brain sciences.

THE EARLY GROUNDWORK for the revolution was laid in the nineteenth century by physicians, who noticed that injuries to certain parts of the brain result in special kinds of disability. Perhaps the most famous case was that of Phineas P. Gage, who in 1848 was a young construction foreman in charge of a crew laying railroad track across Vermont. Part of the job was to blast away outcrops of hard rock in order to straighten out turns in the advancing path. As Gage pressed powder into a newly drilled hole, a premature explosion fired the iron tamping bar like a missile toward his head. It entered his left cheek and exited the top of his skull, carrying with it a good part of the prefrontal lobe of his cerebral cortex, then arced away more than a hundred feet before coming to earth. Gage fell to the ground, miraculously still alive. To the amazement of all, he was able within minutes to sit up and even walk with assistance. He never lost consciousness. "Wonderful accident" was the later headline in the *Vermont Mercury*. In time his external injuries healed, and he retained the ability to speak and reason. But his personality had changed drastically. Where previously he had been cheerful, responsible, and well-mannered, a valued employee of the Rutland & Burlington Railroad, now he was a habitual liar, unreliable at work, and given to vagrant, self-destructive behavior. Studies on other patients with injuries to the same part of the brain over many years have confirmed the general conclusion suggested by Gage's misfortune: The prefrontal lobe houses centers important for initiative and emotional balance.

For two centuries the medical archives have filled with such anecdotes on the effects of localized brain damage. The data have made it possible for neurologists to piece together a map of functions performed by different parts of the brain. The injuries, which occur throughout the brain, include physical traumas, strokes, tumors, infections, and poisoning. They vary in extent from barely detectable pinpoints to deletions and transections of large parts of the brain. Depending on location and magnitude, they have multifarious effects on thought and behavior.

The most celebrated recent case is that of Karen Ann Quinlan. On April 14, 1975, the young New Jersey woman, while dosed with the tranquilizer Valium and painkiller drug Darvon, made the mistake of drinking gin and tonic. Although the combination does not sound dangerous, it essentially killed Karen Ann Quinlan. She fell into a coma that lasted until her death from massive infections ten years later. An autopsy revealed that her brain was largely intact, which explains why her body survived and even continued its daily cycle of waking and sleep. It lived on even when Quinlan's parents arranged, in the midst of national controversy, to have her ventilator removed. The autopsy revealed that Quinlan's brain damage was local but very severe: The thalamus had been obliterated as though burned out with a laser. Why that particular center deteriorated is unknown. A brain injured by a heavy blow or certain forms of poisoning usually responds by widespread swelling. If the reaction is intense, it presses on centers that control heartbeat and respiration, shutting down blood circulation and soon ending in death of the whole body.

The result of thalamus excision alone is brain death, or, more precisely put, mind death. The thalamus comprises twin egg-shaped masses of nerve cells near the center of the brain. It functions as a relay center through which all sensory information other than smell is transmitted to the cerebral cortex, and therefore to the conscious mind. Even dreams are triggered by impulses that pass through thalamic circuits. Quinlan's drug accident was the equivalent of blowing up a power station: All her lights downline went out, and she entered a sleep from which she had no chance of wakening. Her cerebral cortex lived on, waiting to be activated. But consciousness, even in dreams, was no longer possible.

Such research on brain damage, while enormously informative, is nevertheless dependent on chance occurrence. Over the years it has been greatly enhanced by experimental brain surgery. Neurosurgeons routinely keep patients conscious to test their response to electrical stimulation of the cortex, in order to locate healthy tissue and avoid excising it. The procedure is not uncomfortable: Brain tissue, while processing impulses from all over the body, has no receptors of its own. Instead of pain, the roving probes evoke a medley of sensations and muscular contractions. When certain sites on the surface of the cortex are stimulated, patients experience images, melodies, incoherent sounds, and a gamut of other impressions. Sometimes they involuntarily move fingers and other body parts.

Beginning with experiments in brain surgery by Wilder Penfield and other pioneers in the 1920s and 1930s, researchers have mapped sensory and motor functions over all parts of the cerebral cortex. The method is nevertheless limited in two important respects. It is not easily extended beneath the cortex into the dark nether regions of the brain, and it cannot be used to observe neural activity through time. To reach those objectives—to create motion pictures of the whole brain in action—scientists have adopted a broad range of sophisticated techniques borrowed from physics and chemistry. Since its inception in the 1970s, brain imaging, as the methods are collectively called, has followed a trajectory similar to that of microscopy, toward ever finer resolution in snapshots separated by shrinking intervals of time. The scientists hope eventually to monitor the activity of entire networks of individual nerve cells, both continuously and throughout the living brain.

GRANTED, the brain's machinery remains forbiddingly alien and scientists have traced only a minute fraction of its circuitry. Still, the major anatomical features of the brain are known, and a great deal has been learned of their various functions. Before addressing the nature of mind as a product of these operations, I wish to provide a quick look at the physical groundwork.

The surest way to grasp complexity in the brain, as in any other biological system, is to think of it as an engineering problem. What are the broad principles needed to create a brain from scratch? Whether contrived by advance planning or by blind natural selection, the key features of architecture can be expected to be very broadly predictable. Researchers in biomechanics have discovered time and again that organic structures evolved by natural selection conform to high levels of efficiency when judged by engineering criteria. And at a more microscopic level, biochemists marvel at the exactitude and power of the enzyme molecules controlling the actions of the cells. Like the mills of God, the processes of evolution grind slowly—yet, as the poet said, they grind exceeding fine.

So let us spread the specification sheets out and consider the brain as a solution to a set of physical problems. It is best to start with simple geometry. Because a huge amount of circuitry is required, and the wiring elements must be built from living cells, a relatively huge mass of new tissue needs to be manufactured and housed in the brain case. The ideal brain case will be spherical or close to it. One compelling reason is that a sphere has the smallest surface relative to volume of any geometric form and hence provides the least access to its vulnerable interior. Another reason is that a sphere allows more circuits to be placed close together. The average length of circuits can thus be minimized, raising the speed of transmission while lowering the energy cost for their construction and maintenance.

Because the basic units of the brain-machine must be made of cells, it is best to stretch these elements out into string-shaped forms that serve simultaneously as receiving stations and coaxial cables. The dual-purpose cells created by evolution are in fact the neurons, also called nerve cells or nerve fibers. It is further practical to design the neurons so that their main bodies serve as the receiving sites for impulses from other cells. The neurons can send their own signals out along axons, cablelike extensions of the cell bodies.

For speed, make the transmission an electric discharge by depolarization of the cell membrane. The neurons are then said to "fire." For accuracy during neuron firing, surround the axons with insulating sheaths. These in fact exist as white fatty myelin membranes that together give the brain its light color.

To achieve a higher level of integration, the brain must be very intricately and precisely wired. Given again that its elements are living cells, the number of neuron connections are best multiplied by growing threadlike extensions from the tips of the axons, which reach out and transmit individually to the bodies of many other cells. The discharge of the axon travels to these multiple terminal extensions all the way to their tips, which then make contact with the receptor cells. The receptor cells accept some of the tips of the terminal axon branches on the surface of their main cell bodies. They accept other tips on their dendrites, which are threadlike receptor branches growing out from the cell bodies.

Now visualize the entire nerve cell as a miniature squid. From its body sprouts a cluster of tentacles (the dendrites). One tentacle (the axon) is much longer than the others, and from its tip it sprouts more tentacles. The message is received on the body and short tentacles of the squid and travels along the long tentacle to other squids. The brain comprises the equivalent of one hundred billion squids linked together.

The cell-to-cell connections—more precisely, the points of connection and the ultramicroscopic spaces separating them—are called synapses. When an electric discharge reaches a synapse, it induces the tip of the terminal branch to release a neurotransmitter, a chemical that either excites an electric discharge in the receiving cell or prevents one from occurring. Each nerve cell sends signals to hundreds or thousands of other cells through its synapses at the end of its axon, and it receives input from a similar myriad of synapses on its main cell body and dendrites. In each instant a nerve cell either fires an impulse along its axon to other cells or falls silent. Which of the two responses it makes depends on the summation of the neurotransmissions received from all the cells that feed stimuli into it.

The activity of the brain as a whole, hence the wakefulness and moods experienced by the conscious mind, is profoundly affected by the levels of the neurotransmitters that wash its trillions of synapses. Among the most important of the neurotransmitters are acetylcholine and the amines norepinephrine, serotonin, and dopamine. Others include the amino acid GABA (gamma aminobutyric acid) and, surprisingly, the elementary gas nitrous oxide. Some neurotransmitters excite the neurons they contact, while others inhibit them. Still others can exert either effect depending on the location of the circuit within the nervous system.

During development of the nervous system in the fetus and infant, the neurons extend their axons and dendrites into the cellular environment—like growing tentacles of squids. The connections they make are precisely programmed and guided to their destinations by chemical cues. Once in place each neuron is poised to play a special role in signal transmission. Its axon may stretch only a few millionths of a meter or thousands of times longer. Its dendrites and terminal axon branches can take any of a number of forms, coming to resemble, say, the leafless crown of a tree in winter or a dense, feltlike mat. Possessing the aesthetic inherent to pure function, and riveting to behold, they invite us to imagine their powers. Concerning them, Santiago Ramón y Cajal, the great Spanish histologist, wrote of his own experience, after receiving the 1906 Nobel Prize for his research on the subject: "Like the entomologist in pursuit of brightly colored butterflies, my attention hunted, in the flower garden of the gray matter, cells with delicate and elegant forms, the mysterious butterflies of the soul, the beatings of whose wings may some day—who knows?—clarify the secret of mental life."

The meaning of the neuron shape, which so pleases the biologist, is this: Neuron systems are directed networks, receiving and broadcasting signals. They cross-talk with other complexes to form systems of systems, in places forming a circle, like a snake catching its own tail, to create reverberating circuits. Each neuron is touched by the terminal axon branches of many other neurons, established by a kind of democratic vote whether it is to be active or silent. Using a Morselike code of staccato firing, the cell sends its own messages outward to others. The number of connections made by the cell, their pattern of spread, and the code they use determine the role the cell plays in the overall activity of the brain.

Now to complete the engineering metaphor. When you're setting out to design a hominid brain, it is important to observe another optimum design principle: Information transfer is improved when neuron circuits filling specialized functions are placed together in clusters. Examples of such aggregates in the real brain are the sensory relay stations, integrative centers, memory modules, and emotional control centers identified thus far by neurobiologists. Nerve cell bodies are gathered in flat assemblages called layers and rounded ones called nuclei. Most are placed at or near the surface of the brain. They are interconnected both by their own axons and by intervening neurons that course through the deeper brain tissues. One result is the gray or light-brown color of the surface due to the massing of the cell bodies—the "gray matter" of the brain—and a white color from the myelin sheaths of axons in the interior of the brain.

Human beings may possess the most voluminous brain in proportion to body size of any large animal species that has ever lived. For a primate species the human brain is evidently at or close to its physical limit. If it were much larger in the newborn, the passage of its protecting skull through the birth canal would be dangerous to both mother and child. Even the adult brain size is mechanically risky: The head is a fragile, internally liquescent globe balanced on a delicate bone-and-muscle stem, within which the brain is vulnerable and the mind easily stunned and disabled. Human beings are innately disposed to avoid violent physical contact. Because our evolving ancestors traded brute strength for intelligence, we no longer need to seize and rip enemies with fanged jaws.

Given this intrinsic limit in brain volume, some way must be found to fit in the memory banks and higher-order integrating systems needed to generate conscious thought. The only means available is to increase surface area: Spread the cells out into a broad sheet and crumple it up into a ball. The human cerebral cortex is such a sheet about one thousand square inches in area, packed with millions of cell bodies per square inch, folded and wadded precisely like an origami into many winding ridges and fissures, neatly stuffed in turn into the quart-sized cranial cavity.

WHAT MORE CAN be said of brain structure? If a Divine Engineer designed it, unconstrained by humanity's biological history, He might have chosen mortal but angelic beings cast in His own image. They would presumably be rational, far-seeing, wise, benevolent, unrebellious, selfless, and guilt-free, and, as such, ready-made stewards of the beautiful planet bequeathed them. But we are nothing like that. We have original sin, which makes us *better* than angels. Whatever good we possess we have earned, during a long and arduous evolutionary history. The human brain bears the stamp of 400 million years of trial and error, traceable by fossils and molecular homology in nearly unbroken sequence from fish to amphibian to reptile to primitive mammal to our immediate primate forerunners. In the final step the brain was catapulted to a radically new level, equipped for language and culture. Because of its ancient pedigree, however, it could not be planted like a new computer into an empty cranial space. The old brain had been assembled there as a vehicle of instinct, and remained vital from one heartbeat to the next as new parts were added. The new brain had to be jury-rigged in steps within and around the old brain. Otherwise the organism could not have survived generation by generation. The result was human nature: genius animated with animal craftiness and emotion, combining the passion of politics and art with rationality, to create a new instrument of survival.

Brain scientists have vindicated the evolutionary view of mind. They have established that passion is inseverably linked to reason. Emotion is not just a perturbation of reason but a vital part of it. This chimeric quality of the mind is what makes it so elusive. The hardest task of brain scientists is to explain the products-tested engineering of the cortical circuits against the background of the species' deep history. Beyond the elements of gross anatomy I have just summarized, the hypothetical role of Divine Engineer is not open to them. Unable to deduce from first principles the optimum balance of instinct and reason, they must ferret out the location and function of the brain's governing circuits one by one. Progress is measured by piecemeal discoveries and cautious inferences. Here are a few of the most important made by researchers to date:

- The human brain preserves the three primitive divisions found throughout the vertebrates from fishes to mammals: hindbrain, midbrain, and forebrain. The first two together, referred to as the brain stem, form the swollen posthead on which the massively enlarged forebrain rests.

• The hindbrain comprises in turn the pons, medulla, and cerebellum. Together they regulate breathing, heartbeat, and coordination of body movements. The midbrain controls sleep and arousal. It also partly regulates auditory reflexes and perception.

• A major part of the forebrain is composed of the limbic system, the master traffic-control complex that regulates emotional response as well as the integration and transfer of sensory information. Its key centers are the amygdala (emotion), hippocampus (memory, especially short-term memory), hypothalamus (memory, temperature control, sexual drive, hunger, and thirst), and thalamus (awareness of temperature and all other senses except smell, awareness of pain, and the mediation of some processes of memory).

• The forebrain also includes the cerebral cortex, which has grown and expanded during evolution to cover the rest of the brain. As the primary seat of consciousness, it stores and collates information from the senses. It also directs voluntary motor activity and integrates higher functions, including speech and motivation.

• The key functions of the three successive divisions—hind- plus midbrain, limbic system, and cerebral cortex—can be neatly summarized in this sequence: *heartbeat, heartstrings, heartless.*

• No single part of the forebrain is the site of conscious experience. Higher levels of mental activity sweep through circuits that embrace a large part of the forebrain. When we see and speak of color, for example, visual information passes from the cones and interneurons of the retina through the thalamus to the visual cortex at the rear of the brain. After the information is codified and integrated anew at each step, through patterns of neuron firing, it then spreads forward to the speech centers of the lateral cortex. As a result, we first see red and then say "red." Thinking about the phenomenon consists of adding more and more connections of pattern and meaning, and thus activating additional areas of the brain. The more novel and complicated the connections, the greater the amount of this spreading activation. The better the connections are learned by such experience, the more they are put on autopilot. When the same stimulus is applied later, new activation is diminished and the circuits are more predictable. The procedure becomes a "habit." In one such inferred pathway of memory formation, sensory information is conveyed from the cerebral cortex to the amygdala and hippocampus, then to the thalamus, then to the prefrontal cortex (just behind the brow), and back to the original sensory regions of the cortex for storage. Along the way codes are interpreted and altered according to inputs from other parts of the brain.

• Because of the microscopic size of the nerve cells, a large amount of circuitry can be packed into a very small space. The hypothalamus, a major relay and control center at the base of the brain, is about the size of a lima bean. (The nervous systems of animals are even more impressively miniaturized. The entire brains of gnats and other extremely small insects, which carry instructions for a series of complex instinctive acts, from flight to mating, are barely visible to the naked eye.)

• Disturbance of particular circuits of the human brain often produce bizarre results. Injuries to certain sites of the undersurface of the parietal and occipital lobes, which occupy the side and rear of the cerebral cortex, cause the rare condition called prosopagnosia. The patient can no longer recognize other persons by their faces, but he can still remember them by their voices. Just as oddly, he retains the ability to recognize objects other than faces by sight alone.

• There may be centers in the brain that are especially active in the organization and perception of free will. One appears to be located within or at least close to the anterior cingulate sulcus, on the inside of a fold of the cerebral cortex. Patients who have sustained damage to the region lose initiative and concern for their own welfare. From one moment to the next they focus on nothing in particular, yet remain capable of reasoned responses when pressed.

• Other complex mental operations, while engaging regions over large parts of the brain, are vulnerable to localized perturbation. Patients with temporal lobe epilepsy often develop hyperreligiosity, the tendency to charge all events, large and small, with cosmic significance. They are also prone to hypergraphia, a compulsion to express their visions in an undisciplined stream of poems, letters, or stories.

• The neural pathways used in sensory integration are also highly specialized. When subjects name pictures of animals during PET (positron emission tomography) imaging, a method that reveals patterns of nerve-cell firing, their visual cortices light up in the same pattern seen when they sort out subtle differences in the appearance of objects. When, on the other hand, they silently name pictures of tools, neural activity shifts to parts of the cortex concerned with hand movements and action words, such as "write" for pencil.

I HAVE SPOKEN so far about the physical processes that produce the mind. Now, to come to the heart of the matter, what *is* the mind? Brain scientists understandably dance around this question. Wisely, they rarely commit themselves to a simple declarative definition. Most believe that the fundamental properties of the elements responsible for mind—neurons, neurotransmitters, and hormones—are reasonably well known. What is lacking is a sufficient grasp of the emergent, holistic properties of the neuron circuits, and of cognition, the way the circuits process information to create perception and knowledge. Although dispatches from the research front grow yearly in number and sophistication, it is hard to judge how much we know in comparison with what we need to know in order to create a powerful and enduring theory of mind production by the brain. The grand synthesis could come quickly, or it could come with painful slowness over a period of decades.

Still, the experts cannot resist speculation on the essential nature of mind. While it is very risky to speak of consensus, and while I have no great trust in my own biases as interpreter, I believe I have been able to piece together enough of their overlapping opinions to forecast a probable outline of the eventual theory, as follows.

Mind is a stream of conscious and subconscious experience. It is at root the coded representation of sensory impressions and the memory and imagination of sensory impressions. The information composing it is most likely sorted and retrieved by vector coding, which denotes direction and magnitude. For example, a particular taste might be partly classified by the combined activity of nerve cells responding to different degrees of sweetness, saltiness, and sourness. If the brain were designed to distinguish ten increments in each of these taste dimensions, the coding could discriminate 10 × 10 × 10, or 1,000 substances.

Consciousness consists of the parallel processing of vast numbers of such coding networks. Many are linked by the synchronized firing of the nerve cells at forty cycles per second, allowing the simultaneous internal mapping of multiple sensory impressions. Some of the impressions are real, fed by ongoing stimulation from outside the nervous system, while others are recalled from the memory banks of the cortex. All together they create scenarios that flow realistically back and forth through time. The scenarios are a virtual reality. They can either closely match pieces of the external world or depart indefinitely far from it. They re-create the past and cast up alternative futures that serve as choices for future thought and bodily action. The scenarios comprise dense and finely differentiated patterns in the brain circuits. When fully open to input from the outside, they correspond well to all the parts of the environment, including activity of the body parts, monitored by the sense organs.

Who or what within the brain monitors all this activity? No one. Nothing. The scenarios are not seen by some other part of the brain. They just *are*. Consciousness is the virtual world composed by the scenarios. There is not even a Cartesian theater, to use Daniel Dennett's dismissive phrase, no single locus of the brain where the scenarios are played out in coherent form. Instead, there are interlacing patterns of neural activity within and among particular sites throughout the forebrain, from cerebral cortex to other specialized centers of cognition such as the thalamus, amygdala, and hippocampus. There is no single stream of consciousness in which all information is brought together by an executive ego. There are instead multiple streams of activity, some of which contribute momentarily to conscious thought and then phase out. Consciousness is the massive coupled aggregates of such participating circuits. The mind is a self-organizing republic of scenarios that individually germinate, grow, evolve, disappear, and occasionally linger to spawn additional thought and physical activity.

The neural circuits do not turn on and off like parts of an electrical grid. In many sectors of the forebrain at least, they are arranged in parallel relays stepping from one neuron level to the next, integrating more and more coded information with each step. The energy of light striking the retina, to expand the example I gave earlier, is transduced into patterns of neuron firing. The patterns are relayed through a sequence of intermediate neuron systems out of the retinal fields through the lateral geniculate nuclei of the thalamus back to the primary visual cortex at the rear of the brain. Cells in the visual cortex fed by the integrated stimuli sum up the information from different parts of the retina. They recognize and by their own pattern of firing specify spots or lines. Further systems of these higher-order cells integrate the information from multiple feeder cells to map the shape and movement of objects. In ways still not understood, this pattern is coupled with simultaneous input from other parts of the brain to create the full scenarios of consciousness. The biologist S. J. Singer has drily expressed the matter thus: I link, therefore I am.

Because just to generate consciousness requires an astronomically large population of cells, the brain is sharply limited in its capacity to create and hold complex moving imagery. A key measure of that capacity lies in the distinction made by psychologists between short-term and long-term memory. Short-term memory is the ready state of the conscious mind. It composes all of the current and remembered parts of the virtual scenarios. It can handle only about seven words or other symbols simultaneously. The brain takes about one second to scan these symbols fully, and it forgets most of the information within thirty seconds. Long-term memory takes much longer to acquire, but it has an almost unlimited capacity, and a large fraction of it is retained for life. By spreading activation, the conscious mind summons information from the store of long-term memory and holds it for a brief interval in short-term memory. During this time it processes the information, at a rate of about one symbol per 25 milliseconds, while scenarios arising from the information compete for dominance.

Long-term memory recalls specific events by drawing particular persons, objects, and actions into the conscious mind through a time sequence. For example, it easily re-creates an Olympic moment: the lighting of the torch, a running athlete, the cheering of the crowd. It also re-creates not just moving images and sound but *meaning* in the form of linked concepts simultaneously experienced. Fire is connected to hot, red, dangerous, cooked, the passion of sex, and the creative act, and on out through multitudinous hypertext pathways selected by context, sometimes building new associations in memory for future recall. The concepts are the nodes or reference points in long-term memory. Many are labeled by words in ordinary language, but others are not. Recall of images from the long-term banks with little or no linkage is just memory. Recall with linkages, and especially when tinged by the resonance of emotional circuits, is remembrance.

The capacity for remembrance by the manipulation of symbols is a transcendent achievement for an organic machine. It has authored all of culture. But it still falls far short of the demands placed by the body on the nervous system. Hundreds of organs must be regulated continuously and precisely; any serious perturbation is followed by illness or death. A heart forgetful for ten seconds can drop you like a stone. The proper functioning of the organs is under the control of hard-wired autopilots in the brain and spinal cord, whose neuron circuits are our inheritance from hundreds of millions of years of vertebrate evolution prior to the origin of human consciousness. The autopilot circuits are shorter and simpler than those of the higher cerebral centers and only marginally communicate with them. Only by intense meditative training can they occasionally be brought under conscious control.

Under automatic control, and specifically through balance of the antagonistic elements of the autonomic nervous system, pupils of the eye constrict or dilate, saliva pours out or is contained, the stomach churns or quietens, the heart pounds or calms, and so on through alternative states in all the organs. The sympathetic nerves of the autonomic nervous system pump the body up for action. They arise from the middle sections of the spinal cord, and typically regulate target organs by release of the neurotransmitter norepinephrine. The parasympathetic nerves relax the body as a whole while intensifying the processes of digestion. They rise from the brain stem and lowermost segment of the spinal cord, and the neurotransmitter they release to the target organs is acetylcholine—also the agent of sleep.

Reflexes are swift automatic responses mediated by short circuits of neurons through the spinal cord and lower brain. The most complex is the startle response, which prepares the body for an imminent blow or collision. Imagine that you are surprised by a loud noise close by—a car horn blasts, someone shouts, a dog charges in a fury of barking. You react without thinking. Your eyes close, your head sags, your mouth opens, your knees buckle slightly. All are reactions that prepare you for the violent contact that might follow an instant later. The startle response occurs in a split second, faster than the conscious mind can follow, faster than can be imitated by conscious effort even with long practice.

Automatic responses, true to their primal role, are relatively impervious to the conscious will. This principle of archaism extends even to the facial expressions that communicate emotion. A spontaneous and genuine smile, which originates in the limbic system and is emotion-driven, is unmistakable to the practiced observer. A contrived smile is constructed from the conscious processes of the cerebrum and is betrayed by telltale nuances: a slightly different configuration of facial muscle contraction and a tendency toward lopsidedness of the upward curving mouth. A natural smile can be closely imitated by an experienced actor. It can also be evoked by artificially inducing the appropriate emotion—the basic technique of method acting. In ordinary usage it is modified deliberately in accordance with local culture, to convey irony (the pursed smile), restrained politeness (the thin smile), threat (the wolfish smile), and other refined presentations of self.

Much of the input to the brain does not come from the outside world but from internal body sensors that monitor the state of respiration, heartbeat, digestion, and other physiological activities. The flood of "gut feeling" that results is blended with rational thought, feeding it, and being fed by it through reflexes of internal organs and neurohormonal loops.

As the scenarios of consciousness fly by, driven by stimuli and drawing upon memories of prior scenarios, they are weighted and modified by emotion. What is emotion? It is the modification of neural activity that animates and focuses mental activity. It is created by physiological activity that selects certain streams of information over others, shifting the body and mind to higher or lower degrees of activity, agitating the circuits that create scenarios, and selecting ones that end in certain ways. The winning scenarios are those that match goals preprogrammed by instinct and the satisfactions of prior experience. Current experience and memory continually perturb the states of mind and body. By thought and action the states are then moved backward to the original condition or forward to conditions conceived in new scenarios. The dynamism of the process provokes labeling by words that denote the basic categories of emotion—anger, disgust, fear, pleasure, surprise. It breaks the categories into many degrees and joins them to create myriad subtle compounds. Thus we experience feelings that are variously weak, strong, mixed, and new.

Without the stimulus and guidance of emotion, rational thought slows and disintegrates. The rational mind does not float above the irrational; it cannot free itself to engage in pure reason. There are pure theorems in mathematics but no pure thoughts that discover them. In the brain-in-the-vat fantasy of neurobiological theory and science fiction, the organ in its nutrient bath has been detached from the impediments of the body and liberated to explore the inner universe of the mind. But that is not what would ensue in reality. All the evidence from the brain sciences points in the opposite direction, to a waiting coffin-bound hell of the wakened dead, where the remembered and imagined world decays until chaos mercifully grants oblivion.

Consciousness satisfies emotion by the physical actions it selects in the midst of turbulent sensation. It is the specialized part of the mind that creates and sorts scenarios, the means by which the future is guessed and courses of action chosen. Consciousness is not a remote command center but part of the system, intimately wired to all the neural and hormonal circuits regulating physiology. Consciousness acts and reacts to achieve a dynamic steady state. It perturbs the body in precise ways with each changing circumstance, as required for well-being and response to opportunity, and helps return it to the original condition when challenge and opportunity have been met.

The reciprocity of mind and body can be visualized in the following scenario, which I have adapted from an account by the neurologist Antonio R. Damasio. Imagine that you are strolling along a deserted city street at night. Your reverie is interrupted by quick footsteps drawing close behind. Your brain focuses instantly and churns out alternative scenarios—ignore, freeze, turn and confront, or escape. The last scenario prevails and you act. You run toward a lighted storefront further down the street. In the space of a few seconds, the conscious response triggers automatic changes in your physiology.

The catecholamine hormones epinephrine ("adrenaline") and norepinephrine pour into the bloodstream from the adrenal medulla and travel to all parts of the body, increasing the basal metabolic rate, breaking down glycogen in the liver and skeletal muscles to glucose for a quick energy feed. The heart races. The bronchioles of the lungs dilate to admit more air. Digestion slows. The bladder and colon prepare to void their contents, disencumbering the body to prepare for violent action and possible injury.

A few seconds more pass. Time slows in the crisis: The event span seems like minutes. Signals arising from all the changes are relayed back to the brain by more nerve fibers and the rise of hormone titers in the bloodstream. As further seconds tick away, the body and brain shift together in precisely programmed ways. Emotional circuits of the limbic system kick in—the new scenarios flooding the mind are charged with fright, then anger that sharply focuses the attention of the cerebral cortex, closing out all other thought not relevant to immediate survival.

The storefront is reached, the race won. People are inside, the pursuer is gone. Was the follower really in pursuit? No matter. The republic of bodily systems, informed by reassuring signals from the conscious brain, begins its slow stand-down to the original calm state.

Damasio, in depicting the mind holistically in such episodes, has suggested the existence of two broad categories of emotion. The first, primary emotion, comprises the responses ordinarily called inborn or instinctive. Primary emotion requires little conscious activity beyond the recognition of certain elementary stimuli, the kind that students of instinctive behavior in animals call releasers—they are said to "release" the preprogrammed behavior. For human beings such stimuli include sexual enticement, loud noises, the sudden appearance of large shapes, the writhing movements of snakes or serpentine objects, and the particular configurations of pain associated with heart attacks or broken bones. The primary emotions have been passed down with little change from the vertebrate forebears of the human line. They are activated by circuits of the limbic system, among which the amygdala appears to be the master integrating and relay center.

Secondary emotions arise from personalized events of life. To meet an old friend, fall in love, win a promotion, or suffer an insult is to fire the limbic circuits of primary emotion, but only after the highest integrative processes of the cerebral cortex have been engaged. We must know who is friend or enemy, and why they are behaving a certain way. By this interpretation, the emperor's rage and poet's rapture are cultural elaborations retrofitted to the same machinery that drives the prehuman primates. Nature, Damasio observes, "with its tinkerish knack for economy, did not select independent mechanisms for expressing primary and secondary emotions. It simply allowed secondary emotions to be expressed by the same channel already prepared to convey primary emotions."

Ordinary words used to denote emotion and other processes of mental activity make only a crude fit to the models used by the brain scientists in their attempts at rigorous explanation. But the ordinary and conventional conceptions—what some philosophers call folk psychology—are necessary if we are to make better sense of thousands of years of literate history, and thereby join the cultures of the past with those of the future. To that end I offer the following neuroscience-accented definitions of several of the most important concepts of mental activity.

What we call *meaning* is the linkage among the neural networks created by the spreading excitation that enlarges imagery and engages emotion. The competitive selection among scenarios is what we call *decision making*. The outcome, in terms of the match of the winning scenario to instinctive or learned favorable states, sets the kind and intensity of subsequent emotion. The persistent form and intensity of emotions is called *mood*. The ability of the brain to generate novel scenarios and settle on the most effective among them is called *creativity*. The persistent production of scenarios lacking reality and survival value is called *insanity*.

The explicit material constructions I have put upon mental life will be disputed by some brain scientists, and reckoned inadequate by others. That is the unavoidable fate of synthesis. In choosing certain hypotheses over others, I have tried to serve as an honest broker searching for the gravitational center of opinion, where by and large the supporting data are most persuasive and mutually consistent. To include all models and hypotheses deserving respect in this tumultuous discipline, and then to clarify the distinctions among them, would require a full-dress textbook. Undoubtedly events will prove that in places I chose badly. For that eventuality I apologize now to the slighted scientists, a concession I comfortably make, knowing that the recognition they deserve and will inevitably receive cannot be blunted by premature omission on the part of any one observer.

THE SUBJECT thus qualified, I will next describe the deeper problems that must be resolved before the physical basis of mind can be said to be truly solved. The one universally judged to be the most difficult of all is the nature of subjective experience. The Australian philosopher David Chalmers recently put the matter in perspective by contrasting the "easy problems" of general consciousness with the "hard problem" of subjective experience. In the first group (easy, I suppose, in the sense that Mont Blanc is more readily climbed in beachwear than Everest) are the classical problems of mind research: how the brain responds to sensory stimuli, how it incorporates information into patterns, and how it converts the patterns into words. Each of these steps of cognition are the subjects of vigorous contemporary research.

The hard problem is more elusive: how physical processes in the brain addressed in the easy problems give rise to subjective feeling. What exactly does it mean when we say we *experience* a color such as red or blue? Or experience, in Chalmers' words, "the ineffable sound of a distant oboe, the agony of an intense pain, the sparkle of happiness or the meditative quality of a moment lost in thought. All are part of what I am calling consciousness. It is these phenomena that compose the real mystery of the mind."

An imaginary experiment proposed by the philosopher Frank Jackson in 1983 illustrates the supposed unattainability of subjective thought by the natural sciences. Consider a neurobiologist two centuries hence who understands all the physics of color and all the brain's circuitry giving rise to color vision. But the scientist (call her Mary) has never experienced color; she has been cloistered all her life in a black-and-white room. She does not know what it is like for another person to see red or blue; she cannot imagine how they feel about color. According to Jackson and Chalmers, it follows that there are qualities of conscious experience that cannot be deduced from knowledge of the physical functioning of the brain.

Although it is the nature of philosophers to imagine impasses and expatiate upon them at book length with schoolmasterish dedication, the hard problem is conceptually easy to solve. What material description might explain subjective experience? The answer must begin by conceding that Mary cannot know what it feels like to see color. The chromatic nuances of a westering sun are not hers to enjoy. And for the same reason she and all her fellow human beings *a fortiori* cannot know how a honeybee feels when it senses magnetism or what an electric fish thinks as it orients by an electric field. We can translate the energies of magnetism and electricity into sight and sound, the sensory modalities we biologically possess. We can read the active neural circuits of bees and fish by scanning their sense organs and brains. But we cannot feel as they do—ever. Even the most imaginative and expert observers cannot think as animals, however they may wish or deceive themselves otherwise.

But incapacity is not the point. The distinction that illuminates subjective experience lies elsewhere, in the respective roles of science and art. Science perceives who can feel blue and other sensations and who cannot feel them, and explains why that difference exists. Art in contrast transmits feelings among persons of the same capacity. In other words, science explains feeling, while art transmits it. The majority of human beings, unlike Mary, see a full color spectrum, and they feel its productions in reverberating pathways through the forebrain. The basic patterns are demonstrably similar across all color-sighted human beings. Variations exist, owing to remembrances that arise from the personal memories and cultural biases of different people. But in theory these variations can also be read in the patterns of their brain activity. The physical explanations derived from the patterns would be understandable to Mary the confined scientist. She might say, "Yes, that is the wavelength span classified by others as blue, and there is the pattern of neural activity by which it is recognized and named." The explanations would be equally clear to bee and fish scientists if their species could somehow be raised to human levels of intelligence.

Art is the means by which people of similar cognition reach out to others in order to transmit feeling. But how can we know for sure that art communicates this way with accuracy, that people really, truly *feel* the same in the presence of art? We know it intuitively by the sheer weight of our cumulative responses through the many media of art. We know it by detailed verbal descriptions of emotion, by critical analyses, and in fact through data from all the vast, nuanced, and interlocking armamentaria of the humanities. That vital role in the sharing of culture is what the humanities are all about. Nevertheless, fundamental new information will come from science by studying the dynamic patterns of the sensory and brain systems during episodes when commonly shared feelings are evoked and experienced through art.

But surely, skeptics will say, that is impossible. Scientific fact and art can never be translated one into the other. Such a response is indeed the conventional wisdom. But I believe it is wrong. The crucial link exists: The common property of science and art is the transmission of information, and in one sense the respective modes of transmission in science and art can be made logically equivalent. Imagine the following experiment: A team of scholars—led perhaps by color-challenged Mary—has constructed an iconic language from the visual patterns of brain activity. The result resembles a stream of Chinese ideograms, each one representing an entity, process, or concept. The new writing—call it "mind script"—is translated into other languages. As the fluency of its readers increases, the mind script can be read directly by brain imaging.

In the silent recesses of the mind, volunteer subjects recount episodes, summon adventure in dreams, recite poems, solve equations, recall melodies, and while they are doing this the fiery play of their neuronal circuitry is made visible by the techniques of neurobiology. The observer reads the script unfolding not as ink on paper but as electric patterns in live tissue. At least some of the thinker's subjective experience—his feeling—is transferred. The observer reflects, he laughs or weeps. And from his own mind patterns he is able to transmit the subjective responses back. The two brains are linked by perception of brain activity.

Whether seated across from one another at a table, or alone in separate rooms or even in separate cities, the communicants can perform feats that resemble extrasensory perception (ESP). But only superficially. The first thinker glances at a playing card he holds cupped in his hand. With no clue other than the neural imagery to guide him, the second thinker reads the face of the card. The first thinker reads a novel; the second thinker follows the narrative.

Accurate transmission of the mind script depends as much as conventional language does on the commonality of the users' culture. When the overlap is slight, the script may be limited in use to a hundred characters; when extensive, the lexicon can expand to thousands. At its most efficient, the script transmits the tones and flourishes indigenous to particular cultures and individual minds.

Mind script would resemble Chinese calligraphy, not only a medium employed for the communication of factual and conceptual information, but also one of the great art forms of Eastern civilization. The ideograms contain subtle variations with aesthetic and other subjective meanings of their own shared by writer and reader. Of this property the Sinologist Simon Leys has written, "The silk or paper used for calligraphy has an absorbent quality: the lightest touch of the brush, the slightest drop of ink, registers at once—irretrievably and indelibly. The brush acts like a seismograph of the mind, answering every pressure, every turn of the wrist. Like painting, Chinese calligraphy addresses the eye and is an art of space; like music, it unfolds in time; like dance, it develops a dynamic sequence of movements, pulsating in rhythm."

AN OLD IMPASSE nonetheless remains: If the mind is bound by the laws of physics, and if it can conceivably be read like calligraphy, how can there be free will? I do not mean free will in the trivial sense, the ability to choose one's thoughts and behavior free of the will of others and the rest of the world all around. I mean, instead, freedom from the constraints imposed by the physiochemical states of one's own body and mind. In the naturalistic view, free will in this deeper sense is the outcome of competition among the scenarios that compose the conscious mind. The dominant scenarios are those that rouse the emotion circuits and engage them to greatest effect during reverie. They energize and focus the mind as a whole and direct the body in particular courses of action. The self is the entity that seems to make such choices. But what is the self?

The self is not an ineffable being living apart within the brain. Rather, it is the key dramatic character of the scenarios. It must exist, and play on center stage, because the senses are located in the body and the body creates the mind to represent the governance of all conscious actions. The self and body are therefore inseparably fused: The self, despite the illusion of its independence created in the scenarios, cannot exist apart from the body, and the body cannot survive for long without the self. So close is this union that it is almost impossible to envision souls in heaven and hell without at least the fantastical equivalent of corporeal existence. Even Christ, we have been instructed, and Mary soon afterward, ascended to heaven in bodies—supernal in quality, but bodies nonetheless. If the naturalistic view of mind is correct, as all the empirical evidence suggests, and if there is also such a thing as the soul, theology has a new Mystery to solve. The soul is immaterial, this Mystery goes, it exists apart from the mind, yet it cannot be separated from the body.

The self, an actor in a perpetually changing drama, lacks full command of its own actions. It does not make decisions solely by conscious, purely rational choice. Much of the computation in decision making is unconscious—strings dancing the puppet ego. Circuits and determining molecular processes exist outside conscious thought. They consolidate certain memories and delete others, bias connections and analogies, and reinforce the neurohormonal loops that regulate subsequent emotional response. Before the curtain is drawn and the play unfolds, the stage has already been partly set and much of the script written.

The hidden preparation of mental activity gives the illusion of free will. We make decisions for reasons we often sense only vaguely, and seldom if ever understand fully. Ignorance of this kind is conceived by the conscious mind as uncertainty to be resolved; hence freedom of choice is ensured. An omniscient mind with total commitment to pure reason and fixed goals would lack free will. Even the gods, who grant that freedom to men and show displeasure when they choose foolishly, avoid assuming such nightmarish power.

Free will as a side product of illusion would seem to be free will enough to drive human progress and offer happiness. Shall we leave it at that? No, we cannot. The philosophers won't let us. They will say: Suppose that with the aid of science we knew all the hidden processes in detail. Would it then be correct to claim that the mind of a particular individual is predictable, and therefore truly, fundamentally determined and lacking in free will? We must concede that much in principle, but only in the following, very peculiar sense. If within the interval of a microsecond the active networks composing the thought were known down to every neuron, molecule, and ion, their exact state in the next microsecond might be predicted. But to pursue this line of reasoning into the ordinary realm of conscious thought is futile in pragmatic terms, for this reason: If the operations of a brain are to be seized and mastered, they must also be altered. In addition, the principles of mathematical chaos hold. The body and brain comprise noisy legions of cells, shifting microscopically in discordant patterns that unaided consciousness cannot even begin to imagine. The cells are bombarded every instant by outside stimuli unknowable by human intelligence in advance. Any one of the events can entrain a cascade of microscopic episodes leading to new neural patterns. The computer needed to track the consequences would have to be of stupendous proportions, with operations conceivably far more complex than those of the thinking brain itself. Furthermore, scenarios of the mind are all but infinite in detail, their content evolving in accordance with the unique history and physiology of the individual. How are we to feed that into a computer?

So there can be no simple determinism of human thought, at least not in obedience to causation in the way physical laws describe the motion of bodies and the atomic assembly of molecules. Because the individual mind cannot be fully known and predicted, the self can go on passionately believing in its own free will. And that is a fortunate circumstance. Confidence in free will is biologically adaptive. Without it the mind, imprisoned by fatalism, would slow and deteriorate. Thus in organismic time and space, in every operational sense that applies to the knowable self, the mind *does* have free will.

FINALLY, given that conscious experience is a physical and not a supernatural phenomenon, might it be possible to create an artificial human mind? I believe the answer to this philosophically troubling question to be yes in principle, but no in practice, at least not as a prospect for many decades or even centuries to come.

Descartes, in first conceiving the question over three centuries ago, declared artificial human intelligence to be impossible. Two absolutely certain criteria, he said, would always distinguish the machine from a real mind. It could never "modify its phrases to reply to the sense of whatever was said in its presence, as even the most stupid men can do," and it could never "behave in all the occurrences of life as our reason makes us behave." The test was recast in operational terms by the English mathematician Alan Turing in 1950. In the Turing test, as it is now generally called, a human interpreter is invited to ask any question of a hidden computer. All he is told is that either another person or a computer will answer. If, after a respectable period of time, the questioner is unable to tell whether the interlocutor is human or machine, he loses the game; and the mind of the machine is accorded human status. Mortimer Adler, the American philosopher and educator, proposed essentially the same criterion in order to challenge not just the feasibility of humanoids but also the entire philosophy of materialism. We cannot accept a thoroughly material basis for human existence, he said, until such an artificial being is created. Turing thought the humanoid could be built within a few years. Adler, a devout Christian, arrived at the same conclusion as Descartes: No such machine will ever be possible.

Scientists, when told something is impossible, as a habit set out to do it. It is not, however, their purpose to search for the ultimate meaning of existence in their experiments. Their response to cosmic inquiry is most likely to be: "What you suggest is not a productive question." Their occupation is instead exploration of the universe in concrete steps, one at a time. Their greatest reward is occasionally to reach the summit of some improbable peak and from there, like Keats' Cortez at Darien, look in "wild surmise" upon the vastness beyond. In their ethos it is better to have begun a great journey than to have finished it, better to make a seminal discovery than to put the final touches on a theory.

The scientific field of artificial intelligence, AI for short, was inaugurated in the 1950s hard upon the invention of the first electronic computers. It is defined by its practitioners as the study of computation needed for intelligent behavior and the attempt to duplicate that behavior using computers. A half century of work has yielded some impressive results. Programs are available that recognize objects and faces from a few select features and at different angles, drawing on rules of geometric symmetry in the manner of human cognition. Others can translate languages, albeit crudely, or generalize and classify novel objects on the basis of cumulative experience—much in the manner of the human mind.

Some programs can scan and choose options for particular courses of action according to preselected goals. In 1996 Deep Blue, an advanced chess-playing computer, earned grand master status by narrowly losing a six-game match to Gary Kasparov, the reigning human world champion. Deep Blue works by brute force, using thirty-two microprocessors to examine two hundred million chess positions each second. It finally lost because it lacked Kasparov's ability to assess an opponent's weakness and plan long-term strategy based in part on deception. In 1997 a reprogrammed and improved Deep Blue narrowly defeated Kasparov: the first game to Kasparov, the second to Deep Blue, then three ties and the final game to Deep Blue.

The search is on for quantum advances in the simulation of all domains of human thought. In evolutionary computation, AI programmers have incorporated an organismlike procedure in the evolution of design. They provide the computers with a range of options in solving problems, then let them select and modify the available procedures to be followed. By this means the machines have come to resemble bacteria and other simple one-celled organisms. A truly Darwinian twist can be added by placing elements in the machines that mutate at random to change the available procedures. The programs then compete to solve problems, such as gaining access to food and space. Which mutated programs will be born and which among the neonates will succeed are not always predictable, so the "species" of machines as a whole can evolve in ways not anticipated by the human designer. It is within the reach of computer scientists to create mutable robots that travel about the laboratory, learn and classify real resources, and thwart other robots in attaining their goals. At this level their programs would be close to the instinctive repertories not of bacteria but of simple multicellular animals such as flatworms and snails. In fifty years the computer scientists—if successful—will have traversed the equivalent of hundreds of millions of years of organic evolution.

But for all that advance, no AI enthusiast claims to have a road map from flatworm instinct to the human mind. How might such an immense gap be closed? There are two schools of thought. One, represented by Rodney Brooks of the Massachusetts Institute of Technology, takes a bottom-up approach. In this version, the designers would follow the Darwinian robot model to higher and higher levels, gaining new insights and elaborating technology along the way. It is possible that in time, humanoid capability might emerge. The other approach is top-down. Favored by Marvin Minsky, a founding father of AI and colleague of Brooks at MIT, it concentrates directly on the highest-order phenomena of learning and intelligence as they might be conceived and built into a machine without intervening evolutionary steps.

In the teeth of all pessimistic assessments of human limitation likely to be raised, human genius is unpredictable and capable of stunning advances. In the near future a capacity for at least a crude simulation of the human mind might be attained, comprising a level of brain sciences sophisticated enough to understand the basic operations of the mind fully, with computer technology advanced enough to imitate it. We might wake up one morning to find such a triumph announced in the *New York Times*, perhaps along with a generic cure for cancer or the discovery of living organisms on Mars. But I seriously doubt that any such event will ever occur, and I believe a great majority of AI experts are inclined to agree. There are two reasons, which can be called respectively the functional obstacle and the evolutionary obstacle.

The functional obstacle is the overwhelming complexity of inputs of information to and through the human mind. Rational thought emerges from continuous exchanges between body and brain through nerve discharges and blood-borne flow of hormones, influenced in turn by emotional controls that regulate mental set, attention, and the selection of goals. In order to duplicate the mind in a machine, it will not be nearly enough to perfect the brain sciences and AI technology, because the simulation pioneers must also invent and install an entirely new form of computation—artificial emotion, or AE.

The second, or evolutionary, obstacle to the creation of a humanoid mind is the unique genetic history of the human species. Generic human nature—the psychic unity of mankind—is the product of millions of years of evolution in environments now mostly forgotten. Without detailed attention to the hereditary blueprint of human nature, the simulated mind might be awesome in power, but it would be more nearly that of some alien visitor, not of a human.

And even if the blueprint were known, and even if it could be followed, it would serve only as a beginning. To be human, the artificial mind must imitate that of an individual person, with its memory banks filled by a lifetime's experience—visual, auditory, chemoreceptive, tactile, and kinesthetic, all freighted with nuances of emotion. And social: There must be intellectual and emotional exposure to countless human contacts. And with these memories, there must be meaning, the expansive connections made to each and every word and bit of sensory information given the programs. Without all these tasks completed, the artificial mind is fated to fail Turing's test. Any human jury could tear away the pretense of the machine in minutes. Either that, or certifiably commit it to a psychiatric institution.

CHAPTER 7

From Genes to Culture

The natural sciences have constructed a webwork of causal explanation that runs all the way from quantum physics to the brain sciences and evolutionary biology. There are gaps in this fabric of unknown breadth, and many of the strands composing it are as delicate as spider's silk. Predictive syntheses, the ultimate goal of science, are still in an early stage, and especially so in biology. Yet I think it fair to say that enough is known to justify confidence in the principle of universal rational consilience across all the natural sciences.

The explanatory network now touches the edge of culture itself. It has reached the boundary that separates the natural sciences on one side from the humanities and humanistic social sciences on the other. Granted, for most scholars the two domains, commonly called the scientific and literary cultures, still have a look of permanence about them. From Apollonian law to Dionysian spirit, prose to poetry, left cortical hemisphere to right, the line between the two domains can be easily crossed back and forth, but no one knows how to translate the tongue of one into that of the other. Should we even try? I believe so, and for the best of reasons: The goal is both important and attainable. The time has come to reassess the boundary.

Even if that perception is disputed—and it will be—few can deny that the division between the two cultures is a perennial source of misunderstanding and conflict. "This polarisation is sheer loss to us all," wrote C. P. Snow in his defining 1959 essay *The Two Cultures and the Scientific Revolution*. "To us as people, and to our society. It is at the same time practical and intellectual and creative loss."

The polarization promotes, for one thing, the perpetual recycling of the nature-nurture controversy, spinning off mostly sterile debates on gender, sexual preferences, ethnicity, and human nature itself. The root cause of the problem is as obvious today as it was when Snow ruminated on it at Christ College high table: the overspecialization of the educated elite. Public intellectuals, and trailing close behind them the media professionals, have been trained almost without exception in the social sciences and humanities. They consider human nature to be their province and have difficulty conceiving the relevance of the natural sciences to social behavior and policy. Natural scientists, whose expertise is diced into narrow compartments with little connection to human affairs, are indeed ill prepared to engage the same subjects. What does a biochemist know of legal theory and the China trade? It is not enough to repeat the old nostrum that all scholars, natural and social scientists and humanists alike, are animated by a common creative spirit. They are indeed creative siblings, but they lack a common language.

There is only one way to unite the great branches of learning and end the culture wars. It is to view the boundary between the scientific and literary cultures not as a territorial line but as a broad and mostly unexplored terrain awaiting cooperative entry from both sides. The misunderstandings arise from ignorance of the terrain, not from a fundamental difference in mentality. The two cultures share the following challenge. We know that virtually all of human behavior is transmitted by culture. We also know that biology has an important effect on the origin of culture and its transmission. The question remaining is how biology and culture interact, and in particular how they interact across all societies to create the commonalities of human nature. What, in final analysis, joins the deep, mostly genetic history of the species as a whole to the more recent cultural histories of its far-flung societies? That, in my opinion, is the nub of the relationship between the two cultures. It can be stated as a problem to be solved, the central problem of the social sciences and the humanities, and simultaneously one of the great remaining problems of the natural sciences.

At the present time no one has a solution. But in the sense that no one in 1842 knew the true cause of evolution and in 1952 no one knew the nature of the genetic code, the way to solve the problem may lie within our grasp. A few researchers, and I am one of them, even think they know the approximate form the answer will take. From diverse vantage points in biology, psychology, and anthropology, they have conceived a process called *gene-culture coevolution*. In essence, the conception observes, first, that to genetic evolution the human lineage has added the parallel track of cultural evolution, and, second, that the two forms of evolution are linked. I believe the majority of contributors to the theory during the past twenty years would agree to the following outline of its principles:

Culture is created by the communal mind, and each mind in turn is the product of the genetically structured human brain. Genes and culture are therefore inseverably linked. But the linkage is flexible, to a degree still mostly unmeasured. The linkage is also tortuous: Genes prescribe epigenetic rules, which are the neural pathways and regularities in cognitive development by which the individual mind assembles itself. The mind grows from birth to death by absorbing parts of the existing culture available to it, with selections guided through epigenetic rules inherited by the individual brain.

To visualize gene-culture coevolution more concretely, consider the example of snakes and dream serpents, which I used earlier to argue the plausibility of complete consilience. The innate tendency to react with both fear and fascination toward snakes is the epigenetic rule. The culture draws on that fear and fascination to create metaphors and narratives. The process is thus:

As part of gene-culture coevolution, culture is reconstructed each generation collectively in the minds of individuals. When oral tradition is supplemented by writing and art, culture can grow indefinitely large and it can even skip generations. But the fundamental biasing influence of the epigenetic rules, being genetic and ineradicable, stays constant.

Hence the prominence of dream serpents in the legends and art of the Amazonian shamans enriches their culture across generations under the guidance of the serpentine epigenetic rule.

Some individuals inherit epigenetic rules enabling them to survive and reproduce better in the surrounding environment and culture than individuals who lack those rules, or at least possess them in weaker valence. By this means, over many generations, the more successful epigenetic rules have spread through the population along with the genes that prescribe the rules. As a consequence the human species has evolved genetically by natural selection in behavior, just as it has in the anatomy and physiology of the brain.

Poisonous snakes have been an important source of mortality in almost all societies throughout human evolution. Close attention to them, enhanced by dream serpents and the symbols of culture, undoubtedly improves the chances of survival.

The nature of the genetic leash and the role of culture can now be better understood, as follows. Certain cultural norms also survive and reproduce better than competing norms, causing culture to evolve in a track parallel to and usually much faster than genetic evolution. The quicker the pace of cultural evolution, the looser the connection between genes and culture, although the connection is never completely broken. Culture allows a rapid adjustment to changes in the environment through finely tuned adaptations invented and transmitted without correspondingly precise genetic prescription. In this respect human beings differ fundamentally from all other animal species.

Finally, to complete the example of gene-culture coevolution, the frequency with which dream serpents and serpent symbols inhabit a culture is seen to be adjusted to the abundance of real poisonous snakes in the environment. But owing to the power of fear and fascination given them by the epigenetic rule, they easily acquire additional mythic meaning; they serve in different cultures variously as healers, messengers, demons, and gods.

Gene-culture coevolution is a special extension of the more general process of evolution by natural selection. Biologists generally agree that the primary force behind evolution in human beings and all other organisms is natural selection. That is what created *Homo sapiens* during the five or six million years after the ancestral hominid species split off from a primitive chimpanzeelike stock. Evolution by natural selection is not an idle hypothesis. The genetic variation on which selection acts is well understood in principle all the way down to the molecular level. "Evolution watchers" among field biologists have monitored evolution by natural selection, generation by generation, in natural populations of animals and plants. The result can often be reproduced in the laboratory, even up to the creation of new species, for example by hybridization and the breeding of reproductively isolated strains. The manner in which traits of anatomy, physiology, and behavior adapt organisms to their environment has been massively documented. The fossil hominid record, from man-apes to modern humans, while still lacking many details, is solid in main outline, with a well established chronology.

In simplest terms, evolution by natural selection proceeds, as the French biologist Jacques Monod once put it (rephrasing Democritus), by chance and necessity. Different forms of the same gene, called alleles, originate by mutations, which are random changes in the long sequences of DNA (deoxyribonucleic acid) that compose the gene. In addition to such point-by-point scrambling of the DNA, new mixes of alleles are created each generation by the recombining processes of sexual reproduction. The alleles that enhance survival and reproduction of the carrier organisms spread through the population, while those that do not, disappear. Chance mutations are the raw material of evolution. Environmental challenge, deciding which mutants and their combinations will survive, is the necessity that molds us further from this protean genetic clay.

If given enough generations, mutations and recombination can generate a nearly infinite amount of hereditary variation among individuals in a population. For example, if even a mere thousand genes out of the fifty thousand to a hundred thousand in the human genome were to exist in two forms in the population, the number of genetic combinations conceivable is 10^{500}, more than all the atoms in the visible universe. So except for identical siblings the probability that any two human beings share identical genes, or have ever shared them throughout the history of the hominid line, is vanishingly small.

With each generation the chromosomes and genes of the parents are scrambled to produce new mixes. But this perpetual shearing and reconfiguration does not of itself cause evolution. The consistent guiding force is natural selection. Genes that confer higher survival and reproductive success on the organisms bearing them, through the prescribed traits of anatomy, physiology, and behavior, increase in the population from one generation to the next. Those that do not, decrease. Similarly, populations or even entire species with higher survival and reproductive success prevail over competing populations or species, to the same general end in evolution.

Such is the impersonal force that evidently made us what we are today. All of biology, from molecular to evolutionary, points that way. At the risk of seeming defensive, I am obliged to acknowledge that many people, some very well educated, prefer special creation as an explanation for the origin of life. According to a poll conducted by the National Opinion Research Center in 1994, 23 percent of Americans reject the idea of human evolution, and a third more are undecided. This pattern is unlikely to change radically in the years immediately ahead. Because I was raised in a predominantly anti-evolutionist culture in the Protestant southern United States, I am inclined to be empathetic to these feelings, and conciliatory. Anything is possible, it can be said, if you believe in miracles. Perhaps God did create all organisms, including human beings, in finished form, in one stroke, and maybe it all happened several thousand years ago. But if that is true, He also salted the earth with false evidence in such endless and exquisite detail, and so thoroughly from pole to pole, as to make us conclude first that life evolved, and second that the process took billions of years. Surely Scripture tells us He would not do that. The Prime Mover of the Old and New Testaments is variously loving, magisterial, denying, thunderously angry, and mysterious, but never tricky.

Virtually all biologists closely familiar with the details find the evidence for human evolution compelling, and give natural selection the commanding role. There is at least one other force, however, that must be mentioned in any account of evolution. By chance alone, the biologists agree, substitutions are occurring through long stretches of time in some of the DNA letters and the proteins they encode. The continuity of change is often smooth enough to measure the age of different evolving lines of organisms. But this genetic drift, as it is called, adds very little to evolution at the level of cells, organisms, and societies. The reason is that the mutants involved in drift have proven to be neutral, or nearly so: They have little or no effect on the higher levels of biological organization manifest in cells and organisms.

TO GENETIC EVOLUTION, putting the matter as concisely as possible, natural selection has added the parallel track of cultural evolution, and the two forms of evolution are somehow linked. We are trapped, we sometimes think, for ultimate good or evil, not just by our genes but also by our culture. What precisely is this superorganism, this strange creature called culture? To anthropologists, who have analyzed thousands of examples, should go the privilege of response. For them, a culture is the total way of life of a discrete society—its religion, myths, art, technology, sports, and all the other systematic knowledge transmitted across generations. In 1952 Alfred Kroeber and Clyde Kluckhohn melded 164 prior definitions pertaining to all cultures into one, as follows: "Culture is a product; is historical; includes ideas, patterns, and values; is selective; is learned; is based upon symbols; and is an abstraction from behavior and the products of behavior." As Kroeber had earlier declared, it is also holistic, "an accommodation of discrete parts, largely inflowing parts, into a more or less workable fit." Among the parts are artifacts, but these physical objects have no significance except when addressed as concepts in living minds.

In the extreme nurturist view, which has prevailed in social theory for most of the twentieth century, culture has departed from the genes and become a thing unto itself. Possessing a life of its own, growing like wildfire ignited by the strike of a match, it has acquired emergent properties no longer connected to the genetic and psychological processes that initiated it. Hence, *omnis cultura ex cultura*. All culture comes from culture.

Whether that metaphor is accepted or not, the undeniable truth is that each society creates culture and is created by it. Through constant grooming, decorating, exchange of gifts, sharing of food and fermented beverages, music, and storytelling, the symbolic communal life of the mind takes form, unifying the group into a dreamworld that masters the external reality into which the group has been thrust, whether in forest, grassland, desert, ice field, or city, spinning from it the webs of moral consensus and ritual that bind each tribal member to the common fate.

Culture is constructed with language that is productive, comprising arbitrary words and symbols invented purely to convey information. In this respect *Homo sapiens* is unique. Animals have communication systems that are sometimes impressively sophisticated, but they neither invent them nor teach them to others. With a few exceptions, such as bird song dialects, they are instinctive, hence unchanging across generations. The waggle dance of the honeybee and the odor trails of ants contain symbolic elements, but the performances and meanings are tightly prescribed by genes and cannot be altered by learning.

Among animals true linguistic capacity is most closely approached by the great apes. Chimpanzees and gorillas can learn the meanings of arbitrary symbols when trained to use signaling keyboards. Their champion is Kanzi, a bonobo, or pygmy chimpanzee *(Pan paniscus)*, arguably the smartest animal ever observed in captivity. I met this primate genius when he was a precocious youngster at the Yerkes Regional Primate Center of Emory University in Atlanta. He had been studied intensively since birth by Sue Savage-Rumbaugh and her colleagues. As I played games and shared a cup of grape juice with him, I was more than a bit disoriented by his general demeanor, which I found uncannily close to that of a human two-year-old. More than a decade later, as I write, the adult Kanzi has acquired a large vocabulary, with which he signals his wishes and intentions on a picture-symbol keyboard. He constructs sentences that are lexically if not grammatically correct. On one occasion, for example, *Ice water go* ("Bring me some ice water") got him the drink. He has even managed to pick up about 150 spoken English words spontaneously, listening to conversation among humans, without the kind of training needed by border collies and other smart breeds of dogs to go through their many tricks. On another occasion Savage-Rumbaugh, pointing to a companion chimpanzee nearby, said, "Kanzi, if you give Austin your mask, I'll let you have some of Austin's cereal." Kanzi promptly gave Austin the mask and pointed to the cereal box. He has acted upon words in a focused and specific manner too frequently for the connection to be due to chance alone. Even so, Kanzi uses only words and symbols supplied him by human beings. His linguistic powers have not yet risen to the level of early human childhood.

Bonobos and other great apes possess high levels of intelligence by animal standards but lack the singular human capacity to invent rather than merely to use symbolic language. It is further true that common chimpanzees are humanlike in guile and deception, the animal masters of "Machiavellian intelligence." As Frans de Waal and his fellow primatologists have observed in the African wild and the Arnhem zoo in the Netherlands, they form and break coalitions, manipulate friends, and outwit enemies. Their intentions are conveyed by voiced signals and postures, body movements, facial expressions, and the bristling of fur. But in spite of the great advantage a productive, humanlike language would bestow, chimpanzees never create anything resembling it, or any other form of free-ranging symbolic language.

In fact, the great apes are completely silent most of the time. The primatologist Allen Gardner described his experience in Tanzania as follows: "A group of ten wild chimpanzees of assorted ages and sexes feeding peacefully in a fig tree at Gombe may make so little sound that an inexperienced observer passing below can altogether fail to detect them."

Homo sapiens, by contrast, can rightfully be called the babbling ape. Humans communicate vocally all the time; it is far easier to start them talking than to shut them up. They begin in infancy during exchanges with adults, who urge them on with the slow, vowel-heavy, emotionally exaggerated singsong called motherese. Left alone, they continue with "crib speech," composed of squeaks, coos, and nonsense monosyllables, which evolve over a few months into a complex play of words and phrases. These early verbal repertories, conforming more or less to adult vocabularies, are repeated *ad nauseam*, modified, and combined in experimental mixtures. By the age of four the average child has mastered syntax. By six, in the United States at least, he has a vocabulary of about fourteen thousand words. In contrast, young bonobos play and experiment freely with movements and sounds and sometimes with symbols, but so far progress toward the Kanzi level depends on the rich linguistic environment provided by human trainers.

Even if the great apes lack true language, is it possible they possess culture? From evidence in the field it appears they do, and many expert observers have so concluded. Wild chimps regularly invent and use tools. And the particular kinds of artifacts they invent, just as in human culture, are often limited to local populations. Where one group breaks nuts open with a rock, another cracks them against tree trunks. Where some groups use twigs to fish ants and termites from the nests for food, others do not. Among those that fish, a minority first peel the bark off the twigs. One chimp group has been observed using long hooked branches to pull down branches of fig trees to obtain fruit.

It is natural to conclude from such observations that chimpanzees have the rudiments of culture, and to suppose that their capability differs from human culture by degree alone. But that perception needs to be accepted with caution: Chimpanzee inventions may not be culture in any sense. The still scanty evidence on the subject suggests that while chimps pick up the use of a tool more quickly when they see others using one, they seldom imitate the precise movements employed or show any clear sign of understanding the purpose of the activity. Some observers have gone so far as to claim that they are merely stirred into greater activity by watching others. This kind of response, which zoologists call social facilitation, is common in many kinds of social animals, from ants to birds and mammals. Although the evidence is inconclusive, social facilitation alone, combined with trial-and-error manipulation of materials conveniently at hand, might guide the chimps to tool-using behavior in the free-ranging African populations.

Human infants, on the other hand, do engage in precise imitation and with astonishing precocity. As early as forty minutes after birth, to cite the ultimate example, they stick out their tongues and move their heads from side to side in close concert with adults. By twelve days they imitate complex facial expressions and hand gestures. By two years they can be verbally instructed in the use of simple tools.

In summary, the language instinct consists of precise mimicry, compulsive loquacity, near-automatic mastery of syntax, and the swift acquisition of a large vocabulary. The instinct is a diagnostic and evidently unique human trait, based upon a mental power beyond the reach of any animal species, and it is the precondition for true culture. To learn how language originated during evolution would be a discovery of surpassing importance. Unfortunately, the evidences of behavior rarely fossilize. All the millennia of campsite chattering and gesticulation, and with them all the linguistic steps up from our chimplike ancestors, have vanished without trace.

What paleontologists have instead are fossil bones, which tell of the downward migration and lengthening of the voice box, as well as possible changes in the linguistic regions of the brain impressed upon the inner cranial case. They also have steadily improving evidence of the evolution of artifacts, from the controlled use of fire 450,000 years ago, presumably by the ancestral species *Homo erectus*, to the construction of well-wrought tools by early *Homo sapiens* 250,000 years ago in Kenya, then elaborate spearheads and daggers 160,000 years later in the Congo, and finally elaborate painting and the accouterments of ritual 30,000 and 20,000 years ago in southern Europe.

This pace in the evolution of artifactual culture is intriguing. We know that the modern *Homo sapiens* brain was anatomically fully formed by no later than 100,000 years before the present. From that time forward the material culture at first evolved slowly, later expanded, and then exploded. It passed from a handful of stone and bone tools at the beginning of the interval to agricultural fields and villages at the 90 percent mark, and then—in a virtual eyeblink—to prodigiously elaborate technologies (example: five million patents so far in the United States alone). In essence, cultural evolution has followed an exponential trajectory. It leaves us with a mystery: When did symbolic language arise, and exactly how did it ignite the exponentiation of cultural evolution?

TOO BAD, but this great puzzle of human paleontology seems insoluble, at least for the time being. To pick up the trail of gene-culture coevolution, it is better to defer reconstruction of the prehistoric record and proceed to the production of culture by the contemporary human brain. The next best approach, I believe, is to search for the basic unit of culture. Although no such element has yet been identified, at least to the general satisfaction of experts, its existence and some of its characteristics can be reasonably inferred.

Such a focus may seem at first contrived and artificial, but it has many worthy precedents. The great success of the natural sciences has been achieved substantially by the reduction of each physical phenomenon to its constituent elements, followed by the use of the elements to reconstitute the holistic properties of the phenomenon. Advances in the chemistry of macromolecules, for example, led to the exact characterization of genes, and the study of population biology based on genes has refined our understanding of biological species.

What then, if anything, is the basic unit of culture? Why should it be supposed even to exist? Consider first the distinction made by the Canadian neuroscientist Endel Tulving in 1972 between episodic and semantic memory. Episodic memory recalls the direct *perception* of people and other concrete entities through time, like images seen in a motion picture. Semantic memory, on the other hand, recalls *meaning* by the connection of objects and ideas to other objects and ideas, either directly by their images held in episodic memory or by the symbols denoting the images. Of course, semantic memory originates in episodes and almost invariably causes the brain to recall other episodes. But the brain has a strong tendency to condense repeated episodes of a kind into concepts, which are then represented by symbols. Thus, "Proceed to the airport this way" yields to a silhouette of an airplane and arrow, and "This substance is poisonous" becomes a skull and crossbones on the side of a container.

With the two forms of memory distinguished, the next step in the search for the unit of culture is to envision concepts as "nodes," or reference points, in semantic memory that ultimately can be associated with neural activity in the brain. Concepts and their symbols are usually labeled by words. Complex information is thus organized and transmitted by language composed of words. Nodes are almost always linked to other nodes, so that to recall one node is to summon others. This linkage, with all the emotional coloring pulled up with it, is the essence of what we refer to as meaning. The linkage of nodes is assembled as a hierarchy to organize information with more and more meaning. "Hound," "hare" and "chasing" are nodes, each symbolizing collectively a class of more or less similar images. A hound chasing a hare is called a proposition, the next order of complexity in information. The higher order above the proposition is the schema. A typical schema is Ovid's telling of Apollo's courtship of Daphne, like an unstoppable hound in pursuit of an unattainable hare, wherein the dilemma is resolved when Daphne, the hare and a concept, turns into a laurel tree, another concept reached by a proposition.

I have faith that the unstoppable neuroscientists will encounter no such dilemma. In due course they will capture the physical basis of mental concepts through the mapping of neural activity patterns. They already have direct evidence of "spreading activation" of different parts of the brain during memory search. In the prevailing view of the researchers, new information is classified and stored in a similar manner. When new episodes and concepts are added to memory, they are processed by a spreading search through the limbic and cortical systems, which establishes links with previously created nodes. The nodes are not spatially isolated centers connected to other isolated centers. They are typically complex circuits of large numbers of nerve cells deployed over wide, overlapping areas of the brain.

Suppose, for example, you are handed an unfamiliar piece of fruit. You automatically classify it by its physical appearance, smell, taste, and the circumstances under which it is given. A large amount of information is activated within seconds, not just the comparison of the fruit in hand with other kinds but also the emotional feelings, recollections of previous discoveries of similar nature, and memories of dietary customs that seem appropriate. The fruit—all its characteristics compounded—is given a name. Consider the durian of Southeast Asia, regarded by aficionados as the greatest of all tropical fruits. It looks like a spiny grapefruit, tastes sweet with a transient custardlike nuance, and when held away from the mouth smells like a sewer. The experience of a single piece establishes, I assure you, the concept "durian" for a lifetime.

The natural elements of culture can be reasonably supposed to be the hierarchically arranged components of semantic memory, encoded by discrete neural circuits awaiting identification. The notion of a culture unit, the most basic element of all, has been around for over thirty years, and has been dubbed by different authors variously as mnemotype, idea, idene, meme, sociogene, concept, culturgen, and culture type. The one label that has caught on the most, and for which I now vote to be winner, is meme, introduced by Richard Dawkins in his influential work *The Selfish Gene* in 1976.

The definition of meme I suggest is nevertheless more focused and somewhat different from that of Dawkins. It is the one posed by the theoretical biologist Charles J. Lumsden and myself in 1981, when we outlined the first full theory of gene-culture coevolution. We recommended that the unit of culture—now called meme—be the same as the node of semantic memory and its correlates in brain activity. The level of the node, whether concept (the simplest recognizable unit), proposition, or schema, determines the complexity of the idea, behavior, or artifact that it helps to sustain in the culture at large.

I realize that with advances in the neurosciences and psychology the notion of node-as-meme, and perhaps even the distinction between episodic and semantic memory, are likely to give way to more sophisticated and complex taxonomies. I realize also that the assignment of the unit of culture to neuroscience might seem at first an attempt to short-circuit semiotics, the formal study of all forms of communication. That objection would be unjustified. My purpose in this exposition is the opposite, to establish the plausibility of the central program of consilience, in this instance the causal connections between semiotics and biology. If the connections can be established empirically, then future discoveries concerning the nodes of semantic memory will correspondingly sharpen the definition of memes. Such an advance will enrich, not replace, semiotics.

I CONCEDE that the very expression "genes to culture," as the conceptual keystone of the bridge between science and the humanities, has an ethereal feel to it. How can anyone presume to speak of a gene that prescribes culture? The answer is that no serious scientist ever has. The web of causal events comprising gene-culture coevolution is more complicated—and immensely more interesting. Thousands of genes prescribe the brain, the sensory system, and all the other physiological processes that interact with the physical and social environment to produce the holistic properties of mind and culture. Through natural selection, the environment ultimately selects which genes will do the prescribing.

For its implications throughout biology and the social sciences, no subject is intellectually more important. All biologists speak of the interaction between heredity and environment. They do not, except in laboratory shorthand, speak of a gene "causing" a particular behavior, and they never mean it literally. That would make no more sense than its converse, the idea of behavior arising from culture without the intervention of brain activity. The accepted explanation of causation from genes to culture, as from genes to any other product of life, is not heredity alone. It is not environment alone. It is interaction between the two.

Of course it is interaction. But we need more information about interaction in order to encompass gene-culture coevolution. The central clarifying concept of interactionism is the *norm of reaction*. The idea is easily grasped as follows. Choose a species of organism, whether animal, plant, or microorganism. Select either one gene or a group of genes that act together to affect a particular trait. Then list all the environments in which the species can survive. The different environments may or may not cause variation in the trait prescribed by the selected gene or group of genes. The total variation in the trait in all the survivable environments is the norm of reaction of that gene or group of genes in that species.

The textbook case of a norm of reaction is leaf shape in the arrowleaf, an amphibious plant. When an individual of the species grows on the land, its leaves resemble arrowheads. When it grows in shallow water, the leaves at the surface are shaped like lily pads; and when submerged in deeper water, the leaves develop as eelgrasslike ribbons that sway back and forth in the surrounding current. No known genetic differences among the plants underlie this extraordinary variation. The three basic types are variations in the expression of the same group of genes caused by different environments. Together they compose the norm of reaction of the genes prescribing leaf form. They embrace, in other words, the full variation in expression of the genes in all known survivable environments.

When some of the variation within a species is due to differences in genes possessed by separate members of the species, and not just different environments, norms of reaction can still in principle be defined for each of the genes or set of genes in turn. The relation of variation in a trait to variation in genes and their norms of reaction is illustrated by human body weight. There is abundant evidence that body form is influenced by heredity. A person genetically predisposed to obesity by heredity can diet to moderate slimness, although he is prone to slide back when off the diet. A hereditarily slender person, on the other hand, is likely to stay that way, and only persistent overeating or endocrine imbalance can push him into obesity. The relevant genes of the two individuals have different norms of reaction. They produce different results when both individuals occupy identical environments, including diet and exercise. The more familiar way to express the matter is in reverse, noting that hereditarily distinct individuals require different environments, in particular different diets and regimes of exercise, in order to produce the same result.

The same kind of interaction between genes and environment occurs in every category of human biology, including social behavior. In his important 1996 work *Born to Rebel*, the American social historian Frank J. Sulloway has demonstrated that people respond powerfully during personality development to the order in which they were born and thus the roles they assume in family dynamics. Later-borns, who identify least with the roles and beliefs of the parents, tend to become more innovative and accepting of political and scientific revolutions than do first-borns. As a result they have, on average, contributed more than first-borns have to cultural change throughout history. They do it by gravitating toward independent, often rebellious roles, first within the family and then within society at large. Because first- and later-borns do not differ genetically in any way correlated with their birth order, the genes influencing development can be said to spread their effects among various niches available in the environment. The birth-order effect documented by Sulloway is their norm of reaction.

In some categories of biology, such as the most elementary molecular processes and properties of gross anatomy, almost everyone has the same genes affecting traits in these categories and hence the same norms of reaction. Long ago in geological time, when the truly universal traits were evolving, there probably was variation in the prescribing genes, but natural selection has since narrowed the variation almost to zero. All primates, for example, have ten fingers and ten toes, and there is virtually no variation due to environment; so the norm of reaction is exactly the single state, of ten fingers and ten toes. In most categories, however, people differ genetically to a considerable degree, even in traits consistent enough to be regarded as cultural universals. In order to make the most of the variation, to cultivate health and talent and realize human potential, it is necessary to understand the roles of both heredity and environment.

By environment I do not mean merely the immediate circumstances in which people find themselves. A snapshot will not suffice. The required meaning is that used by developmental biologists and psychologists. It is nothing less than the myriad influences that shape body and mind step by step throughout every stage of life.

Because human beings cannot be bred and reared under controlled conditions like animals, information about the interaction of genes and environment comes hard. Relatively few genes affecting behavior (some of which I will describe later) have been located on chromosomes, and the exact pathways of development they influence have seldom been traced. In the interim the preferred measure of interaction is *heritability*, the percentage of variation in the trait due to heredity. Heritability does not apply to individuals; it is used only for populations. It is incongruous to say, "This marathoner's athletic ability is 20 percent due to his genes and 80 percent to his environment." It *is* correct to make a statement such as, to use an imaginary example, "Twenty percent of the variation in performance of Kenyan marathoners is due to their heredity and 80 percent to their environment." For the reader who would like more precise definitions of heritability and variance, the measure of variation used by statisticians and geneticists, I will add them here:

> Heritability, minus mathematical refinements, is estimated as follows. In a sample of individuals from the population, measure the trait in a standardized way, say aerobic performance on a treadmill to represent endurance. Take the variation in the measure among the individuals in the sample, and estimate the amount of the variation due to heredity. That fraction is the heritability. The measure of variation used is the variance. To get it, first take the average score obtained from individuals in the sample. Subtract each individual's score in turn from the average and square the difference. The variance is the average of all the squared differences.

The principal method of estimating the fraction of variation due to the genes—the heritability—is by studies of twins. Identical twins, which have exactly the same genes, are compared with fraternal twins, which on average share only the same number of genes as the number shared by siblings born at different times. Fraternal twins are consistently less alike than identical twins, and the difference between pairs of fraternal twins and pairs of identical twins serves as an approximate measure of the contribution of heredity to the overall variation in the trait. The method can be considerably enhanced by studies of those special pairs of identical twins who were separated in infancy and adopted by different families, thus possessing the same heredity but reared in different environments. It is further improved by multiple correlation studies, in which the key environmental influences are identified and their contributions to the overall variation individually assessed.

Heritability has been a standard measure for decades in plant and animal breeding. It has gained recent controversial attention for its human applications through *The Bell Curve*, the 1994 book by Richard J. Herrnstein and Charles Murray, and other popular works on the heredity of intelligence and personality. The measure has considerable merit, and in fact is the backbone of human behavioral genetics. But it contains oddities that deserve close attention with reference to the consilience between genetics and the social sciences. The first is the peculiar twist called "genotype-environment correlation," which serves to increase human diversity beyond the ambit of its immediate biological origins. The twist works as follows. People do not merely select roles suited to their native talents and personalities. They also gravitate to environments that reward their hereditary inclinations. Their parents, who possess similar inborn traits, are also likely to create a family atmosphere nurturing development in the same direction. The genes, in other words, help to create a particular environment in which they will find greater expression than would otherwise occur. The overall result is a greater divergence of roles within societies due to the interaction of genes and environment. For example, a musically gifted child, receiving encouragement from adults, may take up an instrument early and spend long hours practicing. His classmate, innately thrill-seeking, persistently impulsive and aggressive, is drawn to fast cars. The first child grows up to be a professional musician, the second (if he stays out of trouble) a successful racing-car driver. The hereditary differences in talent and personality between the classmates may be small, but their effects have been amplified by the diverging pathways into which they were guided by the differences. To put genotype-environment correlation in a phrase, heritability measured at the level of biology reacts with the environment to increase heritability measured at the level of behavior.

Understanding genotype-environment correlation clarifies a second principle of the relation of genes to culture. There is no gene for playing the piano well, or even a particular "Rubinstein gene" for playing it extremely well. There is instead a large ensemble of genes whose effects enhance manual dexterity, creativity, emotive expression, focus, attention span, and control of pitch, rhythm, and timbre. All of these together compose the special human ability that the American psychologist Howard Gardner calls musical intelligence. The combination also inclines the gifted child to seize the right opportunity at the right time. He tries a musical instrument, likely provided by musically gifted parents, is then reinforced by deserved praise, repeats, is reinforced again, and soon embraces what is to be the central preoccupation of his life.

Another important peculiarity of heritability is its flexibility. By simply changing the environment, the percentage of variation due to heredity can be increased or decreased. Scores for heritability in IQ and measurable personality traits in white Americans, a segment of population typically chosen for convenience and in order to increase statistical reliability by making the sample more uniform, mostly fall around the 50 percent mark, at least closer to it than to zero or 100 percent.

Do we wish to change these numbers? I think not, at least not as a primary goal. Imagine the result if a society became truly egalitarian, so that all children were raised in nearly identical circumstances and encouraged to enter any occupation they chose within reach of their abilities. Variation in environment would thus be drastically reduced, while the original innate abilities and personality traits endured. Heritability in such a society would increase. Any socioeconomic class divisions that persisted would come to reflect heredity as never before.

Suppose instead that all children were tested for ability at an early age and put on educational tracks that reflected their scores, with the aim of directing them to occupations most appropriate to their gifts. Environmental variation in this Brave New World would rise and innate ability would stay the same. If the scores and hence environments reflected the genes, heritability would increase. Finally, imagine a society with the reverse policy: uniformity of outcome is valued above all else. Gifted children are discouraged and slow children provided with intensive personal training in an effort to bring everyone to the same level in abilities and achievement. Because a wide range of tailored environments is required to approach this goal, heritability would fall.

These idealized societies are posed not to recommend any one of them—all have a totalitarian stench—but to clarify the social meaning of this important phase of genetic research. Heritability is a sound measure of the influence of genes on variation in existing environments. It is invaluable in establishing the presence of the genes in the first place. Until the 1960s, for example, schizophrenia was thought to be a result of what parents, especially mothers, do to their children in the first three years of their lives. Until the 1970s autism was also thought to be an environmental disorder. Now, thanks to heritability studies, we know that in both disabilities genes play a significant role. In the reverse direction, alcoholism was once assumed to be largely inherited, so much so that careful heritability studies were not conducted until the 1990s. Now we know that alcoholism is only moderately heritable in males and scarcely at all in females.

Still, except for the rare behavioral conditions approaching total genetic determination, heritabilities are at best risky predictors of personal capacity in existing and future environments. The examples I have cited also illustrate the danger of using them as measures of the worth of either individuals or societies. The message from geneticists to intellectuals and policy-makers is this: Choose the society you want to promote, then prepare to live with its heritabilities. Never favor the reverse, of promoting social policies just to change heritabilities. For best results, cultivate individuals, not groups.

I HAVE PUT these ideas from genetics in play so as to clarify the vexing differences between nurturists and hereditarians, and to try to establish a common ground between them. Until that much is accomplished, the search for consilience risks being sidetracked by endless ideological bickering, with adversaries who promote different political and social agendas talking past one another. Nurturists traditionally emphasize the contributions of the environment to behavior, while hereditarians emphasize the genes. (Nurturists are sometimes called environmentalists, but that label has been preempted by protectors of the environment; and hereditarians cannot be called naturists, unless they hold their conferences in the nude.) Redefined with the more precise concepts of genetics, nurturists can now be seen to believe that human behavioral genes have very broad norms of reaction, while hereditarians think the norms are relatively narrow. In this sense the difference between the two opinions is thus one of degree, not of kind. It becomes a matter that can be settled and agreed upon empirically, should the adversaries agree to take an objective approach.

Nurturists have also traditionally thought that the heritability of intelligence and personality traits is low, while hereditarians have considered it to be high. That disagreement has been largely resolved. In contemporary Caucasians of Europe and the United States at least, heritability is usually in mid-range, with its exact value varying from one trait to another.

Nurturists think that culture is held on a very long genetic leash, if held at all, so that the cultures of different societies can diverge from one another indefinitely. Hereditarians believe the leash is short, causing cultures to evolve major features in common. This problem is technically less tractable than the first two, but it is also empirical in nature, and in principle can be solved. I will take it up again shortly, and give several examples that illustrate how a resolution can in fact be reached.

There is already at least some common ground to build upon. Nurturists and hereditarians generally agree that almost all the differences between cultures are likely to be the product of history and environment. While individuals *within* a particular society vary greatly in behavioral genes, the differences mostly wash out statistically *between* societies. The culture of the Kalahari hunter-gatherers is very distinct from that of Parisians, but the differences between them are primarily a result of divergence in history and environment, and are not genetic in origin.

THE CLARIFICATION OF norms of reaction and heritability, while admittedly a bit technical and dry, is the crucial first step toward unbraiding the roles of heredity and environment in human behavior, and hence important for the attainment of consilience of biology with the social sciences. The logical next step is the location of the genes that affect behavior. Once genes have been mapped on chromosomes and their pathways of expression identified, their interaction with the environment can be more precisely traced. When many such interactions have been defined, the whole can be braided back again to attempt a more complete picture of mental development.

The state of the art in human behavioral genetics, including its still formidable difficulties in gene mapping, is illustrated by the study of schizophrenia. This most common of psychoses afflicts just under 1 percent of people in populations around the world. The symptoms of schizophrenia are highly variable from person to person. But they share one diagnostic trait: mental activity that consistently breaks with reality. In some cases the patient believes he is a great personage (the Messiah is a popular choice) or the target of a clever and pervasive conspiracy. In others, he hallucinates voices or visions, often bizarre, as in a dream while fully awake.

In 1995 independent groups of scientists achieved three breakthroughs while probing the physical origins of schizophrenia. Neurobiologists at the

University of California in Irvine discovered that during fetal development some nerve cells in the prefrontal cortex of future schizophrenics fail to communicate with other cells required for normal exchanges with the rest of the brain. In particular, the cells are unable to manufacture messenger RNA molecules that guide synthesis of the neurotransmitter GABA, or gamma aminobutyric acid. With GABA missing, the nerve cells cannot function, even though they look normal. In some manner still unknown, the impairment promotes internal mental constructions with no connection to external stimuli or ordinary rational thought. The brain creates a world of its own, as though closed off in sleep.

In the same year a second team from Cornell University and two medical research centers in England reported the first direct observation of brain activity in hallucinating schizophrenic patients. Using positron emission tomography (PET) imaging, the investigators monitored active sites in the cortex and limbic systems of patients during periods of both normal and psychotic activity. In one case, they watched a male patient's brain light up while (according to his testimony) disembodied heads rolled through his mind barking orders. The region responsible for the most abnormal events is the anterior cingulate cortex, a region thought to regulate other portions of the cerebral cortex. Its malfunction evidently diminishes the integration of external information and provokes erratic, dreamlike confabulation by the wakened brain.

What is the ultimate cause of schizophrenia? For years data from twin and family-history studies have suggested that the malfunction has at least a partially genetic origin. Early attempts to locate the responsible genes misfired; particular chromosomes were tentatively identified as sites of schizophrenia genes, but then further studies failed to duplicate the results. In 1995, four independent research groups, using advanced chromosome mapping techniques on large samples of subjects, placed at least one gene responsible for schizophrenia on the short arm of chromosome 6. (Humans have 22 pairs of chromosomes in addition to the sex chromosomes X and Y; each of the pairs of chromosomes is arbitrarily assigned a different number for easy reference.) Two other groups failed to confirm the result, but as I write two years later the weight of evidence from the four combined positive tests has led to wide acceptance of their conclusion as to the probable placement of at least one of the schizophrenia genes.

These recent advances and others have cleared the way toward an eventual understanding, not merely of one of the most important mental diseases but of a complex piece of human behavior. Although the behavior can in no way be called normal, it affects the evolution of culture. From the delusions and visions of madmen have come some of the world's despotisms, religious cults, and great works of art. The codified responses of societies to extreme strangeness have furthermore been part of the culture of the many societies that regard schizophrenics as either blessed by gods or inhabited by demons.

But surely, you may respond, culture is still based mainly on normal responses, not insanity. Why has so little progress been made on love, altruism, competitiveness, and other elements of everyday social behavior? The answer lies in the pragmatic bias of genetic research. Geneticists who study inheritance and development first look for big effects caused by single mutations, those easy to detect and analyze. In the classical period of Mendelian genetics, for example, they began with instantly recognizable traits, such as vestigial wings in drosophila fruit flies and wrinkled seed coats in garden peas. It so happens that big mutations are also harmful mutations, for the same reason that large random changes in an automobile engine are more likely to stall it than small random changes. Big mutations almost always reduce survival rates and reproductive capacity. Much of pioneering human genetics has therefore been medical genetics, as exemplified by the studies of schizophrenia.

The practical value of the approach is beyond question. The use of large effects has been parlayed many times into important advances in medical research. Over 1,200 physical and psychological disorders have been tied to single genes. They range (alphabetically) from Aarskog-Scott syndrome to Zellweger syndrome. The result is the OGOD principle: One Gene, One Disease. So successful is the OGOD approach that researchers joke about the Disease of the Month reported in scientific journals and mainstream media. Consider this diverse set of examples: color blindness, cystic fibrosis, hemophilia, Huntington's chorea, hypercholesterolemia, Lesch-Nyhan syndrome, retinoblastoma, sickle-cell anemia. And so pervasive is the evidence of the origin of pathologies in single and multiple gene deviations—even cigarette smoking has a discernible heritability—that biomedical scientists like to quote the maxim that "all disease is genetic."

Researchers and practicing physicians are especially pleased with the OGOD discoveries, because a single gene mutation invariably has a biochemical signature that can be used to simplify diagnosis. Because the signature is a defect somewhere in the sequence of molecular events entrained by transcription off the affected gene, it can often be disclosed with a simple biochemical test. Hope also rises that genetic disease can be corrected with magic-bullet therapy, by which one elegant and noninvasive procedure corrects the biochemical defect and erases the symptoms of the disease.

For all its early success, however, the OGOD principle can be profoundly misleading when applied to human behavior. While it is true that a mutation in a single gene often causes a significant change in a trait, it does not at all follow that the gene *determines* the organ or process affected. Typically, many genes contribute to the prescription of each complex biological phenomenon. How many? For that kind of information it is necessary to turn from human beings to the house mouse, which, being a prime laboratory animal with a short life span, is genetically the best known of all the mammals. Even here knowledge is sketchy. In the mouse, genes contributing to the texture of the hairs and skin are known from no fewer than seventy-two chromosome sites. At least forty-one other genes have variants that cause defects in the organ of balance in the inner ear, resulting in abnormal head shaking and circling behavior.

The complexity of mouse heredity is a clue to the difficulties still facing human genetics. Whole organs and processes, as well as narrowly defined features within them, are commonly prescribed by ensembles of genes, each of which occupies a different array of positions on the chromosomes. The difference in skin pigmentation between people of African and European ancestry is believed to be determined by three to six such "polygenes." The estimates for this and other such systems may be on the low side. In addition to the more potent genes, which are easier to detect, there can be many others that contribute small portions of the variation observed and thus remain undiscovered.

It follows that a mutation in any one of the polygenes might produce a large, overriding OGOD effect, or it may prescribe a much smaller quantitative deviation from the average. The common occurrence of mutations of the second type is one reason that genes predisposing the development of chronic depression, manic-depressive syndrome, and other disorders have proven so elusive. Clinical depression in Ireland, for example, may have at least a partially different gene-based predisposition from clinical depression in Denmark. In such a case, careful research in one laboratory that locates a gene site on one chromosome will fail to find confirmation by equally careful research conducted in a second laboratory.

Subtle differences in environment can also distort the classic patterns of Mendelian inheritance. One common effect is the condition called incomplete penetrance. The trait appears in one person but not another, even though both possess the same enabling genes. When one identical twin develops schizophrenia, for example, the chance that the other twin will follow suit is only 50 percent, despite the fact that exactly the same genes are found in both. Another consequence is variable expressivity. Those who develop schizophrenia have it in greatly varying form and intensity.

To summarize, human behavioral genetics provides one of the crucial links in the track from genes to culture. The discipline is still in its infancy, and hampered by formidable theoretical and technical difficulties. Its principal methods are classical twin studies and family-tree analysis, gene mapping, and, most recently, DNA sequence identification. These approaches have so far been but crudely joined. As their synthesis proceeds and is supplemented by studies of psychological development, a clearer picture of the foundations of human nature will emerge.

MEANWHILE, what we know or (to be completely forthright) what we *think* we know, about the hereditary basis of human nature can be expressed by linking together three determining levels of biological organization. I will present them from the top down, in a sequence that begins with the universals of culture, proceeds to epigenetic rules of social behavior, and ends in a second look at behavioral genetics.

In a classic 1945 compendium, the American anthropologist George P. Murdock listed the universals of culture, which he defined as the social behaviors and institutions recorded in the Human Relations Area Files for every one of the hundreds of societies studied to that time. There are sixty-seven universals in the list: age-grading, athletic sports, bodily adornment, calendar, cleanliness training, community organization, cooking, cooperative labor, cosmology, courtship, dancing, decorative art, divination, division of labor, dream interpretation, education, eschatology, ethics, ethnobotany, etiquette, faith healing, family feasting, fire-making, folklore, food taboos, funeral rites, games, gestures, gift-giving, government, greetings, hair styles, hospitality, housing, hygiene, incest taboos, inheritance rules, joking, kin groups, kinship nomenclature, language, law, luck superstitions, magic, marriage, mealtimes, medicine, obstetrics, penal sanctions, personal names, population policy, postnatal care, pregnancy usages, property rights, propitiation of supernatural beings, puberty customs, religious ritual, residence rules, sexual restrictions, soul concepts, status differentiation, surgery, tool-making, trade, visiting, weather control, and weaving.

It is tempting to dismiss these traits as not truly diagnostic for human beings, not really genetic, but inevitable in the evolution of *any* species that attains complex societies based on high intelligence and complex language, regardless of their hereditary predispositions. But that interpretation is easily refuted. Imagine a termite species that evolved a civilization from the social level of a living species. Take for the purpose the mound-building termites *Macrotermes bellicosus* of Africa, whose citylike nests beneath the soil each contain millions of inhabitants. Elevate the basic qualities of their social organization in their present-day insectile condition to a culture that is guided, as in human culture, by heredity-based epigenetic rules. The "termite nature" at the foundation of this hexapod civilization would include celibacy and nonreproduction by the workers, the exchange of symbiotic bacteria by the eating of one another's feces, the use of chemical secretions (pheromones) to communicate, and the routine cannibalism of shed skins and dead or injured family members. I have composed the following state-of-the-colony speech for a termite leader to deliver to the multitude, in her attempt to reinforce the supertermite ethical code:

Ever since our ancestors, the macrotermitine termites, achieved ten-kilogram weight and larger brains during their rapid evolution through the late Tertiary Period, and learned to write with pheromonal script, termitic scholarship has elevated and refined ethical philosophy. It is now possible to express the imperatives of moral behavior with precision. These imperatives are self-evident and universal. They are the very essence of termitity. They include the love of darkness and of the deep, saprophytic, basidiomycetic penetralia of the soil; the centrality of colony life amidst the richness of war and trade with other colonies; the sanctity of the physiological caste system; the evil of personal rights (the colony is ALL!); our deep love for the royal siblings allowed to reproduce; the joy of chemical song; the aesthetic pleasure and deep social satisfaction of eating feces from nestmates' anuses after the shedding of our skins; and the ecstasy of cannibalism and surrender of our own bodies when we are sick or injured (it is more blessed to be eaten than to eat).

FURTHER EVIDENCE of human cultural universals is the dual origin of civilization in the Old and New Worlds, evolved in mutual isolation yet remarkably convergent in broad detail. The second part of "the grand experiment" began twelve thousand or more years ago, when the New World was invaded by nomadic tribes from Siberia. The colonists were at that time Paleolithic hunter-gatherers who most likely lived in groups of a hundred or fewer. In the centuries to follow they spread south through the length of the

New World, from the Arctic tundra to the icy forests of Tierra del Fuego ten thousand miles distant, splitting as they went into local tribes that adapted to each of the land environments they encountered. Along the way, here and there, some of the societies evolved into chiefdoms and imperial states remarkably similar in their basic structure to those in the Old World.

In 1940 the American archaeologist Alfred V. Kidder, a pioneer student of early North American settlements and Mayan cities, summarized the independent histories of civilization in the Old and New Worlds to make the case for a hereditary human nature. In both hemispheres, he said, people started from the same base as stone-age primitives. First they brought wild plants under cultivation, allowing their populations to increase and form villages. While this was happening they elaborated social groupings and evolved sophisticated arts and religions, with priests and rulers receiving special powers from the gods. They invented pottery, and wove plant fibers and wool into cloth. They domesticated local wild animals for food and transport. They worked metal into tools and ornaments, first gold and copper, then bronze, the harder alloy of copper and tin. They invented writing and used it to record their myths, wars, and noble lineages. They created hereditary classes for their nobles, priests, warriors, craftsmen, and peasants. And, Kidder pointed out, "In the New World as well as in the Old, priesthoods grew and, allying themselves with temporal powers, or becoming rulers in their own right, reared to their gods vast temples adorned with painting and sculpture. The priests and chiefs provided for themselves elaborate tombs richly stocked for the future life. In political history it is the same. In both hemispheres group joined group to form tribes; coalitions and conquests brought preeminence; empires grew and assumed the paraphernalia of glory."

IMPRESSIVE AS the universals may be, it is still risky to use them as evidence of the linkage between genes and culture. While the categories listed occur too consistently to be due to chance alone, their finer details differ widely among societies within and between the hemispheres. The hallmarks of civilization are moreover too scattered and recent in origin to have been genetically evolved and somehow carried around the world by hunter-gatherers. It would be absurd to speak of particular genes that prescribe agriculture, writing, the priesthood, and monumental tombs.

In my own writings, from *On Human Nature* in 1978 forward, I have argued that the etiology of culture wends its way tortuously from the genes through the brain and senses to learning and social behavior. What we inherit are neurobiological traits that cause us to see the world in a particular way and to learn certain behaviors in preference to other behaviors. The genetically inherited traits are not memes, not units of culture, but rather the propensity to invent and transmit certain kinds of these elements of memory in preference to others.

As early as 1972 Martin Seligman and other psychologists had defined the bias in development precisely. They called it "prepared learning." By this concept they meant that animals and humans are innately prepared to learn certain behaviors, while being counterprepared against—that is, predisposed to avoid—others. The many documented examples of prepared learning form a subclass of *epigenetic rules*. As recognized in biology, epigenetic rules comprise the full range of inherited regularities of development in anatomy, physiology, cognition, and behavior. They are the algorithms of growth and differentiation that create a fully functioning organism.

A second productive insight, contributed by sociobiology, is that prepared learning of social behavior, like all other classes of epigenesis, is usually adaptive: It confers Darwinian fitness on organisms by improving their survival and reproduction. The adaptiveness of the epigenetic rules of human behavior is not the exclusive result of either biology or culture. It arises from subtle manifestations of both. One of the most efficient ways to study the epigenetic rules of human social behavior is by methods of conventional psychology, informed by the principles of evolutionary process. For this reason the scientists concentrating on the subject often call themselves evolutionary psychologists. Theirs is a hybrid discipline, drawn from both sociobiology—the systematic study of the biological basis of social behavior in all kinds of organisms, including humans—and psychology, the systematic study of the basis of human behavior. Given our growing understanding of gene-culture coevolution, however, and in the interest of simplicity, clarity, and—on occasion—intellectual courage in the face of ideological hostility, evolutionary psychology is best regarded as identical to human sociobiology.

IN THE 1970S, as I stressed in my early syntheses, altruism was the central problem of sociobiology in both animals and humans. That challenge has now been largely met by successful theory and empirical research. In the 1990s attention is beginning to shift in human sociobiology to gene-culture coevolution. In this new phase of research, the definition of epigenetic rules is the best means to make important advances in the understanding of human nature. Such an emphasis seems logically inescapable. The linkage between genes and culture is to be found in the sense organs and programs of the brain. Until this process is better known and taken into account, mathematical models of genetic evolution and cultural evolution will have very limited value.

The epigenetic rules, I believe, operate, like emotion, at two levels. Primary epigenetic rules are the automatic processes that extend from the filtering and coding of stimuli in the sense organs all the way to perception of the stimuli by the brain. The entire sequence is influenced by previous experience only to a minor degree, if at all. Secondary epigenetic rules are regularities in the integration of large amounts of information. Drawing from selected fragments of perception, memory, and emotional coloring, secondary epigenetic rules lead the mind to predisposed decisions through the choice of certain memes and overt responses over others. The division between the two classes of epigenetic rules is subjective, made for convenience only. Intermediate levels of complexity exist, because more complex primary rules grade into simpler secondary rules.

All of the senses impose primary epigenetic rules. Among the most basic properties of such rules is the breaking of otherwise continuous sensations into discrete units. From birth, for example, the cones of the retina and the neurons of the lateral geniculate nuclei of the thalamus classify visible light of differing wavelengths into four basic colors. In similar manner, the hearing apparatus of both children and adults automatically divides continuous speech sounds into phonemes. Series of sounds that run smoothly from *ba* to *ga* are not heard as a continuum but either as *ba* or *ga*; the same is true of the transition from *v* to *s*.

An infant begins life with other built-in acoustic responses that shape later communication and social existence. The newborn can distinguish innately between noise and tone. By four months the infant prefers harmonious tones, sometimes reacting to out-of-tune notes with a facial expression of disgust, the same, it turns out, as elicited by a drop of lemon juice on the tongue. The newborn's response to a loud sound is the Moro reflex: If on its back, the infant first extends its arms forward, brings them slowly together as though in embrace, emits a cry, and then gradually relaxes. In four to six weeks the Moro reflex is replaced by the startle response, which, as I described earlier, is the most complex of the reflexes and lasts for the remainder of life. Within a fraction of a second after an unexpected loud noise is heard, the eyes close, the mouth opens, the head drops, the shoulders and arms sag, the knees buckle slightly. Altogether, the body is positioned as though to absorb a coming blow.

Some preferences in chemical taste also begin at or shortly after birth. Newborns prefer sugar solutions over plain water and in the following fixed order: sucrose, fructose, lactose, glucose. They reject substances that are acid, salty, or bitter, responding to each with the distinctive facial expressions they will use for the rest of their lives.

The primary epigenetic rules gear the human sensory system to process mostly audiovisual information. The predilection is in contrast to that of the vast majority of animal species, which depend mostly on smell and taste. The human audiovisual bias is reflected by the disproportionate weighting of vocabulary. In languages around the world, from English and Japanese to Zulu and Teton Lakota, two-thirds to three-fourths of all the words describing sensory impressions refer to hearing and vision. The remaining minority of words are divided among the other senses, including smell, taste, and touch, as well as sensitivity to temperature, humidity, and electrical fields.

Audiovisual bias also marks the primary epigenetic rules that establish social bonds in infancy and early childhood. Experiments have shown that within ten minutes after birth, infants fixate more on normal facial designs drawn on posters than on abnormal designs. After two days, they prefer to gaze at their mother rather than other, unknown women. Other experiments have revealed an equally remarkable ability to distinguish their mother's voice from voices of other women. For their part, mothers need only a brief contact to distinguish the cry of their newborns, as well as their personal body odor.

The face is the chief arena of visual nonlinguistic communication and the secondary epigenetic rules that bias their psychological development. A few facial expressions have invariant meaning throughout the human species, even though they are modified in different cultures to express particular nuances. In a classic experiment to test the universality of the phenomenon, Paul Ekman of the University of California at San Francisco photographed Americans as they acted out fear, loathing, anger, surprise, and happiness. He also photographed New Guinea highland tribesmen from recently contacted villages as they told stories in which similar feelings were evoked. When individuals were then shown the portraits from the other culture, they interpreted the facial expressions with an accuracy greater than 80 percent.

Within the face the mouth is the principal instrument of visual communication. The smile in particular is a rich site of secondary epigenetic rules. Psychologists and anthropologists have discovered substantial degrees of similar programmed development in the uses of smiling across cultures. The expression is first displayed by infants between the ages of two and four months. It invariably attracts an abundance of affection from attending adults. Environment has little influence on the maturation of smiling. The infants of the !Kung, a hunter-gatherer people of South Africa's Kalahari desert, are nurtured under very different conditions from those in America and Europe. They are delivered by their mothers without assistance or anesthetic, kept in almost constant physical contact with adults, nursed several times an hour, and trained rigorously at the earliest possible age to sit, stand, and walk. Yet their smile is identical in form to that of American and European infants, appears at the same time, and serves the same social function. Smiling also appears on schedule in deaf-blind children and even in thalidomide-deformed children who are not only deaf and blind but also crippled so badly they cannot touch their own faces.

Throughout life smiling is used primarily to signal friendliness and approval, and beyond that to indicate a general sense of pleasure. Each culture molds its meaning into nuances determined by the exact form and the context in which it is displayed. Smiling can be turned into irony and light mockery, or to conceal embarrassment. But even in such cases its messages span only a tiny fraction of those transmitted by all facial expressions taken together.

At the highest levels of mental activity complex secondary epigenetic rules are followed in the process called reification: the telescoping of ideas and complex phenomena into simpler concepts, which are then compared with familiar objects and activities. The Dusun of Borneo—to take one of countless examples from the archives of anthropology—reify each house into a "body" possessing arms, a head, a belly, legs, and other parts. It is believed to "stand" properly only if aligned in a certain direction; it is thought to be "upside down" if built on the slope of a hill. In other dimensions the house is classified as fat or skinny, young or old and worn-out. All its interior details are invested with intense meaning. Every room and piece of furniture is connected to calendric rituals and magical and social beliefs.

Reification is the quick and easy mental algorithm that creates order in a world otherwise overwhelming in flux and detail. One of its manifestations is the dyadic instinct, the proneness to use two-part classifications in treating socially important arrays. Societies everywhere break people into in-group versus out-group, child versus adult, kin versus nonkin, married versus single, and activities into sacred and profane, good and evil. They fortify the boundaries of each division with taboo and ritual. To change from one division to the other requires initiation ceremonies, weddings, blessings, ordinations, and other rites of passage that mark every culture.

The French anthropologist Claude Lévi-Strauss and other writers of the "structuralist" school he helped found have suggested that the binary instinct is governed by the interaction of inborn rules. They posit oppositions such as man:woman, endogamy:exogamy, and earth:heaven as contradictions in the mind that must be met and resolved, often by mythic narrative. Thus the concept of life necessitates the concept of death, which is resolved by the myth of death serving as the gateway to eternal life. Binary oppositions, in the full-dress structuralist version, are linked still further into complex combinations by which cultures are assembled into integrated wholes.

The structuralist approach is potentially consistent with the picture of mind and culture emerging from natural sciences and biological anthropology, but it has been weakened by disagreements within the ranks of the structuralists themselves concerning the best methods of analysis. Their problem is not the basic conception, insofar as I have been able to understand the massive and diffuse literature, but its lack of a realistic connection to biology and cognitive psychology. That may yet be achieved, with potentially fruitful results.

NOW TO THE next step in the search for human nature, the genetic basis of the epigenetic rules. What is that basis, and how much variation is there in the prescribing genes? As a cautionary prelude to an answer, let me again stress the limitations of the genetics of human behavior as a whole. Human behavior genetics is an infant field of study and still vulnerable to ideologues who would be unkind to it in pursuit of their personal agendas. In only one level of analysis, the estimation of heritability, can it be said to be an advanced scientific discipline. With sophisticated statistical techniques, geneticists have calculated the proportionate contributions of genes across a large array of traits in sensory physiology, brain function, personality, and intelligence. They have arrived at this important conclusion: Variation in virtually every aspect of human behavior is heritable to some degree, and thus in some manner influenced by differences in genes among people. The finding should come as no surprise. It is equally true of behavior in all animal species that have been studied to date.

But the measurement of heritability does not identify *particular* genes. Nor does it provide us with a hint of the intricate pathways of physiological development leading from the genes to the epigenetic rules. The principal weakness of contemporary human behavioral genetics and human sociobiology is that only a small number of the relevant genes and epigenetic rules have been identified. This is not to deny that many others exist—quite the contrary—only that they have not yet been identified and pinpointed in genetic maps. The reason is that human behavioral genetics is technically very difficult at this level.

The paucity of examples has another, heightened consequence. Because both the genes affecting epigenetic rules and the rules themselves are usually searched out independently by different teams of researchers, matches between genes and epigenetic rules are even rarer. They come to light mostly by sheer luck. Suppose, at a guess, that 1 percent of the relevant genes and 10 percent of the epigenetic rules have been discovered up to the present time. The number of matches would be as few as the multiple of the two percentages, in this case one tenth of 1 percent. The scarcity of matches is less a failing, however, than an opportunity for scientific discovery waiting to be seized. It is precisely in this domain, on the frontier between biology and the social sciences, that some of the most significant progress in studies of human behavior can be expected to occur.

Among the known gene mutations affecting complex behavior is one that causes dyslexia, a reading disorder produced by impairment of the ability to interpret spatial relationships. Another reduces performance on three psychological tests of spatial ability but not on three other tests that measure verbal skill, speed at perception, and memory. Genes affecting personality have also been discovered. A mutation inducing outbursts of aggressive behavior, still known only in a single Dutch family, has been located on the X chromosome. It evidently causes a deficiency in the enzyme monoamine oxidase, needed to break down neurotransmitters that regulate the fight-or-flight response. Because the neurotransmitters accumulate as a result of this deviation, the brain remains keyed up, prepared to respond violently to low levels of stress. A more normal variant of personality is brought about by a "novelty-seeking gene," which alters the brain's response to the neurotransmitter dopamine. Persons possessing the gene when given standard tests are found to be more impulsive, curiosity-prone, and fickle. The molecules of the gene and the protein receptor it helps prescribe are longer in molecular length than the unmutated forms. They are also widespread, having been detected in different ethnic groups both in Israel and in the United States (but not in a Finnish group). A variety of other gene variants have been discovered that change the metabolism and activity of neurotransmitters, but their effects on behavior await investigation.

I do not mean to suggest by citing these examples that it is only necessary to discover and list genes one by one in order to establish the genetic basis of human behavior. The mapping of genes is just the beginning. Most traits, including even the simplest elements of intelligence and cognition, are influenced by polygenes, which are multiple genes spread across different chromosome sites and acting in concert. In some cases polygenes simply add their effects, so that more genes of a certain array means more of the product—more of a transmitter, say, or a higher concentration of skin pigment. Such additive inheritance, as it is called, typically produces a bell-shaped curve in the distribution of the trait in the population as a whole. Other polygenes add up until they reach a certain threshold number, at which point the trait emerges for the first time. Diabetes and some mental disorders appear to belong to this class. Finally, polygenes can interact epistatically: The presence of a gene at one chromosome site suppresses the action of a gene at another chromosome site. Brain wave patterns as revealed in electroencephalograms (EEGs) are an example of a neurological phenomenon inherited in this manner.

Finally, to complicate matters further, there is pleiotropy, the prescription of multiple effects by a single gene. A classic human example of pleiotropy is provided by the mutant gene that causes phenylketonuria, the symptoms of which include an excess of the amino acid phenylalanine, a deficiency of tyrosine, abnormal metabolic products of phenylalanine, darkening of the urine, lightening of hair color, toxic damage to the central nervous system, and—mental retardation.

The pathways from the genes to the traits they prescribe may seem overwhelmingly convoluted. Still, they can be deciphered. A large part of future human biology will consist of tracing the development of body and mind they influence. In the first two decades of the coming century, if current research stays on track, we will see the complete sequencing of the human genome and a mapping of most of the genes. Furthermore, the modes of inheritance are scientifically manageable. The number of polygenes controlling individual behavioral traits is finite, with those responsible for most of the variation often being fewer than ten. The multiple effects of single genes are also finite. They will be defined more fully as molecular biologists trace the cascades of chemical reactions entrained by groups of genes, and as neuroscientists map the patterns of brain activity that are among the final products of these reactions.

For the immediate future the genetics of human behavior will travel behind two spearheads. The first is research on the heredity of mental disorders, and the second is research on gender difference and sexual preference. Both classes are favored by strong public interest and have the further advantage of entailing processes that are well marked, hence relatively easily isolated and measured. They fit a cardinal principle in the conduct of scientific research: Find a paradigm for which you can raise money and attack with every method of analysis at your disposal.

Gender differences are an especially productive paradigm, even though politically controversial. They are already richly described in the psychological and anthropological literature. Their biological foundations are partly known, having been documented in the corpus callosum and other brain structures; in patterns of brain activity; in smell, taste, and other senses; in spatial and verbal ability; and in innate play behavior during childhood. The hormones that mediate the divergence of the sexes, resulting in statistical differences with overlap in these various traits, are relatively well understood. The major gene that triggers their ultimate manufacture during fetal and childhood development has been located on the Y chromosome. It is called *Sry*, for sex-determining region of Y. In its absence, when the individual has two X chromosomes rather than an X and Y, the fetal gonads develop into ovaries, with all the consequences that follow in endocrine and psychophysiological development. These facts may not satisfy everyone's ideological yearning, but they illustrate in yet another way that, whether we like it or not, *Homo sapiens* is a biological species.

TO THIS POINT I have traced most of the steps of gene-culture coevolution, circling from genes to culture and back around to genes, as evidence allows. These steps can be summed up very briefly as follows:

Genes prescribe epigenetic rules, which are the regularities of sensory perception and mental development that animate and channel the acquisition of culture.

Culture helps to determine which of the prescribing genes survive and multiply from one generation to the next.

Successful new genes alter the epigenetic rules of populations.

The altered epigenetic rules change the direction and effectiveness of the channels of cultural acquisition.

The final step in this series is the most crucial and contentious. It is embodied in the problem of the genetic leash. Throughout prehistory, particularly up to a hundred thousand years ago, by which time the modern *Homo sapiens* brain had evolved, genetic and cultural evolution were closely coupled. With the advent of Neolithic societies, and especially the rise of civilizations, cultural evolution sprinted ahead at a pace that left genetic evolution standing still by comparison. So, in this last exponential phase, how far apart did the epigenetic rules allow different cultures to diverge? How tight was the genetic leash? That is the key question, and it is possible to give only a partial answer.

In general, the epigenetic rules are strong enough to be visibly constraining. They have left an indelible stamp on the behavior of people in even the most sophisticated societies. But to a degree that may prove discomfiting to a diehard hereditarian, cultures have dispersed widely in their evolution under the epigenetic rules so far studied. Particular features of culture have sometimes emerged that reduce Darwinian fitness, at least for a time. Culture can indeed run wild for a while, and even destroy the individuals that foster it.

THE BEST WAY to express our still very imperfect knowledge of the transition from the epigenetic rules to cultural diversity is to describe real cases. I will offer two such examples, one relatively simple, the other complex.

The simple first. If all verbal communication were stripped away, we would still be left with a rich paralanguage that communicates most of our basic needs: body odors, blushing and other telltale reflexes, facial expressions, postures, gesticulations, and nonverbal vocalizations, all of which, in various combinations and often without conscious intent, compose a veritable dictionary of mood and intention. They are our primate heritage, having likely persisted with little change since before the origin of language. Although the signals differ in detail from one culture to the next, they contain invariant elements that reveal their ancient genetic origin. For example:

- Anstrostenol is a male pheromone concentrated in perspiration and fresh urine. Perceived variously as musk or sandalwood, it changes sexual attraction and warmth of mood during social contacts.

- To touch another is a form of greeting regulated by the following innate rules: Touch strangers of the same sex on the arms only, spreading to other parts of the body as familiarity increases, the more so for intimates of the opposite sex.

- Dilation of the pupils is a positive response to others, and one especially prominent in women.

- Pushing the tongue out and spitting are aggressive displays of rejection; flicking the tongue around the lips is a social invitation, used most commonly during flirtation.

• Closing the eyes and wrinkling the nose is another universal sign of rejection.

• Opening the mouth while pulling down the corners of the mouth to expose the lower teeth is to threaten with contempt.

These and other nonverbal signals are ideal subjects for understanding the coevolution of genes and culture. A great deal is already known of their anatomy and physiology; and their genetic prescription and controlling brain activity are likely to prove simple in comparison with verbal communication. The variation in meaning of each signal in turn caused by cultural evolution can be observed by its multiple uses across many societies. Each signal has its own amount of such variation, its own flexibility and resulting scatter of nuance across the cultures of the world. Put another way, each set of genes prescribing the basic structure of particular signals has its own norm of reaction.

The culture of nonverbal signals awaits study from this comparative viewpoint. An instinctive case of moderate dispersion is that of eyebrow flashing, one of many examples provided by the pioneering German ethologist Irenäus Eibl-Eibesfeldt. When a person's attention is caught, he opens his eyes widely to improve vision. When he is surprised, he opens his eyes very widely, while lifting the eyebrows conspicuously. Eyebrow lifting has been universally ritualized, presumably by genetic prescription, into eyebrow flashing, a signal that invites social contact. By ritualization is meant the evolution of a movement with a function in one context, in this case eye opening and eyebrow lifting, into a conspicuous, stereotyped form, in this case eyebrow flashing used for communication. That is the genetic part of the gene-culture coevolution. Eyebrow flashing has also been subjected to moderate dispersion of meaning across societies by the cultural part of gene-culture coevolution. In different societies and contexts it is combined with other forms of body language to signal greeting, flirtation, approval, request for confirmation, thanking, or emphasis of a verbal message. In Polynesia it is used as a factual "yes."

The second case of gene-culture coevolution I wish to present, because it is the most thoroughly researched of the more complex examples to date, is color vocabulary. Scientists have traced it all the way from the genes that prescribe color perception to the final expression of color perception in language.

Color does not exist in nature. At least, it does not exist in nature in the form we think we see. Visible light consists of continuously varying wavelength, with no intrinsic color in it. Color vision is imposed on this variation by the photosensitive cone cells of the retina and the connecting nerve cells of the brain. It begins when light energy is absorbed by three different pigments in the cone cells, which biologists have labeled blue, green, or red cells according to the photosensitive pigments they contain. The molecular reaction triggered by the light energy is transduced into electrical signals that are relayed to the retinal ganglion cells forming the optic nerve. Here the wavelength information is recombined to yield signals distributed along two axes. The brain later interprets one axis as green to red and the other as blue to yellow, with yellow defined as a mixture of green and red. A particular ganglion cell, for example, may be excited by input from red cones and inhibited by input from green cones. How strong an electric signal it then transmits tells the brain how much red or green the retina is receiving. Collective information of this kind from vast numbers of cones and mediating ganglion cells is passed back into the brain, across the optic chiasma to the lateral geniculate nuclei of the thalamus, which are masses of nerve cells composing a relay station near the center of the brain, and finally into arrays of cells in the primary visual cortex at the extreme rear of the brain.

Within milliseconds the visual information, now color-coded, spreads out to different parts of the brain. How the brain responds depends on the input of other kinds of information and the memories they summon. The patterns invoked by many such combinations, for example, may cause the person to think words denoting the patterns, such as: "This is the American flag; its colors are red, white, and blue." Keep the following comparison in mind when pondering the seeming obviousness of human nature: An insect flying by would perceive different wavelengths, and break them into different colors or none at all, depending on its species, and if somehow it could speak, its words would be hard to translate into our own. Its flag would be very different from our flag, thanks to its insect (as opposed to human) nature.

The chemistry of the three cone pigments—the amino acids of which they are composed and the shapes into which their chains are folded—is known. So is the chemistry of the DNA in the genes on the X chromosome that prescribe them, as well as the chemistry of the mutations in the genes that cause color blindness.

So, by inherited and reasonably well understood molecular processes the human sensory system and brain break the continuously varying wavelengths of visible light into the array of more or less discrete units we call the color spectrum. The array is arbitrary in an ultimately biological sense; it is only one of many arrays that might have evolved over the past millions of years. But it is not arbitrary in a cultural sense: Having evolved genetically, it cannot be altered by learning or fiat. All of human culture involving color is derived from this unitary process. As a biological phenomenon color perception exists in contrast to the perception of light intensity, the other primary quality of visible light. When we vary the intensity of light gradually, say by moving a dimmer switch smoothly up or down, we perceive the change as the continuous process it truly is. But if we use monochromatic light—one wavelength only—and change that wavelength gradually, the continuity is not perceived. What we see, in going from the short-wavelength end to the long-wavelength end, is first a broad band of blue (at least one more or less perceived as that color), then green, then yellow, and finally red.

The creation of color vocabularies worldwide is biased by this same biological constraint. In a famous experiment performed in the 1960s at the University of California at Berkeley, Brent Berlin and Paul Kay tested the constraint in native speakers of twenty languages, including Arabic, Bulgarian, Cantonese, Catalan, Hebrew, Ibibio, Thai, Tzeltal, and Urdu. The volunteers were asked to describe their color vocabulary in a direct and precise manner. They were shown a Munsell array, a spread of chips varying across the color spectrum from left to right, and in brightness from the bottom to the top, and asked to place each of the principal color terms of their language on the chips closest to the meaning of the words. Even though the terms vary strikingly from one language to the next in origin and sound, the speakers placed them into clusters on the array that correspond, at least approximately, to the principal colors blue, green, yellow, and red.

The intensity of the learning bias was strikingly revealed by an experiment conducted on color perception during the late 1960s by Eleanor Rosch, also of the University of California at Berkeley. In looking for "natural categories" of cognition, Rosch exploited the fact that the Dani people of New Guinea have no words to denote color; they speak only of *mili* (roughly, "dark") and *mola* ("light"). Rosch considered the following question: If Dani adults set out to learn a color vocabulary, would they do so more readily if the color terms correspond to the principal innate hues? In other words, would cultural innovation be channeled to some extent by the innate genetic constraints? Rosch divided 68 volunteer Dani men into two groups. She taught one a series of newly invented color terms placed on the principal hue categories of the array (blue, green, yellow, red), where most of the natural vocabularies of other cultures are located. She taught a second group of Dani men a series of new terms placed off center, away from the main clusters formed by other languages. The first group of volunteers, following the "natural" propensities of color perception, learned about twice as quickly as those given the competing, less natural color terms. They also selected these terms more readily when allowed a choice.

Now comes the question that must be answered to complete the transit from genes to culture. Given the genetic basis of color vision and its general effect on color vocabulary, how great has been the dispersion of the vocabularies among different cultures? We have at least a partial answer. A few societies are relatively unconcerned with color. They get along with a rudimentary classification. Others make many fine distinctions in hue and intensity within each of the basic colors. They have spaced their vocabularies out.

Has the spacing out been random? Evidently not. In later investigations, Berlin and Kay observed that each society uses from two to eleven basic color terms, which are focal points spread across the four elementary color blocks perceived in the Munsell array. The full complement (to use the English-language terminology) is black, white, red, yellow, green, blue, brown, purple, pink, orange, and gray. The Dani language, for example, uses only two of the terms, the English language all eleven. In passing from societies with simple classifications to those with complicated classifications, the combinations of basic color terms as a rule grow in a hierarchical fashion, as follows:

Languages with only two basic color terms use them to distinguish black and white.

Languages with only three terms have words for black, white, and red.

Languages with only four terms have words for black, white, red, and either green or yellow.

Languages with only five terms have words for black, white, red, green, and yellow.

Languages with only six terms have words for black, white, red, green, yellow, and blue.

Languages with only seven terms have words for black, white, red, green, yellow, blue, and brown.

No such precedence occurs among the remaining four basic colors, purple, pink, orange, and gray, when these have been added on top of the first seven.

If basic color terms were combined at random, which is clearly not the case, human color vocabularies would be drawn helter-skelter from among a mathematically possible 2,036 possibilities. The Berlin-Kay progression suggests that for the most part they are drawn from only twenty-two.

At one level, the twenty-two combinations of basic terms are the dispersion of memes, or cultural units, generated by the epigenetic rules of color vision and semantic memory. In simple language, our genes prescribe that we see different wavelengths of light a certain way. Our additional propensity to break the world into units and label them with words causes us to accumulate up to eleven basic color units in a particular order.

That, however, is not the end of the story. The human mind is much too subtle and productive to stop at eleven words that specify different wavelengths. As the British linguist John Lyons has pointed out, the recognition of a color in the brain does not necessarily lead to a term that denotes only the light wavelength. Color terms are often invented to include other qualities as well, particularly texture, luminosity, freshness, and indelibility. In Hanunóo, a Malayo-Polynesian language of the Philippines, *malatuy* means a brown, wet, shiny surface, the kind seen in freshly cut bamboo, while *marara* is a yellowish, hardened surface, as in aged bamboo. English-language speakers are prone to translate *malatuy* as "brown" and *marara* as "yellow," but they would capture only part of the meaning and perhaps the less important part. Similarly, *chloros* in ancient Greek is usually translated as simply "green" in English, but its original meaning was apparently the freshness or moistness of green foliage.

The brain constantly searches for meaning, for connections between objects and qualities that cross-cut the senses and provide information about external existence. We penetrate that world through the constraining portals of the epigenetic rules. As shown in the elementary cases of paralanguage and color vocabulary, culture has risen from the genes and forever bears their stamp. With the invention of metaphor and new meaning, it has at the same time acquired a life of its own. In order to grasp the human condition, both the genes and culture must be understood, not separately in the traditional manner of science and the humanities, but together, in recognition of the realities of human evolution.

CHAPTER 8

THE FITNESS OF HUMAN NATURE

WHAT IS human nature? It is not the genes, which prescribe it, or culture, its ultimate product. Rather, human nature is something else for which we have only begun to find ready expression. It is the epigenetic rules, the hereditary regularities of mental development that bias cultural evolution in one direction as opposed to another, and thus connect the genes to culture.

Human nature is still an elusive concept because our understanding of the epigenetic rules composing it is rudimentary. The rules I have used as examples in previous chapters are no more than fragments cut from the vast mental landscape. Yet, coming from so many behavioral categories, they offer persuasive testimony of the existence of a genetically based human nature. Consider the variety of examples so far reviewed: the hallucinatory properties of dreams, the mesmerizing fear of snakes, phoneme construction, elementary preferences in the sense of taste, details of mother-infant bonding, the basic facial expressions, the reification of concepts, the personalization of inanimate objects, and the tendency to split continuously varying objects and processes into two discrete classes. One more rule in particular, the breaking of light into the colors of the rainbow, has been placed within a causal sequence running all the way from the genes to the invention of vocabulary. It serves as a prototype for future research aimed at bridging science and the humanities.

Some epigenetic rules, including color vision, are primate traits tens of millions of years old. Others, such as the neural mechanisms of language, are uniquely human and possibly date back no more than several hundred thousand years. The search for human nature can be viewed as the archaeology of the epigenetic rules. It is destined to be a vital part of future interdisciplinary research.

In gene-culture coevolution as now conceived by biologists and social scientists, causal events ripple out from the genes to the cells to tissues and thence to brain and behavior. By interaction with the physical environment and preexisting culture, they bias further evolution of the culture. But this sequence—composing what the genes do to culture by way of epigenesis—is only half the circle. The other half is what culture does to the genes. The question posed by the second half of the coevolutionary circle is how culture helps to select the mutating and recombining genes that underlie human nature.

By expressing gene-culture coevolution in such a simple manner, I have no wish either to overwork the metaphor of the selfish gene or to minimize the creative powers of the mind. After all, the genes prescribing the epigenetic rules of brain and behavior are only segments of giant molecules. They feel nothing, care for nothing, intend nothing. Their role is simply to trigger the sequences of chemical reactions within the highly structured fertilized cell that orchestrate epigenesis. Their writ extends to the levels of molecule, cell, and organ. This early stage of epigenesis, consisting of a series of sequential physicochemical reactions, culminates in the self-assembly of the sensory system and brain. Only then, when the organism is completed, does mental activity appear as an emergent process. The brain is a product of the very highest levels of biological order, which are constrained by epigenetic rules implicit in the organism's anatomy and physiology. Working in a chaotic flood of environmental stimuli, it sees and listens, learns, plans its own future. By that means the brain determines the fate of the genes that prescribed it. Across evolutionary time, the aggregate choices of many brains determine the Darwinian fate of everything human—the genes, the epigenetic rules, the communicating minds, and the culture.

Brains that choose wisely possess superior Darwinian fitness, meaning that statistically they survive longer and leave more offspring than brains that choose badly. That generalization by itself, commonly telescoped into the phrase "survival of the fittest," sounds like a tautology—the fit survive and those who survive are fit—yet it expresses a powerful generative process well documented in nature. During hundreds of millennia of Paleolithic history, the genes prescribing certain human epigenetic rules increased and spread at the expense of others through the species by means of natural selection. By that laborious process human nature assembled itself.

What is truly unique about human evolution, as opposed say to chimpanzee or wolf evolution, is that a large part of the environment shaping it has been cultural. Therefore, construction of a special environment is what culture does to the behavioral genes. Members of past generations who used their culture to best advantage, like foragers gleaning food from a surrounding forest, enjoyed the greatest Darwinian advantage. During prehistory their genes multiplied, changing brain circuitry and behavior traits bit by bit to construct human nature as it exists today. Historical accident played a role in the assembly, and there were many particular expressions of the epigenetic rules that proved self-destructive. But by and large, natural selection, sustained and averaged over long periods of time, was the driving force of human evolution. Human nature is adaptive, or at least was at the time of its genetic origin.

Gene-culture coevolution may seem to create a paradox: At the same time that culture arises from human action, human action arises from culture. The contradiction evaporates, however, if we compare the human condition with the simpler form of reciprocity between environment and behavior widespread in the animal kingdom. African elephants, while consuming the vegetation of large numbers of trees and shrubs, create the open woodlands in which they thrive. Termites, swarming at their feet, consume leftover dead vegetation and build tightly sealed nests from soil and their own excrement, creating moist, high–carbon-dioxide microclimates to which—no surprise—their physiology is closely adapted. To view human beings evolving among elephants and termites in the same habitat during the Pleistocene Epoch, we need only replace environment in part with culture. While it is true that culture, strictly defined as complex socially learned behavior, is evidently limited to humans, and as a consequence the reciprocity between genes and culture-as-environment is also unique, the underlying principle is the same. There is nothing contradictory in saying that culture arises from human action while human action arises from culture.

The general biological imagery of the origin of human nature has repelled some writers, including a few of the most discerning scholars in the social sciences and humanities. They are, I am sure, mistaken. They misunderstand gene-culture coevolution, confusing it with rigid genetic determinism, the discredited idea that genes dictate particular forms of culture. I believe reasonable concerns can be dispelled by the following argument. Genes do not specify elaborate conventions such as totemism, elder councils, and religious ceremonies. To the best of my knowledge no serious scientist or humanities scholar has ever suggested such a thing. Instead, complexes of gene-based epigenetic rules predispose people to invent and adopt such conventions. If the epigenetic rules are powerful enough, they cause the behaviors they affect to evolve convergently across a great many societies. The conventions—evolved by culture, biased by epigenetic rules—are then spoken of as the cultural universals. Rare cultural forms are also possible under the same scenario. The whole matter can be expressed another way by reverting to the imagery of developmental genetics. The norm of reaction of the underwriting genes is greatly narrowed in the case of a cultural universal; in other words, there are few if any environments available to human beings in which the cultural convention does not arise. In contrast, genes that spawn many rare conventions in response to changing environments, thus expanding cultural diversity, are those with broader norms of reaction.

Genetic evolution might have gone the other way by eliminating epigenetic bias altogether, expanding the norm of reaction of the prescribing genes to indefinite degree, and thus causing cultural diversity to explode. That is a theoretical possibility, but the existence of such a phenomenon does not imply that culture can be cut loose from the human genome. It means only that the prescriptive genes can design the brain to learn and respond with equal alacrity to any experience. Bias-free learning, if it exists, is not an erasure of gene-culture coevolution but an extremely specialized product of it, based on a very peculiar kind of epigenetic rule. For the time being, however, the argument is moot, because no example of bias-free mental development has yet been discovered. Some degree of epigenetic bias has been demonstrated in every one of the small number of cultural categories thus far tested for the presence or absence of such bias.

The swiftness of cultural evolution in historical times may by itself seem to imply that humanity has slipped its genetic instructions, or somehow suppressed them. But that is an illusion. The ancient genes and the epigenetic rules of behavior they ordain remain comfortably in place. For most of the evolutionary history of *Homo sapiens* and its antecedent species *Homo habilis*, *Homo erectus*, and *Homo ergaster*, cultural evolution was slow enough to remain tightly coupled to genetic evolution. Both culture and the genes underlying human nature were probably genetically fit throughout that time. For tens of thousands of years during the Pleistocene Epoch the evolution of artifacts remained nearly static, and presumably so did the basic social organization of the hunter-gatherer bands using them. There was time enough, as one millennium passed into another, for the genes and epigenetic rules to evolve in concert with culture. By Upper Paleolithic times, however, from about 40,000 to 10,000 years before the present, the tempo of cultural evolution quickened. During the ensuing Neolithic agricultural advance, the pace accelerated dramatically. According to the theory of population genetics, most of the change was far too fast to be tracked closely by genetic evolution. But there is no evidence that the Paleolithic genes simply disappeared during this "creative revolution." They stayed in place and continued to prescribe the foundational rules of human nature. If they could not keep up with culture, neither could culture expunge them. For better or worse they carried human nature into the chaos of modern history.

TO TAKE behavioral genes into account therefore seems a prudent step when assessing human behavior. Sociobiology (or Darwinian anthropology, or evolutionary psychology, or whatever more politically acceptable term one chooses to call it) offers a key link in the attempt to explain the biological foundation of human nature. By asking questions framed in evolutionary theory, it has already steered research in anthropology and psychology in new directions. Its major research strategy in human studies has been to work from the first principles of population genetics and reproductive biology to predict the forms of social behavior that confer the greatest Darwinian fitness. The predictions are then tested with data taken from ethnographic archives and historical records, as well as from fresh field studies explicitly designed for the purpose. Some of the tests are conducted on preliterate and other traditional societies, whose conservative social practices are likely to resemble most closely those of Paleolithic ancestors. A very few societies in Australia, New Guinea, and South America in fact still have stone-age cultures, which is why anthropologists find them especially interesting. Other tests are conducted with data from modern societies, where fast-evolving cultural norms may no longer be optimally fit. In all these studies a full array of analytic techniques is brought to bear. They include multiple competing hypotheses, mathematical models, statistical analysis, and even the reconstruction of the histories of memes and cultural conventions by the same quantitative procedures used to trace the evolution of genes and species.

In the past quarter-century, human sociobiology has grown into a large and technically complex subject. Nevertheless, it is possible to reduce its primary evolutionary principles to some basic categories, which I will now briefly summarize.

Kin selection is the natural selection of genes based on their effects on individuals carrying them plus the effects the presence of the genes has on all the genetic relatives of the individuals, including parents, children, siblings, cousins, and others who still live and are capable either of reproducing or of affecting the reproduction of blood relatives. Kin selection is especially important in the origin of altruistic behavior. Consider two sisters, who share half their genes by virtue of having the same father and mother. One sacrifices her life, or at least remains childless, in order to help her sister. As a result the sister raises more than twice as many children as she would have otherwise. Since half of her genes are identical to those of her generous sister, the loss in genetic fitness is more than made up by the altruistic nature of the sacrifice. If such actions are predisposed by genes and occur commonly, the genes can spread through the population, even though they induce individuals to surrender personal advantage.

From this simple premise and elaborations of it have come a wealth of predictions about patterns of altruism, patriotism, ethnicity, inheritance rules, adoption practices, and infanticide. Many are novel, and most have held up well under testing.

Parental investment is behavior toward offspring that increases the fitness of the latter at the cost of the parent's ability to invest in other offspring. The different patterns of investment have consequences for the fitness of the genes that predispose individuals to select the patterns. Choose one, and you leave more offspring; choose another, and you leave fewer offspring. The idea has given rise to a biologically based "family theory," spinning off new insights on sex ratios, marriage contracts, parent-offspring conflict, grief at the loss of a child, child abuse, and infanticide. I will take up family theory again in the next chapter, in order to illustrate more fully the relevance of evolutionary reasoning for the social sciences.

Mating strategy is influenced by the cardinal fact that women have more at stake in sexual activity than men, because of the limited age span in which they can reproduce and the heavy investment required of them with each child conceived. One egg, to put the matter in elemental terms, is hugely more valuable than a single sperm, which must compete with millions of other sperm for the egg. The achievement of pregnancy closes off further breeding opportunity of the mother for a substantial fraction of her remaining reproductive life, whereas the father has the physical capacity to inseminate another woman almost immediately. With considerable success, the nuances of this concept have been used by scientists to predict patterns of mate choice and courtship, relative degrees of sexual permissiveness, paternity anxiety, treatment of women as resources, and polygyny (multiple wives, which in the past at least has been an accepted arrangement in three-quarters of societies around the world). The optimum sexual instinct of men, to put the matter in the now familiar formula of popular literature, is to be assertive and ruttish, while that of women is to be coy and selective. Men are expected to be more drawn than women to pornography and prostitution. And in courtship, men are predicted to stress exclusive sexual access and guarantees of paternity, while women consistently emphasize commitment of resources and material security.

Status is central to all complex mammal societies, humanity included. To say that people generally seek status, whether by rank, class, or wealth, is to sum up a large part of the catalogue of human social behavior. In traditional societies genetic fitness of individuals is generally but not universally correlated with status. In chiefdoms and despotic states especially, dominant males have easy access to multiple women and produce more children, often in spectacular disproportion. Throughout history, despots (absolute rulers with arbitrary powers of life and death over their subjects) commanded access to hundreds or even thousands of women. Some states used explicit rules of distribution, as in Inca Peru, where by law petty chiefs were given seven women, governors of a hundred people eight, leaders of a thousand people fifteen, and lords and kings no fewer than seven hundred. Commoners took what was left over. The fathering of children was commensurately lopsided. In modern industrial states, the relationship between status and genetic fitness is more ambiguous. The data show that high male status is correlated with greater longevity and copulation with more women, but not necessarily the fathering of more children.

Territorial expansion and defense by tribes and their modern equivalents the nation states is a cultural universal. The contribution to survival and future reproductive potential, especially of tribal leaders, is overwhelming, and so is the warlike imperative of tribal defense. "Our country!" declared Commodore Stephen Decatur, hard-fighting hero of the War of 1812, "may she always be right; but our country, right or wrong." (Personal aggressiveness has its Darwinian limits, however; Decatur was killed in a duel in 1820.)

Biologists have determined that territoriality is not unavoidable during social evolution. It is apparently entirely absent in many animal species. The territorial instinct arises during evolution when some vital resource serves as a "density-dependent factor." That is, the growth of population density is slowed incrementally by an increasing shortage of food, water, nest sites, or the entire local terrain available to individuals searching for these resources. Death rates increase or birth rates decrease, or both, until the two rates come more or less into balance and population density levels off. Under such circumstances animal species tend to evolve territorial behavior. The theoretical explanation is that individuals hereditarily predisposed to defend private resources for themselves and their social group pass more genes on to the next generation.

In contrast, the growth of other species is not leveled off by limiting resources but by rising amounts of emigration, disease, or predation. When such alternative density-dependent factors are paramount, and resource control is therefore not required, territorial defense usually does not evolve as a hereditary response.

Humanity is decidedly a territorial species. Since the control of limiting resources has been a matter of life and death through millennia of evolutionary time, territorial aggression is widespread and reaction to it often murderous. It is comforting to say that war, being cultural in origin, can be avoided. Unfortunately, that bit of conventional wisdom is only a half truth. It is more nearly correct—and far more prudent—to say that war arises from both genes and culture and can best be avoided by a thorough understanding of the manner in which these two modes of heredity interact within different historical contexts.

Contractual agreement so thoroughly pervades human social behavior, virtually like the air we breathe, that it attracts no special notice—until it goes bad. Yet it deserves focused scientific research for the following reason. All mammals, including humans, form societies based on a conjunction of selfish interests. Unlike the worker castes of ants and other social insects, they resist committing their bodies and services to the common good. Rather, they devote their energies to their own welfare and that of close kin. For mammals, social life is a contrivance to enhance personal survival and reproductive success. As a consequence, societies of nonhuman mammalian species are far less organized than the insect societies. They depend on a combination of dominance hierarchies, rapidly shifting alliances, and blood ties. Human beings have loosened this constraint and improved social organization by extending kinshiplike ties to others through long-term contracts.

Contract formation is more than a cultural universal. It is a human trait as characteristic of our species as language and abstract thought, having been constructed from both instinct and high intelligence. Thanks to groundbreaking experiments by the psychologists Leda Cosmides and John Tooby at the University of California at Santa Barbara, we know that contract formation is not simply the product of a single rational faculty that operates equally across all agreements made among bargaining parties. Instead, one capacity, the detection of cheating, is developed to exceptional levels of sharpness and rapid calculation. Cheater detection stands out in acuity from mere error detection and the assessment of altruistic intent on the part of others. It is furthermore triggered as a computation procedure only when the cost and benefits of a social contract are specified. More than error, more than good deeds, and more even than the margin of profit, the possibility of cheating by others attracts attention. It excites emotion and serves as the principal source of hostile gossip and moralistic aggression by which the integrity of the political economy is maintained.

THE GENETIC FITNESS hypothesis—that the most widely distributed traits of culture confer Darwinian advantage on the genes that predispose them—has been reasonably well borne out by the evidence. Widely distributed traits are usually adaptive, and their existence accords with the first principles of evolution by natural selection. It is further true that by and large people behave in their daily lives as though somehow guided, whether consciously or unconsciously, by these first principles. The value of the genetic fitness hypothesis lies in the insights concerning human nature it provides and the productive new directions in scholarly research it has stimulated.

There are nonetheless many weaknesses in the genetic fitness hypothesis. For the most part the flaws are due not to contradictory evidence but to a scarcity of relevant information. Because human behavioral genetics is still in its infancy, there is a near-absence of direct links between particular genes and behavior underlying the universal culture traits. The observed fit between theory and fact is based mostly on statistical correlation. One of the rare exceptions, described in the previous chapter, is the connection successfully made between the genetics and vocabulary of color vision.

The epigenetic rules that guide behavioral development are also largely unexplored, and as a result the exact nature of gene-culture coevolution can in most cases only be guessed. It makes all the difference in the world whether epigenetic rules are rigid, specialized functions of the brain, and thus resemble animal instinct, or whether they are more generalized rational algorithms that function across a wide range of behavioral categories. The evidence to date shows that both kinds of epigenetic rules, narrow and broad, exist. For example, the use of the smile is narrowly channeled by one set of rules, while territorial response is broadly channeled by another. But until such rules are better documented and disentangled, along with the manner in which they guide mental development, it will be difficult to account for the wide cultural variation that occurs in a majority of behavioral categories.

These shortcomings in behavioral genetics and development are conceptual, technical, and deep. But they are ultimately solvable. Unless new evidence commands otherwise, trust is wisely placed in the natural consilience of the disciplines now addressing the connection between heredity and culture, even if support for it is accumulating slowly and in bits and pieces. The resolution of the difficulties awaits the further expansion of biology and its coalescence with psychology and anthropology.

THE CATEGORY OF human behavior that provides the fullest test of the genetic fitness hypothesis to date is incest avoidance. A large amount of information concerning the phenomenon has become available at different levels of biology and culture. The behavior itself is universal, or nearly so. It is also relatively clear-cut in expression. Sexual activity in all societies is relatively uncommon between siblings and between parents and their offspring; children produced by such activity are rare; and long-term unions made with the consensual purpose of having such children are almost nonexistent.

The current explanation of incest avoidance, which combines genetic and cultural evolution, is a straightforward sociobiological exercise. Inbreeding at the level of siblings and parents and children yields a high percentage of offspring with genetic defects. Humans tend to avoid this risk by unconscious obedience to the following epigenetic rule: If a boy and girl are brought together before one or the other is thirty months of age and then raised in close domestic proximity—use the same potty, so to speak—they are devoid of later sexual interest in each other, and the very thought of it arouses an acute aversion. This emotional incapacity, fortified in many societies by a rational understanding of the consequence of inbreeding, has led to the cultural incest taboos, which prohibit incest by custom and law.

The risk of defective children from incest—inbreeding depression as it is called by geneticists—is now well understood. On average, each person carries somewhere on his twenty-three pairs of chromosomes two sites that contain recessive lethal genes. The sites can be almost anywhere on the chromosomes. They also differ in exact number and location from one person to the next. Only one of the two homologous chromosomes in the affected pair carries lethals at the site; the other homologous chromosome carries a normal gene, which overrides the effects of the lethal gene. The reason is the lethality itself. When both chromosomes carry a lethal gene at a particular site, the fetus is aborted or the child dies in infancy.

Consider a woman with a lethal gene at one such site. If she is impregnated by her brother, and if their parents themselves are unrelated, her child has one chance in eight of dying as a fetus or as an infant. If she has lethal genes at two such sites, her child has about one chance in four of dying. There exist in addition a horde of other recessive genes that cause crippling anatomical and mental defects. The total effect is that early mortality of children born of incest is about twice that of outbred children, and among those that survive, genetic defects such as dwarfism, heart deformities, severe mental retardation, deaf-mutism, enlargement of the colon, and urinary tract abnormalities are ten times more common.

The destructive consequence of incest is a general phenomenon not just in humans but also in plants and animals. Almost all species vulnerable to moderate or severe inbreeding depression use some biologically programmed method to avoid incest. Among the apes, monkeys, and other nonhuman primates the method is two-layered. First, among all nineteen social species whose mating patterns have been studied, young individuals tend to practice the equivalent of human exogamy: Before reaching full adult size they leave the group in which they were born and join another. In the lemurs of Madagascar and in the majority of monkey species from both the Old and New Worlds, it is the males who emigrate. In red colobus monkeys, hamadryas baboons, gorillas, and chimpanzees of Africa, the females leave. In howler monkeys of Central and South America, both sexes depart. The restless young of these diverse primate species are not driven out of the group by aggressive adults. Their departure appears to be entirely voluntary.

Whatever its ultimate evolutionary origin, and however else it affects reproductive success, the emigration of young primates prior to reaching full sexual maturity greatly reduces the potential for inbreeding. But the barrier against inbreeding is reinforced by a second line of resistance. This is the avoidance of sexual activity by even those individuals who remain with their natal group. In all the social nonhuman primate species whose sexual development has been carefully studied, including marmosets and tamarins of South America, Asian macaques, baboons, and chimpanzees, both adult males and females display the "Westermarck effect": They spurn individuals with whom they were closely associated in early life. Mothers and sons almost never copulate, and brothers and sisters kept together mate much less frequently than do more distantly related individuals.

This elemental response was discovered, not in monkeys and apes, but in human beings, by the Finnish anthropologist Edward A. Westermarck and first reported in his 1891 masterwork *The History of Human Marriage*. The existence of the phenomenon has gained increasing support from many sources in the intervening years. None is more persuasive than the study of "minor marriages" in Taiwan by Arthur P. Wolf of Stanford University. Minor marriages, formerly widespread in southern China, are those in which unrelated infant girls are adopted by families, raised with the biological sons in an ordinary brother-sister relationship, and later married to the sons. The motivation for the practice appears to be to insure partners for sons when an unbalanced sex ratio and economic prosperity combine to create a highly competitive marriage market.

Across four decades, from 1957 to 1995, Wolf studied the histories of 14,200 Taiwanese women contracted for minor marriage during the late nineteenth and early twentieth centuries. The statistics were supplemented by personal interviews with many of these "little daughters-in-law," or *sim-pua*, as they are known in the Hokkien language, as well as with their friends and relatives.

What Wolf had hit upon was a controlled—if unintended—experiment in the psychological origins of a major piece of human social behavior. The *sim-pua* and their husbands were not biologically related, thus taking away all of the conceivable factors due to close genetic similarity. Yet they were raised in a proximity as intimate as that experienced by brothers and sisters in Taiwanese households.

The results unequivocally favor the Westermarck hypothesis. When the future wife was adopted before thirty months of age, she usually resisted later marriage with her de facto brother. The parents often had to coerce the couple to consummate the marriage, in some cases by threat of physical punishment. The marriages ended in divorce three times more often than "major marriages" in the same communities. They produced nearly 40 percent fewer children, and a third of the women were reported to have committed adultery, as opposed to about 10 percent of wives in major marriages.

In a meticulous series of cross-analyses, Wolf identified the key inhibiting factor as close coexistence during the first thirty months of life of either or both of the partners. The longer and closer the association during this critical period, the stronger the later effect. Wolf's data allow the reduction or elimination of other imaginable factors that might have played a role, including the experience of adoption, financial status of the host family, health, age at marriage, sibling rivalry, and the natural aversion to incest that could have arisen from confusing the pair with true, genetic siblings.

A parallel unintended experiment has been performed in Israeli kibbutzim, where children are raised in crèches as closely as brothers and sisters in conventional families. The anthropologist Joseph Shepher and his coworkers reported in 1971 that among 2,769 marriages of young adults reared in this environment, none was between members of the same kibbutz peer group who had lived together since birth. There was not even a single known case of heterosexual activity, despite the fact that the kibbutz adults were not especially opposed to it.

From these examples, and a great deal of additional anecdotal evidence gleaned from other societies, it is evident that the human brain is programmed to follow a simple rule of thumb: *Have no sexual interest in those whom you knew intimately during the earliest years of your life.*

The Westermarck effect is also consistent with the principle of graded effect in psychology. The evidence from across many societies shows that the more intimate the association during the critical period of early childhood, the less likely is it that heterosexual activity will occur. Hence mother-son incest, which is inhibited by the intense bonding during the infancy of the son, is by far the rarest kind. Next in scarcity is sibling incest, then sexual abuse of girls by their biological fathers (I say abuse because consent is seldom given freely by the daughters), and finally sexual abuse of girls by their stepfathers.

Yet, while the evidence makes a tidy and persuasive picture, we are still far from a full explanation of incest avoidance. There is no conclusive proof that the Westermarck effect originated from genetic evolution by natural selection. Certainly all the signs point that way. Incest avoidance diminishes inbreeding and thereby increases the production of healthy offspring. Given even a small amount of genetic variability in sexual responsiveness to childhood associates, the differences in fitness based on it would have been strong enough, in population genetics theory at least, to spread the Westermarck effect throughout the population from a very low incidence to widespread occurrence in as few as ten generations. Further evidence is the occurrence of the effect in other primates, including our closest living relatives the chimpanzees, where it is unquestionably genetic, not cultural, in origin. Still, no attempt has been made to measure heritability in the human response or to discover the genes underwriting it.

A second shortcoming on the research front is that we do not know the exact psychological source of the Westermarck effect. The stimuli from childmates that trigger the inhibition have not been pinpointed. It is not known whether they occur during play, eating together, unavoidable aggressive exchanges, or other events more subtle and perhaps only subliminally sensed. The critical stimuli could be anything, large or small, visual, auditory, or olfactory, and not necessarily understood in any ordinary adult sense. The essence of instinct as interpreted by biologists is that it is evoked by simple cues that need only be associated in real life with the object to which it is directed. A scent or a single touch at a critical moment can unleash complex behavior, or inhibit it.

A further complication in the story of human incest avoidance is the existence of a third barrier, incest taboos, the culturally transmitted sets of rules that prohibit sexual activity among very close relatives. Many societies permit or even encourage marriages between first cousins, especially when the bonding serves group cohesion and consolidates wealth, but forbid it between siblings and half siblings.

The taboos, being conscious inventions and not simple instinctive responses, vary enormously in detail from one society to the next. In many cultures they are interwoven with the strictures of kinship classification and exogamous marriage contracts. In preliterate societies incest is commonly thought to be connected with cannibalism, vampirism, and malign witchcraft, each of which is punishable on its own account. Modern societies enact laws to discourage incest. During the Commonwealth and Protectorate period of England, from 1650 to the Restoration a decade later, it was punishable by death. In Scotland until 1887, it was nominally a capital offense, although transgressions seldom drew more than life imprisonment. In the United States incest has been generally treated as a felony punishable by fine, imprisonment, or both. The sexual abuse of children is considered all the more abhorrent when it is in addition incestuous.

History, as ever true for human mores generally, records exceptions. Societies with some degree of permissiveness have included the Incas, Hawaiians, Thais, ancient Egyptians, Nkole (Uganda), Bunyoro (Uganda), Ganda (Uganda), Zande (Sudan), and Dahomeyans of West Africa. In each case the practice is (or in most instances was, having been discontinued) surrounded by ritual and limited to royalty or other groups of high status. In all the incestuous arrangements the male also consorted with other women, fathering outbred children in addition to "pure" progeny. The ruling families are or were patrilineal. The strategy yielding maximum genetic fitness for a high-ranking male is to mate with his own sister, producing children who share with him 75 percent of their genes by common descent, instead of the usual 50 percent, and also to mate with women who are genetically unrelated and more likely to give birth to normal children. Less easily explained are the common and well-documented cases of brother-sister marriages among commoners in Roman Egypt, from about 30 B.C. to A.D. 324. Papyrus texts from the period reveal beyond reasonable doubt that at least some of the siblings engaged in full and unabashed sexual relations.

Incest taboos have led us, once again, to the borderland between the natural and social sciences. The question they raise is as follows: What is the relation between the Westermarck effect, which is biological, and the incest taboos, which are cultural?

The issue can be drawn more sharply by distinguishing the two principal hypotheses that compete for the explanation of human incest avoidance. The first is Westermarck's, which I will now summarize in updated language: People avoid incest because of a hereditary epigenetic rule of human nature that they have translated into taboos. The opposing hypothesis is that of Sigmund Freud. There is no Westermarck effect, the great theoretician insisted when he learned of it. Just the opposite: Heterosexual lust among members of the same family is primal and compelling, and not forestalled by any instinctive inhibition. In order to prevent such incest, and the consequent disastrous ripping apart of family bonds, societies invent taboos. One result, which Freud developed as part of his grand scheme for psychology, is the Oedipus complex, the unresolved desire of a son for sexual gratification with his mother and his simultaneous hatred for the father, who is seen as a rival. "The first choice of object in mankind," he wrote in 1917, "is regularly an incestuous one, directed to the mother and sister of men, and the most stringent prohibitions are required to prevent this sustained infantile tendency from being carried into effect."

Labeling the idea of the Westermarck effect "preposterous," Freud carried the day from the very start. The findings of psychoanalysis, he asserted, make the phenomenon untenable. He also drew heavily on a rebuttal by James Frazer, the British anthropologist, classicist, and author of *The Golden Bough*. If the Westermarck effect really existed, Frazer reasoned, no taboos would be required. "It is not easy to see why any deep human instinct should need to be reinforced by law." That logic prevailed in textbooks and scholarly reviews for most of the rest of the twentieth century.

Westermarck's response to Frazer was simple, equally logical, and supported by growing amounts of evidence, but ignored in the triumphant onrush of psychoanalytic theory. Individual humans, Westermarck said, reason as follows: *I am sexually indifferent to my parents and siblings. Yet occasionally I wonder what it would be like to have sex with them. The thought is repugnant! Incest is forced and unnatural. It would alter or break other bonds I have formed with them and must maintain on a day-to-day basis for my own welfare. Incest by others is by extension also repugnant to my mind, and evidently to that of others too, and so the rare cases in which it occurs should be condemned as immoral.*

Reasonable as that explanation may be, and supported by evidence, it is nevertheless easy to see why Freud and a host of other influential social theorists reacted so vehemently to the Westermarck effect. It imperiled a foundation piece of modernist thought, calling into question what had come to be regarded as a major intellectual advance of the era. Wolf has expressed the difficulty with precision: "Freud saw all too clearly that if Westermarck was right, *he* was wrong. The possibility that early childhood association suppressed sexual attraction had to be denied lest the basis of the Oedipus complex crumble and with it his conception of personality dynamics, his explanation of neuroses, and his grand view of the origins of law, art, and civilization."

The Westermarck effect rocks other boats as well. There is the matter of whether social regulation in general exists to repress human nature or to express it. And from that comes the not so trivial question of what incest taboos imply about the origins of morality. Orthodox social theory holds that morality is largely a convention of obligation and duty constructed from mode and custom. The alternative view, favored by Westermarck in his writings on ethics, is that moral concepts are derived from innate emotions.

In the clash of ethical theory at least, the matter of incest avoidance can be settled empirically. Either Westermarck or Freud was factually right. The evidence now leans strongly to Westermarck. Yet there is more to incest taboos than the mere grafting of cultural conventions onto personal preference. It is also possible for people to observe the effects of inbreeding directly. They are capable of recognizing in at least a vague way that deformed children are a frequent product of incestuous unions. William H. Durham, a colleague of Arthur Wolf's at Stanford University, searched the ethnographic records of sixty societies chosen at random from around the world for references to any form of understanding of the consequences of incest. He found that twenty showed some degree of such awareness. The Tlingit Amerindians of the Pacific Northwest, for example, grasped in a straightforward manner that defective children are often produced from matings of very close kin. Other societies not only knew that much, but also developed folk theories to explain it. The Lapps of Scandinavia spoke of "bad blood" created by incest. The Tikopian Polynesians thought that *mara*, the doom generated by partners in incest, is transmitted to their young. The Kapauku of New Guinea, in a similar theory, believed that the act of incest causes a deterioration of the vital substances of the transgressors, which is then passed on to their children. The Toradja of Sulawesi, Indonesia, were more cosmic in their interpretation. They said that whenever people mate who have certain conflicting characteristics, as between close kin, nature is thrown into confusion.

Curiously, while fifty-six of Durham's sixty societies had incest motifs in one or more of their myths, only five contained accounts of evil effects. A somewhat larger number ascribed beneficial results, in particular the creation of giants and heroes. But even here incest was viewed as something special if not abnormal.

In summary, the factual picture emerging from research on human incest avoidance is one of multiple, successive barriers. Up front is the Westermarck effect, the ancient sexual desensitization found in all other primates thus far, and thus likely to be universal in humans. Next there is the dispersal of the young at sexual maturity, also a universal primate trait, manifested in humans by adolescent restlessness and the formal practices of exogamous marriage. The deeper psychological motivations of the dispersal behaviors and the epigenetic rules composing them remain unknown. Finally, there are the cultural incest taboos, which enhance the Westermarck effect and dispersal. The taboos seem likely to have arisen from the Westermarck effect but also, in a minority of societies, from a direct perception of the destructive effects of inbreeding.

By translating the Westermarck effect into incest taboos, humans appear to pass from pure instinct to pure rational choice. But do they really? What is rational choice anyway? I suggest that rational choice is the casting about among alternative mental scenarios to hit upon the ones which, in a given context, satisfy the strongest epigenetic rules. It is these rules and this hierarchy of their relative strengths by which human beings have successfully survived and reproduced for hundreds of millennia. The incest avoidance case may illustrate the manner in which the coevolution of genes and culture has woven not just part but all of the rich fabric of human social behavior.

CHAPTER 9

THE SOCIAL SCIENCES

PEOPLE EXPECT from the social sciences—anthropology, sociology, economics, and political science—the knowledge to understand their lives and control their future. They want the power to predict, not the preordained unfolding of events, which does not exist, but what will happen if society selects one course of action over another.

Political life and the economy are already pivoted upon the presumed existence of such a predictive capacity. The social sciences are striving to achieve it, and to do so largely without linkage to the natural sciences. How well are they doing on their own? Not very well, considering their track record in comparison with the resources placed at their command.

The current status of the social sciences can be put in perspective by comparing them with the medical sciences. Both have been entrusted with big, urgent problems. Medical scientists are paid, for example, to cure cancer, correct genetic birth defects, and repair severed nerve cords. Social scientists are expected to tell us how to moderate ethnic conflict, convert developing countries into prosperous democracies, and optimize world trade. In both spheres the problems have been intractably complex, partly because the root causes are poorly understood.

The medical sciences are nevertheless progressing dramatically. Breakthroughs have been achieved in basic research and others are expected at any time, perhaps leading to more and more noninvasive, magic-bullet cures.

Excitement runs high through global information networks connecting thousands of well-funded research groups. Neurobiologists, virologists, and molecular geneticists understand and encourage one another even as they compete in the race for discovery.

There is also progress in the social sciences, but it is much slower, and not at all animated by the same information flow and optimistic spirit. Cooperation is sluggish at best; even genuine discoveries are often obscured by bitter ideological disputes. For the most part, anthropologists, economists, sociologists, and political scientists fail to understand and encourage one another.

The crucial difference between the two domains is consilience: The medical sciences have it and the social sciences do not. Medical scientists build upon a coherent foundation of molecular and cell biology. They pursue elements of health and illness all the way down to the level of biophysical chemistry. The success of their individual projects depends on the fidelity of their experimental design to fundamental principles, which the researchers endeavor to make consistent across all levels of biological organization from the whole organism down, step by step, to the molecule.

Social scientists, like medical scientists, have a vast store of factual information and an arsenal of sophisticated statistical techniques for its analysis. They are intellectually capable. Many of their leading thinkers will tell you, if asked, that all is well, that the disciplines are on track—sort of, more or less. Still, it is obvious to even casual inspection that the efforts of social scientists are snarled by disunity and a failure of vision. And the reasons for the confusion are becoming increasingly clear. Social scientists by and large spurn the idea of the hierarchical ordering of knowledge that unites and drives the natural sciences. Split into independent cadres, they stress precision in words within their specialty but seldom speak the same technical language from one specialty to the next. A great many even enjoy the resulting overall atmosphere of chaos, mistaking it for creative ferment. Some favor partisan social activism, directing theory into the service of their personal political philosophies. In past decades, social scientists have endorsed Marxism-Leninism, or—as much as the misguided biologists who usually receive the blame—the worst excesses of Social Darwinism. Today various factions favor ideological positions ranging from laissez-faire capitalism to radical socialism, while a few promote versions of postmodernist relativism that question the very idea of objective knowledge itself.

They are easily shackled by tribal loyalty. Much of what passes for social theory is still in thrall to the original grand masters—a bad sign, given the principle that progress in a scientific discipline can be measured by how quickly its founders are forgotten. Simon Blackburn, in *The Oxford Dictionary of Philosophy*, provides an instructive example: "The tradition of semiotics that follows Saussure is sometimes referred to as semiology. Confusingly, in the work of Kristeva, the term is appropriated for the nonrational effluxes of the infantile part of the self." And so on through the byways of critical theory, functionalism, historicism, antihistoricism, structuralism, poststructuralism, and—if the mind is not steeled to resist—thence into the pits of Marxism and psychoanalytic theory where so much of academia disappeared in the twentieth century.

Each of these enterprises has contributed something to understanding the human condition. The best of the insights, if pieced together, explain the broad sweep of social behavior, at least in the same elementary sense that preliterate creation myths explain the universe, that is, with conviction and a certain internal consistency. But never—I do not think that too strong a word—have social scientists been able to embed their narratives in the physical realities of human biology and psychology, even though it is surely there and not some astral plane from which culture has arisen.

I grant that a large measure of humility is in order for any critic. Everyone knows that the social sciences are hypercomplex. They are inherently far more difficult than physics and chemistry, and as a result they, not physics and chemistry, should be called the hard sciences. They just seem easier, because we can talk with other human beings but not with photons, gluons, and sulfide radicals. Consequently, too many social-science textbooks are a scandal of banality.

Such is the paradox of the social sciences. Familiarity bestows comfort, and comfort breeds carelessness and error. Most people believe they know how they themselves think, how others think too, and even how institutions evolve. But they are wrong. Their understanding is based on folk psychology, the grasp of human nature by common sense—defined (by Einstein) as everything learned to the age of eighteen—shot through with misconceptions, and only slightly advanced over ideas employed by the Greek philosophers. Advanced social theorists, including those who spin out sophisticated mathematical models, are equally happy with folk psychology. As a rule they ignore the findings of scientific psychology and biology. That is part of the reason, for example, why social scientists overestimated the strength of communist rule and underestimated the strength of ethnic hostility. They were genuinely startled when the Soviet empire collapsed, popping the cap off the superpower pressure cooker, and were surprised again when one result of this release of energies was the breakout of ethnic strife and nationalistic wars in the spheres of diminished Russian influence. The theorists have consistently misjudged Muslim fundamentalism, which is religion inflamed by ethnicity. At home in America, they not only failed to foresee the collapse of the welfare state, but still cannot agree on its causes. In short, social scientists as a whole have paid little attention to the foundations of human nature, and they have had almost no interest in its deep origins.

The social sciences are hampered in this last regard by the residue of strong historical precedent. Ignorance of the natural sciences by design was a strategy fashioned by the founders, most notably Émile Durkheim, Karl Marx, Franz Boas, and Sigmund Freud, and their immediate followers. They aimed to isolate their nascent disciplines from the foundational sciences of biology and psychology, which at the inception of the social sciences were in any case too primitive to be of clear relevance. This stance was fruitful at first. It allowed scholars to search widely for patterns in culture and social organization unencumbered by the patronage of the natural sciences, and to compose such laws of social action as the prima facie evidence demanded. But once the pioneering era ended, the theorists were mistaken not to include biology and psychology. It was no longer a virtue to avoid the roots of human nature.

The theorists were inhibited from probing in that direction by another problem endemic to the social sciences: political ideology. Its effects have been especially clear in American anthropology. Franz Boas, aided by his famous students Ruth Benedict and Margaret Mead, led a crusade against what they perceived (correctly) to be the eugenics and racism implicit in Social Darwinism. With caution swept aside by moral zeal, they turned opposition into the new ideology of cultural relativism. The logic of the ideology, still shared in varying degree by most professional anthropologists, can be expressed as follows:

It is wrong to suppose that "civilized" peoples are the winners over "primitive" peoples in a Darwinian struggle for existence, hence superior; it is wrong to think that the differences between them are due to their genes rather than a product of historical circumstance. Furthermore, culture is wondrously complex and tuned to the environment in which it has evolved. Therefore, it is misleading to think of cultures as evolving from a lower to a higher status, and it is wrong to entertain biological explanations of cultural diversity.

Believing it a virtue to declare that all cultures are equal but in different ways, Boas and other influential anthropologists nailed their flag of cultural relativism to the mast. During the 1960s and 1970s this scientific belief lent strength in the United States and other Western societies to political multiculturalism. Also known as identity politics, it holds that ethnics, women, and homosexuals possess subcultures deserving equal standing with those of the "majority," even if the doctrine demotes the idea of a unifying national culture. The United States motto, *E pluribus unum*, "out of the many, one," was turned around to "out of the one, many"; and those who wished it so asked this question with a good measure of reasonableness: What can be wrong with identity politics if it increases the civil rights of individuals? Many anthropologists, their instincts fortified by humanitarian purpose, grew stronger in their support of cultural relativism while stiffening their opposition to biology in any guise.

So, no biology. The reasoning then came full circle with a twist that must have brought a smile to the little gods of irony. Where cultural relativism had been initiated to negate belief in hereditary behavioral differences among ethnic groups—undeniably an unproven and ideologically dangerous conception—it was then turned against the idea of a unified human nature grounded in heredity. A great conundrum of the human condition was created: If neither culture nor a hereditary human nature, what unites humanity? The question cannot be just left hanging, for if ethical standards are molded by culture, and cultures are endlessly diverse and equivalent, what disqualifies theocracy, for example, or colonialism? Or child labor, torture, and slavery?

IN CONFUSED RESPONSE to the question, anthropology is today breaking into two cultures of its own, different but equal (of course) in merit. The biological anthropologists attempt to explain culture as ultimately a product of the genetic history of humanity, renewed each generation by the decisions of individuals influenced by that history. In sharp contrast, the cultural anthropologists, descendants of Boas, see culture as a higher-order phenomenon largely free of genetic history and diverging from one society to the next virtually without limit. The view of the biological anthropologists can be likened to the film series *Star Wars*, whose aliens have different physical anatomies but are rather disconcertingly united by an unshakable human nature. The view of the cultural anthropologists is more that of the film *Invasion of the Body Snatchers*, whose protagonists take human form but retain their alien natures. (The film that got it right is *Independence Day*: If not human, it correctly suggests, *everything* is alien.)

The schismatic state of contemporary anthropology is illustrated by the resolution passed by officers of the American Anthropological Association in

1994, affirming on the one hand an "abiding commitment to biological and cultural variation" and on the other hand a "refusal to biologize or otherwise essentialize diversity." No way was spelled out to reconcile the two contradictory goals.

How then is diversity to be addressed within anthropology? In the absence of a common search for consilient explanation, there is no solution. The schism between the two camps will continue to deepen. While biological anthropologists increasingly focus on heredity and reconstructions of human evolution, cultural anthropologists will drift farther away from the natural sciences. To an increasing degree they already align their scholarship with the humanities, analyzing each culture—say, Kwakiutl, Yanomamö, Kapauku, Japanese—as a unique entity. They see culture overall as neither predictable nor even definable by laws derived from the natural sciences. Some have gone so far as to adopt the extreme postmodernist view that science is just another way of thinking, one respectable intellectual subculture in the company of many.

CONTEMPORARY SOCIOLOGY STANDS even farther apart from the natural sciences than anthropology. As generally practiced, it can be defined as the anthropology of complex societies, especially those to which sociologists themselves belong. Anthropology can be conversely defined as the sociology of simpler, more remote societies, those to which anthropologists do *not* belong. Where a representative sociological topic is the relationship of family income to American divorce rates, a typical anthropological topic is Sudanese bridewealth.

Much of modern sociology features exact measurement and statistical analysis. But apart from scattered heretics, among the most outspoken of whom are Pierre L. van den Berghe of the University of Washington, Lee Ellis of Minot State University, Joseph Lopreato of the University of Texas, and Walter L. Wallace of Princeton University, academic sociologists have remained clustered near the nonbiological end of the culture studies spectrum. Many are, in Ellis' expression, biophobic—fearful of biology and determined to avoid it. Even psychology is treated gingerly. James S. Coleman of the University of Chicago, a distinguished and influential mainstream theorist proficient in the analytic methods of the natural sciences, could say (in 1990) that "the principal task of the social sciences is the explanation of social phenomena, not the behavior of single individuals. In isolated cases the social phenomenon may derive directly, through summation, from the behavior of individuals, but more often this is not so. Consequently, the focus must be on the social system whose behavior is to be explained. This may be as small as a dyad or as large as a society or even a world system, but the essential requirement is that the explanatory focus be on the system as a unit, not on the individuals or other components which make it up."

To appreciate how far removed Coleman's research strategy is from that of the natural sciences, substitute organism for system, cell for individual, and molecules for other components, and his statement becomes, "the essential requirement is that the explanatory focus be on the organism as a unit, not on the cell or molecules which make it up." Biology would have remained stuck around 1850 with such a flat perspective. Instead, biology is a science that traces causation across many levels of organization, from brain and ecosystem down to atom. There is no obvious reason why sociology should not have a similar orientation, guided by a vision sweeping from society to neuron.

A century after the publication of Durkheim's manifesto *The Rules of Sociological Method* (1894), which helped set the ground rules, the narrowly stratal approach of the discipline to the study of industrialized societies remains nearly unchanged. Robert Nisbet of Columbia University, in a revealing interpretation of classical sociology, sees the field as having originated more as an art form than as a science, however grand in conception. Nisbet cites Herbert Read's preferred goal of great art as not just the satisfaction of personal needs, or even the representation of philosophical or religious ideas, but the creation of a synthetic and internally consistent world through images that "tell us something about the universe, something about nature, about man, or about the artist himself."

Sociology did not, in Nisbet's view, grow as a logical extension of the natural sciences, the course its prophets had foretold in the late Enlightenment. Rather it was created whole from the master themes of the Western ethos, among them individualism, freedom, social order, and progressive change. Much of the classic literature of sociology, Nisbet observed, comprises well-wrought vistas of social, economic, and political life in nineteenth- and early-twentieth-century Western Europe. "What Tocqueville and Marx, and then Toennies, Weber, Durkheim, and Simmel, gave us in their greatest works, ranging from *Democracy in America* and *Capital* to Toennies on *Gemeinschaft und Gesellschaft* or Simmel on *Metropolis*, is a series of landscapes, each as distinctive and compelling as any to be found among the greater novels or paintings of their age." The dominant tropes of modern sociology, from community and authority to status and sacrament and finally alienation, have grown luxuriantly in this humanistic soil.

Sociology's chimeric origin, from bits and pieces of science and the humanities, is the reason it remains today the stronghold of the Standard Social Science Model (SSSM), the sovereign doctrine of twentieth-century social theory. The SSSM views culture as a complex system of symbols and meanings that mold individual minds and social institutions. That much is obviously true. But the SSSM also sees culture as an independent phenomenon irreducible to elements of biology and psychology, thus the product of environment and historical antecedents.

In purest form the Standard Social Science Model turns the intuitively obvious sequence of causation upside down: Human minds do not create culture but are themselves the product of culture. This reasoning is based, once again, on the slighting or outright denial of a biologically based human nature. Its polar opposite is the doctrine of genetic determinism, the belief that human behavior is fixed in the genes, and that its most destructive properties, such as racism, war, and class division, are consequently inevitable. Genetic determinism, proponents of the strong form of the SSSM say, must be resisted not only because it is factually incorrect but because it is morally wrong.

To be fair, I have never met a biologist who believes in genetic determinism as just defined. Conversely, although the extreme form of the SSSM was widely held among serious scholars in the social sciences twenty years ago, today it is rare. Still, the clash of antipodean views is a staple of popular culture, and it is unfortunately perpetuated by journalists and college teachers. When the matter is drawn this way, scholars spring to their archaic defensive postures. Confusion continues to reign, and angry emotions flare.

ENOUGH! A century of misunderstanding, the drawn-out Verdun and Somme of Western intellectual history, has run its exhausting course, and the culture wars are an old game turned stale. It is time to call a truce and forge an alliance. Within the broad middle ground between the strong versions of the Standard Social Science Model and genetic determinism, the social sciences are intrinsically compatible with the natural sciences. The two great branches of learning will benefit to the extent that their modes of causal explanation are made consistent.

The first step in the approach to consilience is to recognize that while the social sciences are truly science, when pursued descriptively and analytically, social theory is not yet true theory. The social sciences possess the same general traits as the natural sciences in the early, natural-history or mostly descriptive period of their historical development. From a rich data base they have ordered and classified social phenomena. They have discovered unsuspected patterns of communal behavior and successfully traced interactions of history and cultural evolution. But they have not yet crafted a web of causal explanation that successfully cuts down through the levels of organization from society to mind and brain. Failing to probe this far, they lack what can be called a true scientific theory. Consequently, even though they often speak of "theory" and, moreover, address the same species and the same level of organization, they remain disunited.

One frequently encountered word for natural history in the social sciences is hermeneutics. In its original, restricted usage the expression, drawn from the Greek *hermēneutikós* ("skilled in interpretation"), means the close analysis and interpretation of texts, and especially of the Old and New Testaments. Writers in the social sciences and humanities have expanded it to embrace the systematic exploration of social relations and culture, in which each topic is examined by many scholars of differing viewpoints and cultures. Sound hermeneutics usually takes long periods of time, even entire scholarly generations. Because experiments can seldom be conducted on human relationships, social scientists judge such studies partly by the fullness of the descriptions and analysis, and partly by the reputations of the experts addressing the subject and the degree of consensus they reach. In recent years they have come increasingly to expect statistical treatments of precisely measured replicate samples, wherever circumstances allow this adoption of the standard procedure of the natural sciences.

All these criteria also mark the best of natural history as it is still practiced through large sectors of biology, geology, and other branches of the natural sciences. A respect for fine analysis of factual information by trained intellects is what the social and natural sciences have in common. In this sense the hermeneutics of Balinese religion is comparable to the natural history of the Balinese bird fauna.

But if natural history by whatever name is the foundation of all the sciences, why is it not yet theory? The main reason is that it includes little effort to explain phenomena by webs of causation across adjacent levels of organization. The analysis is lateral, not vertical. In the Balinese examples, natural history travels widely through culture, but not from brain to mind to culture, and it travels across many bird species but not from individual bird to species to ecosystem. Natural history generates scientific theory when it links the best available knowledge across the organizational levels. It creates rigorous scientific theory when scholars propose competing and verifiable hypotheses that capture all of the plausible events operating across the different levels.

If social scientists choose to select rigorous theory as their ultimate goal, as have the natural scientists, they will succeed to the extent they traverse broad stretches of time and space. That means nothing less than aligning their explanations with those of the natural sciences. It also means avoiding, except at cocktail time, playful definitions of the kind proposed by the distinguished philosopher Richard Rorty, who has contrasted hermeneutics with epistemology, the systematic theory of knowledge: "We will be epistemological when we understand perfectly well what is happening but want to codify it in order to extend, or strengthen, or teach, or 'ground' it. We must be hermeneutical where we do not understand what is happening but are honest enough to admit it. . . ." In Rorty's proposal, hermeneutics is not the name for a discipline or a program of research, as I have recognized it, but "an expression of hope that the cultural space left by the demise of epistemology will not be filled—that our culture should become one in which the demand for constraint and confrontation is no longer felt." Discourse among scholars, in short, can proceed without worrying about consilience. About rigor too, it would seem. Although this concession is welcomed by postmodernist scholars, it is a premature surrender that would drain much of the power and joy from scholarly inquiry. Creativity in research can occur unexpectedly in any form of inquiry, of course, but to resist linking discoveries by causal explanation is to diminish their credibility. It waves aside the synthetic scientific method, demonstrably the most powerful instrument hitherto created by the human mind. Lazily, it devalues intellect.

PRECISELY WHAT FORM might the union between the social and natural sciences take? Consider four disciplines in a stack encompassing successively larger spans of space and time, as they might be described by their practitioners.

The *sociologist* says, with justifiable pride, "We are interested in the here and now, the fine analysis of life in particular complex societies, and cause and effect across recent history. We stand close to the fine details, and we ourselves are often part of it, literally swimming in the details. From our perspective variation in human social behavior seems enormous, perhaps indefinitely plastic."

The *anthropologist* responds. "Yes, that's true as far as it goes. But let's stand back and look again. Consider: We anthropologists study thousands of cultures, many preliterate and nonindustrial, and the variation we record is even greater than that encountered by the sociologists. But I grant it is far from infinite in possible range. We have observed clear limits and patterns within them. The information from so many separate experiments in cultural evolution, those conducted separately for many centuries, may allow us to formulate laws of human social action."

The *primatologist*, impatient, joins in. "True enough, comparative information about simple and complex societies is the bone and sinew of the social sciences. Still, your conceptions need to be put in an even broader perspective. The variation in human behavior is enormous, but it doesn't begin to encompass all the social arrangements we have discovered in the apes, monkeys, and other primates, which were created not by millennia but by fifty million years of evolution. It's there, among the more than one hundred species genetically closest to humanity, that we should look for the principles of social evolution if we are to understand the origins of culture."

The *sociobiologist* adds, "Yes, the key is perspective. So why not make it *really* wide? My discipline, which has been developed jointly by biologists and social scientists, examines the biological basis of social behavior in all kinds of organisms. I know that the very idea of a biological influence on human behavior in particular has been controversial, especially in the political arena, but consider this. Human beings may be unique in degree of behavioral plasticity, and they may be alone in the possession of language, self-awareness, and foresightedness, but all of the known human systems taken together form only a small subset of those displayed by the thousands of living species of highly social insects and vertebrates. If we hope to create a true science of social behavior, we will need to trace the divergent evolution of these groups of organisms, through a time scale of hundreds of millions of years. It is also useful to recognize that human social behavior originated ultimately through biological evolution."

Each discipline of the social sciences rules comfortably within its own chosen domain of space and time so long as it stays largely oblivious of the others. But from the lack of a true social theory comes the debilitating failure of the social sciences to communicate with the natural sciences and even with one another. If the social and natural sciences are to be united, the disciplines of both need to be defined by the scales of time and space they individually encompass and not just by subject matter as in past practice, and then they need to be connected.

A convergence has in fact begun. The natural sciences, by their own swift expansion in subject matter during the past several decades, are drawing close to the social sciences. Four bridges across the divide are in place. The first is cognitive neuroscience, or the brain sciences, with elements of cognitive psychology, whose practitioners analyze the physical basis of mental activity and aim to solve the mystery of conscious thought. The second is human behavioral genetics, now in the early stages of teasing apart the hereditary basis of the process, including the biasing influence of the genes on mental development. The third bridging discipline is evolutionary biology, including the hybrid offspring sociobiology, whose researchers have set out to explain the hereditary origins of social behavior. The fourth is the environmental sciences. The connection of the last field to social theory may at first seem tenuous, but is not. The natural environment is the theater in which the human species evolved and to which its physiology and behavior are finely adapted. Neither human biology nor the social sciences can make full sense until their world views take account of that unyielding framework.

IT IS NOT DIFFICULT to visualize how the stepping-stones between the natural and social sciences might be arranged and traversed. Consider a particular macrosocial event such as the decay of families in the American inner city, the implosion of rural populations into Mexico City, or middle-class resistance to the prospective introduction of euro currency in France. Social scientists addressing such issues start at the level of conventional analysis. They bring order to the facts, quantifying them in tables, graphs, and statistical interpretations. They examine the historical background. They draw comparison with similar phenomena in other places, examine the constraints and biases of the surrounding culture, and determine whether the genre to which the event belongs is widespread or instead unique to that time and place. From all this information they intuit the causes of the event and they ask: What does the event mean, will it continue, will it occur again?

Most present-day social scientists stop there, and write their reports. With consilient theory, however, future analysts will probe more deeply and finish with greater understanding and predictive power. In the ideal scenario during the decades to come, they will factor in the principles of psychology, and especially social psychology. By these last two words I do not mean the intuition of a single person or a team, however gifted, or folk beliefs about human behavior, however emotionally satisfying. I mean full knowledge from a mature, exact discipline of psychology. In short, the subject usually ignored by social scientists.

From this point forward let me suggest a full scenario of consilient research. Our future analysts understand very well how social behavior arises from the summation of individual emotion and intention within designated environments. They know how individual behavior in turn originates from the intersection of biology and environment. Their grasp of cultural change is enhanced by insights from evolutionary biology, which interpret the species-wide properties of human behavior as products of genetic evolution. They are careful how they express that idea—avoiding the assumption that genes prescribe behavior in a simple one-on-one manner. Instead, the analysts use a more sophisticated formula that conveys the same meaning more accurately: *Behavior is guided by epigenetic rules.*

Epigenesis, originally a biological concept, means the development of an organism under the joint influence of heredity and environment. Epigenetic rules, to summarize very briefly my account in the previous two chapters, are innate operations in the sensory system and brain. They are rules of thumb that allow organisms to find rapid solutions to problems encountered in the environment. They predispose individuals to view the world in a particular innate way and automatically to make certain choices as opposed to others. With epigenetic rules, we see a rainbow in four basic colors and not in a continuum of light frequencies. We avoid mating with a sibling, speak in grammatically coherent sentences, smile at friends, and when alone fear strangers in first encounters. Typically emotion-driven, epigenetic rules in all categories of behavior direct the individual toward those relatively quick and accurate responses most likely to ensure survival and reproduction. But they leave open the potential generation of an immense array of cultural variations and combinations. Sometimes, especially in complex societies, they no longer contribute to health and well-being. The behavior they direct can go awry and militate against the best interests of the individual and society.

At this point my imagined analysts, by plumbing the irrational in human affairs, will have traced an Ariadne's thread of causal explanation from historical phenomena to the brain sciences and genetics; hence they will have bridged the divide between the social and natural sciences. Such is the optimistic forecast shared nowadays by a small number of scholars on both sides of the divide. It is opposed by at least an equal number of critics who find it philosophically flawed, or if not flawed at least technically too difficult ever to achieve. All my instincts tell me it will happen. If the union can be achieved, the social sciences will span a wider scale of time and space and harvest an abundance of new ideas. Union is the best way for the social sciences to gain in predictive power.

HOW TO EXPAND the scale of time and space? There are many potential entries across the whole range of human behavior, including those entailing art and ethics that I will take up in succeeding chapters. For one immediately relevant to the social sciences, consider the fundamental theory of the family, developed during the past thirty years by evolutionary biologists and psychologists. In 1995 Stephen T. Emlen of Cornell University completed a reworking of the theory with special reference to cooperation and conflict between parents to their grown offspring who form social groups. The basic assumption is evolution by natural selection: Cooperation and conflict have evolved as instincts because they improve the survival and reproduction of the individuals displaying them. The data Emlen used to expand the assumption, and to test the theory built from it, were drawn from studies by many independent investigators of over one hundred species of birds and mammals around the world.

The patterns predicted by the theory were by and large closely matched by the evidence. Although the data were drawn exclusively from the instinctive behavior of animals, the relevance of the patterns to core themes in the social sciences and humanities will become quickly obvious:

In birds and nonhuman mammals, families are basically unstable, but the least so in those controlling high-quality resources. Dynasties, in which one genetic lineage persists over many generations, arise in territories permanently rich in resources.

The closer the genetic relationships of the family members, as for example father-to-son as opposed to uncle-to-nephew, the higher the degree of cooperation.

Due to this cooperativeness and the general instinctive avoidance of incest, the closer the genetic relationship of the family members, the lower the frequency of sexual conflict.

How closely family members are related also affects forms of conflict and commitment. Breeding males invest less in offspring when paternity is uncertain. If the family consists of a single conjugal pair, and one of the parents is lost, the opposite-sex offspring compete with the surviving parent for breeder status. When the father dies, for example, a still fecund mother is likely to enter into conflict with her son over the status of a mate he may newly acquire, and a son is likely to discourage his mother from establishing a new sexual relationship.

One general result of this pattern of conflict and commitment is that step-

families are less stable than biologically intact families. Stepparents invest less in existing offspring than biological parents. In many species they kill current young if such action speeds the success of their own reproduction. This is especially likely when the stepparent belongs to the dominant sex.

Reproduction within a family (using mates acquired from the outside) is increasingly shared when there is an improvement in the alternative option for subordinate members to disperse and start families of their own. Such forbearance is greatest of all when the members are genetically very close, and when the cooperating individuals are siblings rather than parents and offspring.

In applying this documented theory to humans, it is of course ever prudent to remain aware of the massive intervention of cultural change. The resulting variation of conventions is sometimes great enough to include the bizarre and perverse—what else can we call the former eating of the kuru-ridden brains of dead relatives by the Fore people of New Guinea, which condemned them unknowingly to a fatal disease? But experience in behavior such as incest avoidance has shown that the hard instincts of animals are translatable into epigenetic rules of human behavior. Like ancient settlement mounds on the Euphrates plain awaiting the archaeologist's spade, they are where the long history of a culture is most efficiently sought. The practical role of evolutionary theory is to point to the most likely location of the epigenetic rules.

THE ENTERPRISE WITHIN the social sciences best poised to bridge the gap to the natural sciences, the one that most resembles them in style and self-confidence, is economics. This discipline, fortified with mathematical models, garlanded annually by its own Nobel Memorial Prize in Economic Science, and rewarded with power in business and government, deserves the title often given it, Queen of the Social Sciences. But its similarity to "real" science is often superficial and has been purchased at a steep intellectual price.

The potential and price of economic theory can be most clearly understood against a historical background. Jürg Niehans, in his magisterial work *A History of Economic Theory*, recognizes three periods in the evolution of mainstream economics. In the Classical Era of the eighteenth and early nineteenth centuries the founding fathers, including Adam Smith, David Ricardo, and Thomas Malthus, envisioned the economy as a closed system of circulating income. Driven by supply and demand, the economy controls the world's resources and converts them to beneficial ends. The central postulate of free-market economics was introduced during this period by Adam Smith. According to his conception of the invisible hand, individual producers and consumers will, when freed to pursue their own best interests, propel the economy forward and thereby work to the best interests of the society as a whole.

In the Marginalist Era, which began around 1830 and peaked some forty years later, the focus shifted toward the properties of the invisible hand. The imagined inner workings of the economy were broken down into individual decisions by those agents—persons, firms, governments—whose activities could be examined with the aid of mathematical models. Within the framework of abstract, physicslike theory, the analysts could then manipulate the economy as a virtual world, assessing and predicting the effects of shifting levels of production and consumption. Differential calculus was employed to evaluate economic change as the consequence of very small, hence "marginal" shifts in production and consumption. With growing or declining scarcity and demand, each unit of new production—say of gold, oil, or housing—correspondingly rises or falls in price. In aggregate, these shifts, working through complex webs of exchange, drive the economy either toward or away from steady states in supply and demand.

Thus was constructed the foundation of microeconomics, which purports to plot economic change in exact measures: marginal cost, the increase in total cost incurred by the production of one additional unit of product; marginal product, the growth in total output from one additional unit of productive input; marginal revenue, the growth of total revenue from the sale of a unit of output; and marginal utility, the satisfaction added by the consumption of a unit of production. In the manner of the natural sciences, the models of marginalist economics allow the variables to change alone or in combination while holding the remainder constant. When played skillfully, the models make a tidy picture. The macroanalysis of the Classical Era was then combined with the analytic microanalysis of the Marginalist Era, most influentially by Alfred Marshall in his 1890 *Principles of Economics*. The result, in the phrase coined by Thorstein Veblen in 1900, was neoclassical economics.

Neoclassical economics is what we have today, but there was one more overlapping period, the Era of Model Building, that brought it to fruition. Beginning in the 1930s, theorists added linear programming, game theory, and other powerful mathematical and statistical techniques in their efforts to simulate the economic world in ever finer detail. Invigorated by the sense of their own exactitude, they continued to return to the themes of equilibria and perturbations from equilibria. They specified, as faithfully as they could, supply and demand, impulses of firms and consumers, competition, market fluctuations and failures, and the optimal uses of labor and resources.

The cutting edge of economic theory today remains the equilibrium models of neoclassical theory. The emphasis is always on rigor. Analysts heartily agree with Paul Samuelson, one of the most influential economists of the twentieth century, that "economics focuses on concepts that actually can be measured."

Therein lie the strengths and weaknesses of present-day economic theory. Because its strengths have already been abundantly celebrated by legions of textbook writers and journalists, let me dwell on the weaknesses. They can be summarized in two labels: Newtonian and hermetic. Newtonian, because economic theorists aspire to find simple, general laws that cover all possible economic arrangements. Universality is a logical and worthy goal, except that the innate traits of human behavior ensure that only a minute set of such arrangements is probable or even possible. Just as the fundamental laws of physics cannot be used alone to build an airplane, the general constructions of equilibrium theory cannot be used alone to visualize an optimal or even stable economic order. The models also fall short because they are hermetic—that is, sealed off from the complexities of human behavior and the constraints imposed by the environment. As a result, economic theorists, despite the undoubted genius of many, have enjoyed few successes in predicting the economic future, and they have suffered many embarrassing failures.

Among the successes are partial stabilizations of a few national economies. In the United States the Federal Reserve Board now has enough knowledge and legal power to regulate the flow of money and prevent—we trust!—the economy from spinning into catastrophic inflations and depressions. On another front, the driving force of technological innovation on growth is reasonably well understood, at least roughly and in retrospect. On yet another, capital-asset pricing models have a major influence on Wall Street.

We are better off if the economists speak than if they remain silent. But the theorists cannot answer definitively most of the key macroeconomic questions that concern society, including the optimal amount of fiscal regulation, future income distribution within and between nations, optimal population growth and distribution, long-term financial security of individual citizens, the roles of soil, water, biodiversity, and other exhaustible and diminishing resources, and the strength of "externalities" such as the deteriorating global environment. The world economy is a ship speeding through uncharted waters strewn with dangerous shoals. There is no general agreement on how it works. The esteem that economists enjoy arises not so much from their record of successes as from the fact that business and government have nowhere else to turn.

This is not to say that economists would do better to abandon mathematical models in favor of intuition and description. The great merit of models, at least in the natural sciences, is that they force investigators to provide unambiguous definitions of units, such as atoms and genes, as well as processes, such as movement and change. When well conceived, a model leaves no doubt about its assumptions. It lists the important factors and offers educated guesses about their interaction. Within this self-imposed framework, the investigator makes predictions about the real world, and the more precise the prediction, the better. He thus puts the product of his thinking on the line by exposing it to evidential proof or disproof. There is nothing in science more provocative than a cleanly defined and surprising prediction, and nothing held in higher regard than such a prediction confirmed in detail.

To this end, scientists look for four qualities in theory generally and mathematical models in particular. The first is *parsimony*: the fewer the units and processes used to account for the phenomenon, the better. Because of the success of parsimony in the physical sciences, we do not today need an imaginary substance called phlogiston to explain the combustion of firewood, or nonexistent ether to fill the void of space. The second quality is *generality*: the greater the range of phenomena covered by the model, the more likely it is to be true. In reagent chemistry the periodic table precludes a separate theory for each element and compound. One theory works exactly for all.

Next is *consilience*. Units and processes of a discipline that conform with solidly verified knowledge in other disciplines have proven consistently superior in theory and practice to units and processes that do not conform. That is why, in every scrap of data from every level of biology, from the chemistry of DNA to the dating of fossils, it has been the case that organic evolution by natural selection beats Creationism. God may exist, He may be delighted with what we are up to on this minor planet, but His fine hand is not needed to explain the biosphere. And finally, drawing from all of the above virtues, the definitive quality of good theory is *predictiveness*. Those theories endure that are precise in the predictions they make across many phenomena and whose predictions are easiest to test by observation and experiment.

Before evaluating economic theory by these criteria, I think it only fair to assess a branch of biology with a comparable level of technical difficulty. Population genetics addresses the frequencies and distributions of genes and other hereditary units within entire populations (an example of a population is all the members of a species of fish inhabiting a lake). Population genetics, having accumulated, like economic theory, a vast encyclopedia of models and equations, is arguably the most respected discipline within evolutionary biology. Its Ur-model is the Hardy-Weinberg principle, or "law," a simple probability formula based on elementary Mendelian genetics. The Hardy-Weinberg principle tells us that if in a sexually reproducing population there are two forms, or alleles, of the same gene, say each prescribing a different blood type or ear shape, and if we know the percentages of the two alleles in the population, we can accurately predict the percentages of individuals possessing different pairs of the alleles. Conversely, from the known percentage of just one such pair, we can at once state the percentage of the alleles for the whole population. Here is an example to show how it works. The earlobe in different people either hangs free or is attached to the side of the head, and the difference is due to two forms of the same gene. Call the free earlobe allele A and the attached earlobe allele a. Free earlobe is dominant over attached earlobe. Then all individuals in the population have one or the other of the three following combinations:

AA, free earlobe
Aa, free earlobe
aa, attached earlobe

Following convention in genetics, the frequency (ranging from 0 to 1.0, that is, zero to 100 percent) of A is labeled p, and the frequency of a is labeled q. The Hardy-Weinberg principle is the consequence of Mendelian heredity and the randomness with which an allele in an egg is matched with an allele in a sperm at fertilization. It is written as a simple binomial expansion, since by definition $p + q = 1.0$ and therefore $(p + q)^2 = (1.0)^2 = 1.0$, and therefore

$$p + q = (p + q)^2 = p^2 + 2pq + q^2 = 1.0,$$

where p^2 is the frequency of AA, $2pq$ the frequency of Aa, and q^2 the frequency of aa. The rationale of the formula is the following: There is p chance that an egg contains A, and p chance that the sperm fertilizing it is also A, so there is p^2 chance (hence, frequency) that the individual created is AA, and so on through pq and q^2. Suppose that 16 percent (the frequency is 0.16) of the members of a population have an attached earlobe, in other words, their two alleles are aa. Then the Hardy-Weinberg formula predicts that 40 percent

(0.4, the square root of 0.16) of the alleles in the population are a, and 60 percent A. It also predicts that 36 percent (0.36, or 0.60 × 0.60) of the individuals have the combination AA, and 48 percent (0.48, or 2 × 0.4 × 0.6) have Aa.

There are some large conditions attached to the use of the Hardy-Weinberg formula in the real world. But these are not crippling. Instead, they are what make H-W interesting and even more useful. The simple H-W predictions will be exactly correct if natural selection does not favor one of the gene combinations over the others, if all the members of the population mate at random, and if the population is infinitely large. The first two conditions are improbable and the third impossible. In order to get closer to reality, biological theorists "relax" these restrictions one at a time, and then in various combinations. For example, they reduce the number of imagined organisms from infinite to the numbers actually found in real populations, usually somewhere from ten to a million, according to species. They then take into account chance variation in the gene frequencies from one generation to the next. The smaller the population, the greater the variation. The same principle dictates that if a sample of one million unbiased coins are flipped in repeated trials, the result will almost always be very close to half heads, half tails, whereas if only ten coins are flipped at a time, an exact split will be obtained only occasionally; and in one in an average 512 trials all the coins will be either all heads or all tails.

Now think of sexual reproduction as the equivalent of coin flipping, and each generation as a new coin-flipping trial. The change in gene frequency from one generation to the next by chance is evolution by genetic drift. In populations with a hundred individuals or fewer, genetic drift can be a potent force. Its rate can be precisely described by statistical measures that tell us about the fate of large samples of populations of the same size. These measures reveal that the main effect of genetic drift is to reduce variation by eliminating some of the gene forms. That, combined with the randomness of the change, means that genetic drift is a far less creative process than natural selection.

As natural selection is added to the models, it reduces the impact of genetic drift while driving the gene frequencies in one direction or another at predictable velocities. Population geneticists make their models still more complex and presumably closer to nature in various ways. For example, they decree mating to be nonrandom, or break populations into fragments that continue to exchange migrants, or arrange for constellations of genes rather than single genes to prescribe the character traits.

The models of population geneticists yield exact predictions in the virtual worlds bounded by the assumptions selected for evaluation. They can often be matched by carefully managed populations of animals and plants in the laboratory. They are notoriously poor, however, at predicting evolution in nature. The flaw is not in the internal logic of the theory but in the unpredictability of nature itself. The environment constantly shifts, altering the values of the parameters that geneticists put into their models. Climatic change and weather catastrophes break up some populations while freeing others to expand and coalesce. New predators and competitors invade as old ones retreat. Disease sweeps the habitats. Traditional food sources vanish and new ones appear.

Evolutionary biologists, like weather forecasters, are confounded by the turbulence of the real world. They have had some success in predicting changes in small ensembles of genes and traits over a few generations. They can explain retrospectively many of the major twists and turns of long-term evolution from the fossil record and from the logical reconstruction of family trees of living species. But rarely have they been able to predict future events with any degree of accuracy. They have equal difficulty retrodicting past events—that is, predicting the occurrence of such past events before a search is made for traces of the events and reconstructions are performed. They are unlikely to do so until ecology and the other environmental sciences have sufficiently matured to become predictive themselves and thus provide the full and exact context in which evolution occurs.

Economics, at the cutting edge of the social sciences, shares the same difficulties as population genetics and the environmental sciences. It is battered by "exogenous shocks," all the unaccountable events of history and environmental change that push the parameter values up and down. That alone limits the accuracy of economic predictions. Except in the most general and statistical terms, economic models cannot forecast the onset of bull and bear markets, or the decades-long cycles triggered by war and technological innovation. They cannot tell us whether tax cuts or national deficit reduction is the more effective in raising per capita income, or how economic growth will affect income distribution.

Economic theory is impeded by a second, equally fundamental difficulty. Unlike population genetics and the environmental sciences, it lacks a solid foundation of units and processes. It has not acquired or even attempted serious consilience with the natural sciences. All analysts understand that the broad patterns of economic process originate in some fashion or other from vast numbers of decisions made by human beings, whether as individuals or as members of firms and governmental agencies. The most sophisticated models of economic theory attempt to translate such microeconomic behavior into the larger aggregate measures and patterns broadly defined as "the economy." In economics and in the remainder of the social sciences as well, the translation from individual to aggregate behavior is the key analytic problem. Yet in these disciplines the exact nature and sources of individual behavior are rarely considered. Instead, the knowledge used by the modelers is that of folk psychology, based mostly on common perception and unaided intuition, and folk psychology has already been pushed way past its limit.

The flaw is not fatal. Economic theory is not Ptolemaic, not so structurally defective that a revolution in conception is needed. The most advanced of the micro-to-macro models are on the right track. But the theorists have unnecessarily handicapped themselves by closing off their theory from serious biology and psychology, comprising principles drawn from close description, experiments, and statistical analysis. They have done so, I believe, in order to avoid entanglement in the formidable complexities of these foundation sciences. Their strategy has been to solve the micro-to-macro problem with the fewest possible assumptions at the micro level. In other words, they have carried parsimony too far. Economic theories also aim to create models of the widest possible application, often crafting abstractions so extreme as to represent little more than exercises in applied mathematics. That is generality carried too far. The result of such stringency is a body of theory that is internally consistent but little else. Although economics, in my opinion, is headed in the right direction and provides the wedge behind which social theory will wisely follow, it is still mostly irrelevant.

The strengths and weaknesses of economic theory are illustrated in the work of Gary S. Becker of the University of Chicago, awarded the 1992 Nobel Prize in Economic Science for "having extended the domain of economic theory to aspects of human behavior which had previously been dealt with—if at all—by other social science disciplines such as sociology, demography, and criminology." What Becker accomplished was to cut more deeply than previous economists into the sources of human preferences. He recognized that most of economic reasoning is based on the implicit assumption that people are driven by basic biological needs for food, shelter, and recreation. But there are other incentives, he said, such as the type of housing and furniture, the restaurants, and forms of leisure they prefer, that lie outside the elemental imperatives. All these choices and more depend on variations in personal experience and social forces beyond individual control. If human behavior is to be explained fully, the utility of the choices (that is, their value perceived by the consumer) must be entered into economics models.

The inviolable assumption of Becker's thinking is the principle of rational choice. Introduced by earlier economists as the keystone of quantitative modeling, it says simply that people maximize their satisfaction by acts based on calculation. Economic models using this conception had been largely limited to utility based on narrow self-interest. Becker urged his fellow economists to broaden their vista to include the subject matter of the other social sciences. They should consider desires that are variously altruistic, loyal, spiteful, and masochistic. These too, he argued, are forces that govern rational choice.

Extending the reach of formal models, Becker and other economists of like mind have addressed with greater confidence some of industrial society's most vexing problems. Turning to criminology, they have recommended methods of optimal deterrence—economic of course—for different classes of offense, from capital crimes and armed robbery to embezzlement, tax evasion, and the breaking of laws that regulate business and environmental protection. Venturing into sociology, they have assessed the impact of racial discrimination on production and unemployment, and of economic class on marital choice. In public health, they have analyzed the effects of legalization and tax loads on the use of cigarettes and controlled substances.

Their models contain elegant graphical representations and analytic solutions to theoretical problems of equilibria. Yet seen through the established principles of the behavioral sciences, they are simplistic and often misleading. The choices in personal behavior reduce to a small number of options, such as whether to smoke or not, to marry within the same socioeconomic class or not, to risk committing a crime, or to move to a same-race neighborhood. The predictions consist of "more of this, less of that" and they approximate thresholds at which trends will commence, taper off, or reverse direction. Typically the predictions arise from the commonsense intuitions of the modeler, that is, from folk psychology, and following a series of formal analytical steps, confirm commonsense beliefs. We are told in crisp technical language that a permanent increase in the price of cigarettes reduces consumption from the outset more than a temporary increase, that in order to preserve their wealth the rich take measures to avoid meeting and falling in love with the poor, that people gain satisfaction from going to already popular restaurants even if competitors are as good in price and cuisine, and so forth. Seldom are the premises of such models examined closely. Seldom are their conclusions tested to any depth with quantitative field data. Their appeal is in the chrome and roar of the engine, not the velocity or destination.

The goal of psychologically oriented analysts such as Becker, as well as

Jack Hirshleifer, Thomas Schelling, Amartya Sen, George Stigler, and others of similar interests, is to strengthen microeconomics and draw from it more accurate predictions of macroeconomic behavior. That, of course, is admirable. To advance much further, however, they and other social scientists will have to cross the boundary between the social and natural sciences and trade with the biologists and psychologists they find on the other side. Just as, in his Nobel Lecture, Becker stated that his contribution was "to pry economists away from narrow assumptions about self-interest," the next step is for economists to free themselves completely, at long last, from the Standard Social Science Model of behavior and take seriously the biological and psychological foundations of human nature. Amazingly, despite overwhelming evidence against it, the great majority still cling to the view that aside from meeting basic biological needs, people in modern societies make choices, in Becker's words, that "depend on childhood, social interactions, and cultural influences." Not, apparently, the hereditary epigenetic rules of human nature. The impoverishing consequence of this view has been the acceptance of folk psychology in even the most ingenious models.

TO INFUSE PSYCHOLOGY and biology into economic and other social theory, which can only be to its advantage, means teasing out and examining microscopically the delicate concepts of utility, by asking why people ultimately lean toward certain choices, and being so predisposed, why and under what circumstances they act upon them. Beyond this task lies the micro-to-macro problem, the ensemble of processes by which the mass of individual decisions are translated into social patterns. And beyond that, framed by a still wider scale of space and time, is the coevolution problem, the means by which biological evolution influences culture, and the reverse. Together these domains—human nature, micro-to-macro transition, and the coevolution of genes and culture—require the full traverse from the social sciences to psychology and thence to the brain sciences and genetics.

The evidence from scattered studies in psychology and biology already suggest certain generalizations about utility:

- The categories of choice, the major activities in moment-by-moment thought and behavior, are epistatic: Needs and opportunities in one category alter the strength of others. The rank order in dominance among categories such as sex, status protection, and play appears to be genetically programmed.

• Some needs and opportunities are not just epistatic but preemptive. Conditions such as drug addiction and sexual possessiveness can hijack emotions to focus on unitary goals so powerful as to virtually delete activities in many other categories.

• Rational calculation is based on surges of competing emotions, whose interplay is resolved by an interaction of hereditary and environmental factors. Incest avoidance, for example, is underlaid by a strong hereditary epigenetic rule. It can be reinforced by cultural taboos or overcome by special, increasingly well understood personal experiences.

• Rational calculation is often unselfish. For complex, still poorly understood reasons, some of the most powerful emotions are patriotism and altruism. It remains a surprising fact that a substantial percentage of people are willing at a moment's notice to risk their lives to save those of strangers.

• Choices are group-dependent; that much is obvious. But what is less well known is that the power of peer influence varies strikingly from category to category of behavior. Clothing style, for example, is almost wholly dependent on peer influences, while incest avoidance is largely independent. Do these differences have a genetic basis and thus an evolutionary history? Probably they do, and it is time to start examining them for this possibility more carefully.

• Decision-making is shaped category by category by epigenetic rules, which are the innate propensities to learn certain options in the first place and then to select particular ones among them. On average many of the propensities differ according to age and gender.

The psychobiological subtlety of decision-making is nicely illustrated by the r–K continuum of reproductive strategies. When resources are few and unstable, people tend to adopt an r strategy, preferring many children to insure that at least a few will survive. When resources are abundant and stable, they lean toward a K strategy, in which fewer, "high-quality" offspring are carefully protected and educated for entry at an upper socioeconomic level. (The symbol r refers in demography to the rate of population growth, which rises with the r strategy; and the symbol K to the carrying capacity of the environment, the size at which population growth ceases.) Overlying the r–K continuum is the general tendency of socially powerful males to acquire multiple women of reproductive age, thereby boosting their Darwinian edge.

THE FULL UNDERSTANDING of utility will come from biology and psychology by reduction to the elements of human behavior followed by bottom-up synthesis, not from the social sciences by top-down inference and guesswork based on intuitive knowledge. It is in biology and psychology that economists and other social scientists will find the premises needed to fashion more predictive models, just as it was in physics and chemistry that researchers found premises that upgraded biology.

The performance of future social theory also depends on a psychobiological understanding of the process of reason itself. At present the dominant mode of explanation is the aforementioned rational choice theory. First conceived in economics, then spread to political science and other disciplines, its central conception is that above all else human beings are rational in their actions. They examine as best they can all the pertinent factors and weigh the likely outcome of following each potential choice in turn. They add in cost and benefit—investment, risk, and emotional and material return—before deciding. The preferred option is that which maximizes utility.

This is not an adequate picture of how people think. The human brain is not a very swift calculator, and most decisions have to be made rather quickly, in complex settings and with incomplete information. So the question of importance in rational choice theory is, how much information is enough? In other words, at what point do people stop reflecting and make up their minds? One simple strategy that provides a cut-off point is "satisficing," a Scottish term that combines "satisfying" and "sufficing." Introduced to psychology in 1957 by Herbert Simon, an economist at Carnegie Mellon University, satisficing means taking the first satisfactory choice encountered out of those perceived and reasonably available in the short term, as opposed to visualizing the optimum choice in advance and searching until it is found. A young man ready for marriage is more likely, by satisficing, to propose to the most attractive prospect among the available women of his acquaintance than to search at length for a preconceived ideal mate.

An alternative to this and other conceptions of traditional rational choice is that people follow rules of thumb, known more technically as "heuristics." The idea was first advanced by the American psychologists Daniel Kahneman and Amos Tversky in 1974. Rather than calculate costs and benefits, people act upon simple cues and heuristics that work most of the time. By this means the complex tasks of assessing probabilities and predicting outcomes are reduced to a few judgmental operations.

Usually heuristics work, and save a great deal of time and energy, but in many situations they lead to large systematic error. An example is the heuristic used in rapid arithmetical calculation and known as "anchoring." You can see how it works by comparing the two sets of multiplied numbers below for five seconds and guessing the products:

$$8 \times 7 \times 6 \times 5 \times 4 \times 3 \times 2 \times 1$$
$$1 \times 2 \times 3 \times 4 \times 5 \times 6 \times 7 \times 8$$

Most people give the top row the higher value, even though the two sets of numbers are identical. Reading left to right, they anchor their guess on the first numbers encountered. They also underestimate both. High school students tested by Kahneman and Tversky averaged 2,250 for the upper row and 512 for the lower, whereas the correct answer for both is 40,320.

Here is an example of a systematically inaccurate heuristic in the realm of probability. A majority of people, watching a coin being tossed, believe that the following sequence of six alternating heads and tails,

H-T-H-T-T-H

is more likely to occur than one with the same elements repeated in groups, such as

H-H-H-T-T-T

In fact, both are equally probable.

Why are such consistent errors made by minds that can be trained to grasp calculus and statistics? The correct answer may lie in genetic evolution: Over thousands of generations the brain evolved to handle simple numbers and proportions but not complex problems requiring abstract quantitative reasoning. The heuristics illustrated by the two examples above are therefore folk mathematics. Although their solutions are skewed in attempts at complex formal calculation, they may work very well in real life, where most first impressions accurately prefigure events to follow.

The same explanation fits other odd mistakes made by heuristics. For example, a familiar dish with a different taste is likely to be passed over, even though the ingredients are demonstrably fresh and wholesome. Following a plane crash, many intercity travelers switch to automobiles, even though they know the fatality rate per passenger mile is much higher on the road. Irrational choices yes, but perhaps obedient to the superordinate heuristic of risk aversion, which can be translated in these two examples as follows: Take no chances whatsoever with food poisoning, and stay away from places where others have been recently killed, regardless of what the mathematical laws of probability tell you.

Further research may reveal that the brain sometimes operates as a computerlike optimizer and sometimes as a quick decision-maker ruled by powerful and inborn heuristics. Whatever the mix, rational choice theory, though still the light and the way to many social theorists, is a subject of controversy within psychology. It is too dependent, critics say, on analogies with computer algorithms and abstract optimality solutions. It pays too little attention to the properties of the real brain, which is a stone-age organ evolved over hundreds of millennia and only recently thrust into the alien environment of industrialized society. It is thus inconsistent with the evidence of how people in preliterate cultures reason and have likely reasoned throughout evolutionary time. These qualities have been summarized by C. R. Hallpike in *The Foundations of Primitive Thought*, as follows: intuitive and dogmatic, bound up with specific emotional relationships rather than physical causality, preoccupied with essences and metamorphosis, opaque to logical abstraction or arrays of the hypothetically possible, prone to use language for social interaction rather than as a conceptual tool, limited in quantification mostly to rough images of frequency and rarity, and inclined to view mind as stemming partly from the environment and able to project back out into it, so that words become entities with power unto themselves.

It will be at once apparent, and should be a working premise of economists and other social scientists, that the same preliterate traits are commonplace in citizens of modern industrial societies. They are intensified among cult members, the deeply religious, and the less educated. They permeate and enrich the metaphors of art. They are, like it or not, part of modern civilization. Systematic logico-deductive thought, which is very much a specialized product of Western culture, comes hard on the other hand, and is still rare. While perfecting it we will be wise, I think, to discipline the old ways of thought but never to abandon them, never to forget that as part of adaptive human nature they conducted us alive and fecund all the way to the present age.

THE MAGNITUDE OF the technical problems facing the social theorists in particular is, I readily concede, extremely daunting. Some philosophers of science have thrown up their hands, declaring that the borderlands between the natural and social sciences are too complex to be mastered by contemporary imagination and may lie forever beyond reach. Questioning the very idea of consilience from biology to culture, they point to the nonlinearity of the viable equations, to second- and third-order interactions of factors, to stochasticity, and to all the other monsters that dwelleth in the Great Maelstrom Sea, and they sigh, *No hope, no hope*. But that is what philosophers are supposed to do. Their task is to define and explain the limits of science in the larger scheme of things, where the full dimensions of rational process are better left to—well, philosophers. For them to concede that science has no intellectual limits would be unseemly; it would be unprofessional. Their misgivings lend strength to that dwindling number of social theorists who wish to keep the borders of their dominions sealed and the study of culture unroiled by the dreams of biology.

Scientists themselves are fortunately not so bound. If past generations had been so deeply reflective and humble before the unknown, our comprehension of the universe would have stopped growing in the sixteenth century. The chastening sting of the philosopher's tongue is needed but should be taken with the antidote of self-assurance, and never allowed to be fatal. It is the opposite conviction, blind faith if you prefer, that has propelled science and technology into the modern age. Bear in mind that the original Enlightenment died within philosophy but not within science. The more pessimistic philosophers may be right about the social sciences, of course, but it is better to press on as if they were wrong. There is only one way to find out. The more forbidding the task, the greater the prize for those who dare to undertake it.

CHAPTER 10

The Arts and Their Interpretation

In many respects, the most interesting challenge to consilient explanation is the transit from science to the arts. By the latter I mean the creative arts, the personal productions of literature, visual arts, drama, music, and dance marked by those qualities which for lack of better words (and better words may never be coined) we call the true and beautiful.

The arts are sometimes taken to mean all the humanities, which include not only the creative arts but also, following the recommendations of the 1979–80 Commission on the Humanities, the core subjects of history, philosophy, languages, and comparative literature, plus jurisprudence, the comparative study of religions, and "those aspects of the social sciences which have humanistic content and employ humanistic methods." Nevertheless, the arts in the primary and intuitively creative sense, *ars gratia artis*, remain the definition most widely and usefully employed.

Reflection leads us to two questions about the arts: where they come from, in both history and personal experience, and how their essential qualities of truth and beauty are to be described through ordinary language. These matters are the central concern of interpretation, the scholarly analysis and criticism of the arts. Interpretation is itself partly an art, since it expresses not just the factual expertise of the critic but also his character and aesthetic judgment. When of high quality, criticism can be as inspired and idiosyncratic as the work it addresses. Further, as I now hope to show, it can also be part of science, and science part of it. Interpretation will be the more powerful when braided together from history, biography, personal confession—and science.

The profane word now having been spoken on hallowed ground, a quick disclaimer is in order. While it is true that science advances by reducing phenomena to their working elements—by dissecting brains into neurons, for example, and neurons into molecules—it does not aim to diminish the integrity of the whole. On the contrary, synthesis of the elements to re-create their original assembly is the other half of scientific procedure. In fact, it is the ultimate goal of science.

Nor is there any reason to suppose that the arts will decline as science flourishes. They are not, as suggested recently by the distinguished literary critic George Steiner, in a twilight, past high noon in Western civilization, thus unlikely to witness the reappearance of a Dante, a Michelangelo, or a Mozart. I can conceive of no intrinsic limit to future originality and brilliance in the arts as the consequence of the reductionist understanding of the creative process in the arts and science. On the contrary, an alliance is overdue, and can be achieved through the medium of interpretation. Neither science nor the arts can be complete without combining their separate strengths. Science needs the intuition and metaphorical power of the arts, and the arts need the fresh blood of science.

Scholars in the humanities should lift the anathema placed on reductionism. Scientists are not conquistadors out to melt the Inca gold. Science is free and the arts are free, and as I argued in the earlier account of mind, the two domains, despite the similarities in their creative spirit, have radically different goals and methods. The key to the exchange between them is not hybridization, not some unpleasantly self-conscious form of scientific art or artistic science, but reinvigoration of interpretation with the knowledge of science and its proprietary sense of the future. Interpretation is the logical channel of consilient explanation between science and the arts.

FOR A PROMISING EXAMPLE out of many that might be chosen, consider the episode in *Paradise Lost—Book IV*, when, in a riveting narrative, Milton sends Satan to Eden. Upon arrival the arch-felon and grand thief leaps a barrier of impenetrable bramble and a high wall and settles "like a cormorant" in the branches of the Tree of Life. He waits for the fall of night, when he can enter the dreams of innocent Eve. Milton now unleashes his imaginative powers to tell us what humanity is about to lose. All around the roosting schemer is the environment designed by God to aesthetic perfection: "Crispèd brooks, rolling on orient pearl and sands of gold" descend to "a lake, that to the fringèd bank with myrtle crowned her crystal mirror holds." All through the blessed oasis grow "flowers of all hue and without thorn the rose."

Milton, though now blind, has retained a fine sense of biophilia, the innate pleasure from living abundance and diversity, particularly as manifested by the human impulse to imitate Nature with gardens. But he is far from satisfied with the mere dream of natural harmony. In eight lines of astonishing symphonic power he tries to capture the mythic core of paradise:

Not that fair field
Of Enna, where Proserpin gathering flowers,
Herself a fairer flower, by gloomy Dis
Was gathered, which cost Ceres all that pain
To seek her through the world, nor that sweet grove
Of Daphne, by Orontes and the inspired
Castalian spring, might with this Paradise
Of Eden strive.

How can anyone hope to express Creation's heart at the dawn of time? Milton tries. He summons archetypes that have descended undiminished from ancient Greece and Rome to his own time, and thereafter to ours. They are of a kind, as I will suggest later, that are also innate to the human mental process. He shadows beauty with a hint of tragedy, giving us the untrammeled and fertile world awaiting corruption. He transforms the beauty of the garden into that of a young woman, Proserpine, about to be seized and taken away to the underworld by the god Dis. She, as Nature's beauty, will be concealed in darkness because of conflict between gods. Ceres, Proserpine's mother and goddess of agriculture, turns in grief from her duties and the world plunges into famine. The passion of Apollo for beautiful Daphne is unrequited; in order to escape she turns into a tree, a laurel, in a garden of her own.

Milton means to play on the emotions of readers of his own time, the seventeenth century, when Hellenic mythology was second nature to the educated mind. He counterposes emotions to magnify their force. Beauty clashes with darkness, freedom with fate, passion with denial. Building tension, he leads us through lesser paradises to arrive, suddenly, at the mystical prototype of Eden. In yet another well-grounded artifice, reliance on authority, Milton chooses allusions not to his own time, not for example to

Cromwell and Charles II and the Restoration, from which he himself has narrowly escaped death (he had championed revolution and the Commonwealth), but to ancient texts of another civilization, ancient Greece and Rome, robust enough to have survived in remembrance across centuries. He conveys by their use that what we are not told, we must know nevertheless to be true.

The defining quality of the arts is the expression of the human condition by mood and feeling, calling into play all the senses, evoking both order and disorder. From where then does the ability to create art arise? Not cold logic based on fact. Not God's guidance of Milton's thoughts, as the poet himself believed. Nor is there any evidence of a unique spark that ignites such genius as is evident in *Paradise Lost*. Experiments using brain imaging, for example, have failed to disclose singular neurobiological traits in musically gifted people. Instead, they show engagement of a broader area of the same parts of the brain used by those less able. History supports this incremental hypothesis. Behind Shakespeare, Leonardo, Mozart, and others in the foremost rank are a vast legion whose realized powers form a descending continuum to those who are merely competent. What the masters of the Western canon, and those of other high cultures, possessed in common was a combination of exceptional knowledge, technical skill, originality, sensitivity to detail, ambition, boldness, and drive.

They were obsessed; they burned within. But they also had an intuitive grasp of inborn human nature accurate enough to select commanding images from the mostly inferior thoughts that stream through the minds of all of us. The talent they wielded may have been only incrementally greater, but their creations appeared to others to be qualitatively new. They acquired enough influence and longevity to translate into lasting fame, not by magic, not by divine benefaction, but by a quantitative edge in powers shared in smaller degree with those less gifted. They gathered enough lifting speed to soar above the rest.

Artistic inspiration common to everyone in varying degree rises from the artesian wells of human nature. Its creations are meant to be delivered directly to the sensibilities of the beholder without analytic explanation. Creativity is therefore humanistic in the fullest sense. Works of enduring value are those truest to these origins. It follows that even the greatest works of art might be understood fundamentally with knowledge of the biologically evolved epigenetic rules that guided them.

THIS IS NOT the prevailing view of the arts. Academic theorists have paid little attention to biology; consilience is not in their vocabulary. To varying degrees they have been more influenced by postmodernism, the competing hypothesis that denies the existence of a universal human nature. Applied to literary criticism, the extreme manifestation of postmodernism is the deconstructive philosophy formulated most provocatively by Jacques Derrida and Paul de Man. In this view, truth is relative and personal. Each person creates his own inner world by acceptance or rejection of endlessly shifting linguistic signs. There is no privileged point, no lodestar, to guide literary intelligence. And given that science is just another way of looking at the world, there is no scientifically constructible map of human nature from which the deep meaning of texts can be drawn. There is only unlimited opportunity for the reader to invent interpretations and commentaries out of the world he himself constructs. "The author is dead" is a favorite maxim of the deconstructionists.

Deconstructionist scholars search instead for contradictions and ambiguities. They conceive and analyze what is left out by the author. The missing elements allow for personalized commentary in the postmodernist style. Postmodernists who add political ideology to the mix also regard the traditional literary canon as little more than a collection confirming the world view of ruling groups, and in particular that of Western white males.

The postmodernist hypothesis does not conform well to the evidence. It is blissfully free of existing information on how the mind works. Yet there is surely *some* reason for the popularity of postmodernism other than a love of chaos. If the competing biological approach is correct, its widespread appeal must be rooted in human nature. Postmodernism in the arts is more than a School of Resentment—Harold Bloom's indictment in *The Western Canon*—and more than the eunuch's spite, to borrow a phrase from Alexander Pope, and it is sustained by more than the pathetic reverence commonly given Gallic obscurantism by American academics. There is also a surge of revolutionary spirit in postmodernism, generated by the real—not deconstructed—fact that large segments of the population, most notably women, have unique talents and emotional lives that have been relatively neglected for centuries, and are only now beginning to find full expression within the mainstream culture.

If we are to believe evidence from the biological and behavioral sciences gathered especially during the past quarter century, women differ genetically from men in ways other than reproductive anatomy. In aggregate, on average, with wide statistical overlap, and in many venues of social experience, they speak with a different voice. Today it is being heard loud and clear. But I do not read the welcome triumph of feminism, social, economic, and creative, as a brief for postmodernism. The advance, while opening new avenues of expression and liberating deep pools of talent, has not exploded human nature into little pieces. Instead, it has set the stage for a fuller exploration of the universal traits that unite humanity.

Looked at with a different perspective, postmodernism can also be viewed as one extreme in an historical oscillation in literary world view. The great American critic Edmund Wilson noted, in 1926, that Western literature seems "obliged to vibrate" in emphasis between the two poles of neoclassicism and romanticism. Conceived very broadly, the cycle can first be picked up in the Enlightenment with Pope, Racine, and other poets who drew on the scientists' vision of an orderly world. They were replaced in public esteem by the rebellious romantic poets of the nineteenth century, who yielded in turn to Flaubert and others returning to rational order, who gave way to a flow in the opposite direction as embodied in the modernist writings of the French Symbolists, including Mallarmé and Valéry, and of their British peers Yeats, Joyce, and Eliot. Because each of the extremes proved ultimately "unbearable" as a reigning fashion, Wilson said, it guaranteed reversion toward the opposite pole.

The same mood swing can be seen in recent, post-Wilsonian literary criticism. Earlier in this century scholars stressed the personal experiences of the authors and the history of their times. In the 1950s the New Critics insisted on drawing out the full meaning of the text, without much concern for the personal history of the author. They agreed with Joseph Conrad's famous dictum that a work of art "should carry its justification in every line." In the 1980s the New Critics quite suddenly gave way to the postmodernists, who argued the opposite approach. Search, they said, for what the text does not control, and explain the entirety as a social construction on the part of the author. Their stance has been summarized in a pointed manner by the poet and critic Frederick Turner, as follows: Artists and poets should dismiss the constraints of Nature even in a time of ecological crisis, ignore science, abandon the forms and disciplines of the arts and hence their own culture's shamanic tradition, turn away from the idea of a universal human nature, and, having freed themselves from such stifling confinement, favor snideness and rage over hope and other uplifting emotions. According to Turner, a reversal in fashion is already beginning. "The tradition of Homer, Dante, Leonardo, Shakespeare, Beethoven, and Goethe is not dead. It is growing up in the cracks of the postmodern concrete."

Edmund Wilson hoped for a damping of this perpetual cycle in the arts, which he considered a peculiar affliction of the modern mind. Favoring synthesis in principle, he wrote of his admiration for Bertrand Russell and Alfred North Whitehead, the two great culture unifiers of the first half of the twentieth century. We envy the classics, he said, for the equilibrium they appear to have achieved. "Regularity and logic in Sophocles do not exclude either tenderness or violence; and, in Virgil, the sort of thing that Flaubert can do; the exact objective reproduction of things does not exclude the sort of thing that Wordsworth and Shelley can do, the mysterious, the fluid, the pathetic, and the vague." I like to think that Edmund Wilson would have been favorable to the idea of consilience.

CAN THE OPPOSED Apollonian and Dionysian impulses, cool reason against passionate abandonment, which drive the mood swings of the arts and criticism, be reconciled? This is, I believe, an empirical question. Its answer depends on the existence or nonexistence of an inborn human nature. The evidence accumulated to date leaves little room for doubt. Human nature exists, and it is both deep and highly structured.

If that much is granted, the relation of science to interpretation of the arts can be made clearer, as follows. Interpretation has multiple dimensions, namely history, biography, linguistics, and aesthetic judgment. At the foundation of them all lie the material processes of the human mind. Theoretically inclined critics of the past have tried many avenues into that subterranean realm, including most prominently psychoanalysis and postmodernist solipsism. These approaches, which are guided largely by unaided intuition about the way the brain works, have fared badly. In the absence of a compass based on sound material knowledge, they make too many wrong turns into blind ends. If the brain is ever to be charted, and an enduring theory of the arts created as part of the enterprise, it will be by stepwise and consilient contributions from the brain sciences, psychology, and evolutionary biology. And if during this process the creative mind is to be understood, it will need collaboration between scientists and humanities scholars.

The collaboration, now in its early stages, is likely to conclude that innovation is a concrete biological process founded upon an intricacy of nerve circuitry and neurotransmitter release. It is not the outpouring of symbols by an all-purpose generator or any conjuration therein by ethereal agents. To fathom the origin of innovation in the arts will make a great deal of difference in the way we interpret its creations. The natural sciences have begun to form a picture of the mind, including some of the elements of the creative process itself. Although they are still considerably far from the ultimate goal, they cannot help in the end but strengthen interpretation of the arts.

Charles Lumsden and I reached this conclusion in the early 1980s while developing the full theory of gene-culture coevolution, described earlier. A similar position has been reached from different directions by a small but growing circle of artists and theorists of the arts, among whom the more prominent have been Joseph Carroll, Brett Cooke, Ellen Dissanayake, Walter Koch, Robert Storey, and Frederick Turner. Some of these scholars refer to their approach as biopoetics or bioaesthetics. The analyses have been independently bolstered by Irenäus Eibl-Eibesfeldt, the German ethologist, in his global studies of human instinct; by the American anthropologists Robin Fox and Lionel Tiger in their accounts of ritual and folklore; and by numerous researchers in Artificial Intelligence, whose work on artistic innovation is summarized (to take one excellent exposition) by Margaret Boden in *The Creative Mind.*

The body of the research to date can be fitted together into the following narrative of coevolution of genes and culture:

- *During human evolution there was time enough for natural selection to shape the processes of innovation.* For thousands of generations, sufficient for genetic changes in the brain and sensory and endocrine systems, variation among people in thought and behavior caused personal differences in survival and reproductive success.

- *The variation was to some degree heritable.* Individuals differed then, as they do today, not just in what they learned from their culture but also in their hereditary propensity to learn certain things and to respond by statistical preponderance in particular ways.

- *Genetic evolution inevitably ensued.* Natural selection, favoring some of the gene ensembles over others, molded the epigenetic rules, which are the inherited regularities of mental development that compose human nature. Among the most ancient epigenetic rules I have described to this point are the Westermarck effect, which inhibits incest, and the natural aversion to snakes. Those of more recent origin, perhaps no more than a hundred thousand years ago, include the swift programmed steps by which children acquire language and, we may reasonably presume, some of the creative processes of the arts as well.

- *Universals or near-universals emerged in the evolution of culture.* Because of differences in strength among the underlying epigenetic rules, certain thoughts and behavior are more effective than others in the emotional responses they cause and the frequency with which they intrude on reverie and creative thought. They bias cultural evolution toward the invention of archetypes, the widely recurring abstractions and core narratives that are dominant themes in the arts. Examples of archetypes I have already mentioned are Oedipean tragedy (violating the Westermarck effect) and the serpent images of myth and religion.

• *The arts are innately focused toward certain forms and themes but are otherwise freely constructed.* The archetypes spawn legions of metaphors that compose not only a large part of the arts but also of ordinary communication. Metaphors, the consequence of spreading activation of the brain during learning, are the building blocks of creative thought. They connect and synergistically strengthen different spheres of memory.

GENE-CULTURE COEVOLUTION IS, I believe, the underlying process by which the brain evolved and the arts originated. It is the conceivable means most consistent with the joint findings of the brain sciences, psychology, and evolutionary biology. Still, *direct* evidence with reference to the arts is slender. It is possible that new discoveries concerning the brain and evolution will yet change the picture fundamentally. Such is the nature of science. The uncertainty makes the search for the alignment of science and the humanities all the more interesting a prospect.

This much can be said with confidence, however: The growing evidence of an overall structured and powerful human nature, channeling development of the mind, favors a more traditionalist view of the arts. The arts are not solely shaped by errant genius out of historical circumstances and idiosyncratic personal experience. The roots of their inspiration date back in deep history to the genetic origins of the human brain, and are permanent.

While biology has an important part to play in scholarly interpretation, the creative arts themselves can never be locked in by this or any other discipline of science. The reason is that the exclusive role of the arts is the transmission of the intricate details of human experience by artifice to intensify aesthetic and emotional response. Works of art communicate feeling directly from mind to mind, with no intent to explain why the impact occurs. In this defining quality, the arts are the antithesis of science.

When addressing human behavior, science is coarse-grained and encompassing, as opposed to the arts, which are fine-grained and interstitial. That is, science aims to create principles and use them in human biology to define the diagnostic qualities of the species; the arts use fine details to flesh out and make strikingly clear by implication those same qualities. Works of art that prove enduring are intensely humanistic. Born in the imagination of individuals, they nevertheless touch upon what was universally endowed by human evolution. Even when, as part of fantasy, they imagine worlds that cannot possibly exist, they stay anchored to their human origins. As Kurt Vonnegut, Jr., master fantasist, once pointed out, the arts place humanity in the center of the universe, whether we belong there or not.

Several special powers were granted the arts by the genetic evolution of the brain. First is the ability to generate metaphors with ease and move them fluidly from one context to another. Consider the technical language of the arts themselves. A plot first meant a physical site and building plan, then the stage director's plot or blocking plan, then the action or story blocked out. In the sixteenth century a frontispiece was a decorated front of a building, then the title page of a book ornamented with a figure, usually the allegorical representation of a building, and finally the illustrated page that precedes the title page. A stanza, which in Italian is a public room or resting place, has been appropriated in English to mean the roomlike set of four or more lines separated typographically from other similar sets.

In both the arts and sciences the programmed brain seeks elegance, which is the parsimonious and evocative description of pattern to make sense out of a confusion of detail. Edward Rothstein, a critic trained in both mathematics and music, compares their creative processes:

> We begin with objects that look dissimilar. We compare, find patterns, analogies with what we already know. We distance ourselves and create abstractions, laws, systems, using transformations, mappings, and metaphors. This is how mathematics grows increasingly abstract and powerful; it is how music obtains much of its power, with grand structures growing out of small details. This form of comprehension underlies much of Western thought. We pursue knowledge that is universal in its perspective but its powers are grounded in the particular. We use principles that are shared but reveal details that are distinct.

Now compare that insight with the following independent account of creativity in the physical sciences. The writer is Hideki Yukawa, who spent his career working on the nuclear binding forces of the atom, making discoveries for which he became the first Japanese to receive the Nobel Prize in physics.

> Suppose there is something which a person cannot understand. He happens to notice the similarity of this something to some other thing which he understands quite well. By comparing them he may come to understand the thing which he could not understand up to that moment. If his understanding turns out to be appropriate and nobody else has ever come to such an understanding, he can claim that his thinking was really creative.

The arts, like the sciences, start in the real world. They then reach out to all possible worlds, and finally to all conceivable worlds. Throughout they project the human presence on everything in the universe. Given the power of metaphor, perhaps the arts began with what may be called the "Picasso effect." The artist is reported by his photographer and chronicler Brassaï to have said in 1943: "If it occurred to man to create his own images, it's because he discovered them all around him, almost formed, already within his grasp. He saw them in a bone, in the irregular surfaces of cavern walls, in a piece of wood. One form might suggest a woman, another a bison, and still another the head of a demon." They may have come that route by perception of what Gregory Bateson and Tyler Volk have called metapatterns, those circles, spheres, borders and centers, binaries, layers, cycles, breaks, and other geometric configurations that occur repeatedly in nature and provide easily recognized clues to the identity of more complicated objects.

It was a short step not just to see but to re-create images on rock walls with charcoal lines or by etchings on stone, bone, and wood. The first faltering steps were attempts to stimulate and thereby humanize external Nature. The art historian Vincent Scully has observed that in early historical times, people constructed sacred buildings to resemble mountains, rivers, and animals. By so doing they hoped to draw upon the powers of the environment. The greatest ceremonial site of pre-Columbian America, in Scully's opinion, is Teotihuacán in central Mexico. "There the Avenue of the Dead runs directly to the base of the Temple of the Moon, behind which rises the mountain that is called Tenan ('Our Lady of Stone'). That mountain, running with springs, is basically pyramidal and shaped and notched in the center. And the temple imitates the mountain's shape, intensifies it, clarifies it, geometricizes it, and therefore makes it more potent, as if to draw water down from the mountain to the fields below."

Imitate, make it geometrical, intensify: That is not a bad three-part formula for the driving pulse of the arts as a whole. Somehow innovators know how it all is to be done. They select images from nature that are emotionally and aesthetically potent. In the course of history, as techniques grew more sophisticated, the artists projected feelings back out to nature. Those in architecture and the visual arts created designs based on the idealized features of the human body and what they imagined to be gods modeled from the human body. Supplication, reverence, love, grief, triumph, and majesty, all emotion-charged constructions of the human mind, were captured as abstract images and forced onto both living and inanimate landscapes.

Artists, while free-ranging in the details selected, generally remain faithful to the innate universals of aesthetics. In his 1905–08 variations of *The Farm Weltevreden at Duivendrecht*, the young Piet Mondrian depicted a row of spindly trees in front of a shadowy house. The spacing of the tree trunks seems intuitively right, the redundancy in the canopy lacework is close to what (as I will describe shortly) modern EEG monitoring suggests is most arousing to the brain. The arrangement of open space and water nearby are those that recent psychological studies have revealed to be innately among the most attractive out of all such possible arrangements. Unaware of these neurobiological correlates, probably uncaring even if he had been told, Mondrian repeated the tree-row theme many times over a ten-year period as he felt his way toward new forms of expression. With the influences of Vermeer and van Gogh put well behind him, he discovered and experimented with cubism. In *Study of Trees II* (1913) the canopies of several trees are brought forward, dominating fences and other skeletonized and unfocused structures, yet all still balanced in composition and close to optimally complex by measure of brain arousal. Other variations of the same period increasingly abstract the whole into a mazelike configuration of reticulate lines. The interspaces capture patterns of light and color that change from one compartment to the next. The overall effect is not unlike that of a mottled sky viewed upward through a woodland canopy. Other subjects, including buildings, dunes, piers, and the sea, are similarly transformed. In the end Mondrian attained the pure abstract designs for which he was to be celebrated: "nothing human, nothing specific," as he put it. In this sense he liberated his art. But it is not truly free, and I doubt that inwardly he ever wished it to be. It stays true to the ancient hereditary ground rules that define the human aesthetic.

We do not see in the evolution of Mondrian a localized production of Western culture. The same process was at work in the confluence of Asian art and writing. Chinese characters were invented three thousand years ago as crude pictographs resembling the objects they represent. The sun and moon, mountains and rivers, people and animals, dwellings and utensils are all instantly recognizable today in the ancient Chinese script. They too approach the optimum level of complexity by EEG standards. Over centuries the characters evolved into the elegant *karayo* calligraphy of standard script. An early version of *karayo*, after its introduction to Japan, gave rise to new forms, including the flowing *wayo* script unique to that country. As in Western calligraphy and the ornamental initial letters of medieval hand scripts, art imposed on the written word its own aesthetic standards.

BY INTUITION ALONE, and a sensibility that does not submit easily to formulas, artists and writers know how to evoke emotional and aesthetic response. Adding one artifice to another, obedient to the dictum *ars est celare artem*, it is art to conceal art, they steer us away from explanations of their productions. As Louis Armstrong is reported to have said about jazz: If you have to ask, you'll never know. Scientists, in contrast, try to know. They are anxious to tell you everything, and to make it all clear. But they must respectfully wait until the curtain falls or the book covers close.

The arts are eternally discursive. They seek maximum effect with novel imagery. And imagery that burns itself into the memory, so that when recalled it retains some of its original impact. Among examples I especially appreciate is the perfect opening of Nabokov's pedophilic novel. *Lo-lee-ta: the tip of the tongue taking a trip of three steps down the palate to tap, at three, on the teeth. Lo. Lee. Ta.* Thus with anatomical accuracy, alliterative *t*-sounds, and poetic meter Nabokov drenches the name, the book title, and the plot in sensuality.

Surprise, wit, and originality characterize the memorable use of metaphor. In another genre, the poet Elizabeth Spires tells us about a theological lesson given by a nun at St. Joseph's Elementary School in Circleville, Ohio, on a snowy winter morning. The subject was eschatology for beginners.

> *How long will those lost souls pay for their sins? For all eternity.* Eternity. How can we, at eleven years old, she must be thinking, possibly be able to conceive of just how long eternity is? *Imagine the largest mountain in the world, made of solid rock. Once every hundred years, a bird flies past, the tip of its wing brushing lightly against the mountaintop. Eternity is as long as it would take for the bird's wing to wear the mountain down to nothing.* Ever after, I connect hell and eternity not with fire and flames, but with something cold and unchanging, a snowy tundra overshadowed by a huge granite mountain that casts a pall over the landscape.

WHAT CAN WE truly know about the creative powers of the human mind? The explanation of their material basis will be found at the juncture of science and the humanities. The first premise of the scientific contribution is that *Homo sapiens* is a biological species born of natural selection in a biotically rich environment. Its corollary is that the epigenetic rules affecting the human brain were shaped during genetic evolution by the needs of Paleolithic people in this environment.

The premise and corollary have the following consequence. Culture, rising from the productions of many minds that interlace and reinforce one another over many generations, expands like a growing organism into a universe of seemingly infinite possibility. But not all directions are equally likely. Before the scientific revolution, every culture was sharply circumscribed by the primitive state of that culture's empirical knowledge. The culture evolved under the local influence of climate, water distribution, and food resources. Less obviously, its growth was profoundly affected by human nature.

Which brings us back to the arts. The epigenetic rules of human nature bias innovation, learning, and choice. They are gravitational centers that pull the development of mind in certain directions and away from others. Arriving at the centers, artists, composers, and writers over the centuries have built archetypes, the themes most predictably expressed in original works of art.

Although recognizable through their repeated occurrence, archetypes cannot be easily defined by a simple combination of generic traits. They are better understood with examples, collected into groups that share the same prominent features. This method—called definition by specification—works well in elementary biological classification, even when the essential nature of the species as a category remains disputed. In myth and fiction as few as two dozen such subjective groupings cover most of the archetypes usually identified as such. Some of the most frequently cited are the following.

In the beginning, the people are created by gods, or the mating of giants, or the clash of titans; in any case, they begin as special beings at the center of the world.

The tribe emigrates to a promised land (or Arcadia, or the Secret Valley, or the New World).

The tribe meets the forces of evil in a desperate battle for survival; it triumphs against heavy odds.

The hero descends to hell, or is exiled to wilderness, or experiences an iliad in a distant land; he returns in an odyssey against all odds past fearsome obstacles along the way, to complete his destiny.

The world ends in apocalypse, by flood, fire, alien conquerors, or avenging gods; it is restored by a band of heroic survivors.

A source of great power is found in the tree of life, the river of life, philosopher's stone, sacred incantation, forbidden ritual, secret formula.

The nurturing woman is apotheosized as the Great Goddess, the Great Mother, Holy Woman, Divine Queen, Mother Earth, Gaia.

The seer has special knowledge and powers of mind, available to those worthy to receive it; he is the wise old man or woman, the holy man, the magician, the great shaman.

The Virgin has the power of purity, is the vessel of sacred strength, must be protected at all costs, and perhaps surrendered up to propitiate the gods or demonic forces.

Female sexual awakening is bestowed by the unicorn, the gentle beast, the powerful stranger, the magical kiss.

The Trickster disturbs established order and liberates passion as the god of wine, king of the carnival, eternal youth, clown, jester, clever fool.

A monster threatens humanity, appearing as the serpent demon (Satan writhing at the bottom of hell), dragon, gorgon, golem, vampire.

IF THE ARTS are steered by inborn rules of mental development, they are end products not just of conventional history but also of genetic evolution. The question remains: Were the genetic guides mere byproducts—epiphenomena—of that evolution, or were they adaptations that directly improved survival and reproduction? And if adaptations, what exactly were the advantages conferred? The answers, some scholars believe, can be found in artifacts preserved from the dawn of art. They can be tested further with knowledge of the artifacts and customs of present-day hunter-gatherers.

This is the picture of the origin of the arts that appears to be emerging. The most distinctive qualities of the human species are extremely high intelligence, language, culture, and reliance on long-term social contracts. In combination they gave early *Homo sapiens* a decisive edge over all competing animal species, but they also exacted a price we continue to pay, composed of the shocking recognition of the self, of the finiteness of personal existence, and of the chaos of the environment.

These revelations, not disobedience to the gods, are what drove humankind from paradise. *Homo sapiens* is the only species to suffer psychological exile. All animals, while capable of some degree of specialized learning, are instinct-driven, guided by simple cues from the environment that trigger complex behavior patterns. The great apes have the power of self-recognition, but there is no evidence that they can reflect on their own birth and eventual death. Or on the meaning of existence—the complexity of the universe means nothing to them. They and other animals are exquisitely adapted to just those parts of the environment on which their lives depend, and they pay little or no attention to the rest.

The dominating influence that spawned the arts was the need to impose order on the confusion caused by intelligence. In the era prior to mental expansion, the ancestral prehuman populations evolved like any other animal species. They lived by instinctive responses that sustained survival and reproductive success. When *Homo*-level intelligence was attained, it widened that advantage by processing information well beyond the releaser cues. It permitted flexibility of response and the creation of mental scenarios that reached to distant places and far into the future. The evolving brain, nevertheless, could not convert to general intelligence alone; it could not turn into an all-purpose computer. So in the course of evolution the animal instincts of survival and reproduction were transformed into the epigenetic algorithms of human nature. It was necessary to keep in place these inborn programs for the rapid acquisition of language, sexual conduct, and other processes of mental development. Had the algorithms been erased, the species would have faced extinction. The reason is that the lifetime of an individual human being is not long enough to sort out experiences by means of generalized, unchanneled learning. Yet the algorithms were jerry-built: They worked adequately but not superbly well. Because of the slowness of natural selection, which requires tens or hundreds of generations to substitute new genes for old, there was not enough time for human heredity to cope with the vastness of new contingent possibilities revealed by high intelligence. Algorithms could be built, but they weren't numerous and precise enough to respond automatically and optimally to every possible event.

The arts filled the gap. Early humans invented them in an attempt to express and control through magic the abundance of the environment, the power of solidarity, and other forces in their lives that mattered most to survival and reproduction. The arts were the means by which these forces could be ritualized and expressed in a new, simulated reality. They drew consistency from their faithfulness to human nature, to the emotion-guided epigenetic rules—the algorithms—of mental development. They achieved that fidelity by selecting the most evocative words, images, and rhythms, conforming to the emotional guides of the epigenetic rules, making the right moves. The arts still perform this primal function, and in much the same ancient way. Their quality is measured by their humanness, by the precision of their adherence to human nature. To an overwhelming degree that is what we mean when we speak of the true and beautiful in the arts.

ABOUT THIRTY THOUSAND YEARS AGO *Homo sapiens* used the visual arts to bring the representation of large animals into shelters. Some of the oldest and most sophisticated of such works are the wall paintings, engravings, and sculptures found in caverns of the southern half of Ice Age Europe. More than two hundred such caverns containing thousands of images have been found during the past century in Italy, Switzerland, France, and Spain. The most recently discovered, and oldest of all, is the spectacularly painted cave at Chauvet, in the valley of the Ardèche River, a tributary of the Rhône. Chemical tests have established the age of the art at 32,410 ± 720 years. The youngest cave galleries are Magdalenian paintings, etchings, and sculptures created as recently as ten thousand years before the present, near the dawn of the Neolithic era.

The best of the animal drawings are accurate and beautiful even by exacting modern standards. They are rendered with clean, sweeping lines, some of which are shaded to one side as though to convey three-dimensionality. They present a veritable field guide to the largest mammals of the region, from lion to mammoth, bear to horse, rhinoceros to bison, most of which are now extinct. The figures are more than abstract images. Some are clearly male or female, of different ages. A few of the females are swollen with young. Some wear recognizable winter or summer pelages. At Chauvet two rampant male rhinoceros lock horns in battle.

Given the antiquity of Chauvet and the scarcity of even older representational art, it is tempting to conclude that the skills of the cavern artists emerged quickly, perhaps within a few generations. But that would be premature. On the basis of genetic and fossil evidence, it appears that anatomically modern *Homo sapiens* evolved in Africa by about two hundred thousand years before the present, and entered Europe as recently as fifty thousand years ago. In the succeeding interval, up to the time of the Chauvet paintings, they slowly displaced the Neanderthal people, now considered by some anthropologists to be a distinct human species. It is reasonable to suppose that during this era, and before occupying the particular cave sites that today harbor the oldest known works, the artists improved their techniques and style on surfaces now lost. Many of the early paintings might have been applied to outdoor rock walls, a practice still followed by hunter-gatherers in Australia and southern Africa, and as a result failed to survive the harsh climate of Ice Age Europe.

It may never be known whether European cavern art sprang full-blown or was perfected in small steps across millennia, but at least we have strong hints as to *why* it was created. A number of the examples, as many as 28 percent at Cosques near Marseilles, for example, are depicted with arrows or spears flying about the bodies of the animals. A bison at Lascaux has been eviscerated by a spear that enters its anus and emerges through its genitals. The simplest and most persuasive explanation for the embellishment is the one proposed in the early 1900s by Abbé Breuil, the pioneer explorer and interpreter of European Paleolithic art. It is hunting magic, he said, the belief that by re-creating animals and killing their images, the hunters will more readily overcome real prey when the chase begins outdoors.

Art is magic: That has a modern ring, for as we often hear, the purpose of the arts is enchantment. Breuil's hypothesis is supported by an intriguing piece of additional evidence, the repeated depiction of the same animal species on the same rock-surface panels. In one case, chemical tests indicate that the portraits were drawn centuries apart. Duplicates are also commonly drawn—or in some cases etched on bone fragments—on top of the original. Rhinoceros horns are replicated, mammoths bear multiple head domes, lions have two or three complete heads. Although we will never be able to read the minds of the artists, it is a fair guess that they meant the images to be reborn with each duplication in order to serve the purpose of new rituals. Those rituals might have been part of full-blown ceremonies, accompanied by early forms of music and dancing. Flutes made of bone have been discovered in the caves, in good enough condition to be cleaned and played, and the paintings themselves are consistently located in places where the acoustics are excellent.

Hunting sorcery of one form or another has survived in hunter-gatherer societies to the present time. It is a form of sympathetic magic, an expression of the near-universal belief among prescientific peoples that the manipulation of symbols and images can influence the objects they represent. Sticking pins in dolls and other practices of malign voodoo are among the most familiar examples from popular culture. Most religious rituals contain elements of sympathetic magic. Children selected for sacrifice to Tlaloc, the Aztec god of rain and lightning, were first forced to shed tears, in order to bring raindrops to the Valley of Mexico. Christian baptism takes away the sins of the world.

To be cleansed, to be born again, you must be washed in the blood of the Lamb.

Belief in astrology and extrasensory perception, particularly psychokinesis, is built from similar elements in the sorcerer's toolkit. The near-universal faith in sympathetic magic of one form or another is easily explained. In a bewildering and threatening world, people reach out for power by any means they can find. Combining art with sympathetic magic is a quite natural way to make that attempt.

In opposition to the hunting magic hypothesis it can be argued that the cave art images served the much simpler purpose of instructing the young. Perhaps it was indeed only a prehistoric *Peterson's Field Guide to the Large Mammals of Pleistocene Europe*. But with no more than a dozen species to learn, it remains unclear why the portraits were drawn repetitively on the same panels. Or why the skills of hunting could not have been better learned by adolescent apprentices when they accompanied their elders in the field—the method used by hunter-gatherer people today.

The magic hypothesis of animal art is reinforced by other forms of behavior displayed by extant stone-age people. Their hunters are intensely preoccupied with the lives of the big animals around them, especially mammals that can be killed only by tracking or ambush. They are less concerned with smaller species, such as hares and porcupines, that can be snared or dug from burrows. They often impute to their large prey the possession of minds and special powers that project their own fierce human desires. The animals they kill they sometimes propitiate with ceremony. Hunters of many cultures collect skulls, claws, and skins as trophies to memorialize their own prowess. The totemic animals, invested with supernatural qualities and honored with reverential art, are then used as symbols to bind members of the clan together. Their spirits preside over celebrations of victory, and see the people through the dark hours of defeat. They remind each individual of the existence of something greater than himself, something immortal of which he is a part. The totems enforce moderation in dispute, and they soften dissension within the tribe. They are sources of real power. It is not surprising to find that among the few well-rendered human beings in Ice Age art are shamans wearing headgear of stag antlers or the head of a bird or lion. It seems logical that gods in the form of animals ruled the ancient civilizations of the Fertile Crescent and Mesoamerica. Such effects of sympathetic magic radiate out. Not just hunter-gatherer bands but also groups and nations at the level of high civilization are prone to adopt animal species as totems to reflect the qualities they most value in themselves. American football fans, having at last found a way to form their own Paleolithic tribes, cheer for the Detroit Lions, Miami Dolphins, and Chicago Bears.

THE BIOLOGICAL ORIGIN of the arts is a working hypothesis, dependent on the reality of the epigenetic rules and the archetypes they generate. It has been constructed in the spirit of the natural sciences, and as such is meant to be testable, vulnerable, and consilient with the rest of biology.

So how then is the hypothesis to be tested? One way is to predict from evolutionary theory the themes and underlying epigenetic rules most likely to be encountered in the arts. We know that such near-universal themes do exist, and in fact form the scaffolding of most works of fiction and the visual arts. Their generality is the reason Hollywood plays well in Singapore, and why Nobel Prizes in Literature are given to Africans and Asians as well as to Europeans. What we do not understand very well is why this is so, why processes of mental development direct attention so consistently toward certain images and narratives. Evolutionary theory is a potentially powerful means of predicting the underlying epigenetic rules and understanding their origins in genetic history.

Earlier I described one important example of the evolutionary approach, in studies that address incest avoidance and taboos. The inborn inhibitory responses causing these phenomena have reverberated in myth and the arts throughout recorded history. Other responses that can connect biological theory to the arts are parent-infant bonding, family cooperation and conflict, and territorial aggression and defense.

A second, wholly different means of discovering epigenetic rules affecting the arts is simply to scan directly for them with methods from the neurosciences and cognitive psychology. In a pioneering study of "bioaesthetics" published in 1973, the Belgian psychologist Gerda Smets asked subjects to view abstract designs of varying degrees of complexity while she recorded changes in their brain wave patterns. To register arousal she used the desynchronization of alpha waves, a standard neurobiological measure. In general, the more the alpha waves are desynchronized, the greater the psychological arousal subjectively reported by subjects. Smets made a surprising discovery. She found a sharp peak of brain response when the redundancy—repetitiveness of elements—in the designs was about 20 percent. This is the equivalent amount of order found variously in a simple maze, in two complete turns of a logarithmic spiral, or in a cross with asymmetrical arms. The 20 percent redundancy effect appears to be innate. Newborn infants gaze longest at drawings with about the same amount of order.

What does this epigenetic rule have to do with aesthetics and art? The connection is closer than may be immediately apparent. Smets' high-arousal figures, even though generated by a computer, have an intriguing resemblance to abstract designs used worldwide in friezes, grillwork, logos, colophons, and flag designs. They are also close in order and complexity to the pictographs of written Chinese, Japanese, Thai, Tamil, Bengali, and other Asian languages of diverse origin, as well as the glyphs of the ancient Egyptians and Mayans. Finally, it seems likely that some of the most esteemed products of modern abstract art fall near the same optimal level of order, as illustrated in Mondrian's *oeuvre*. Although this connection of neurobiology to the arts is tenuous, it offers a promising cue to the aesthetic instinct, one that has not to my knowledge been explored systematically by either scientists or interpreters of the arts.

Analyzing the beauty of a young woman's face is another way to scan directly for epigenetic rules relevant to aesthetics. For more than a century it has been known that photographic composites of many faces blended together are considered more attractive than most of the individual faces viewed separately. The phenomenon has led to the belief that ideal facial beauty is simply the average condition for the population as a whole. That entirely reasonable conclusion turns out to be only half true. In 1994 new studies revealed that a blend of individual faces considered attractive at the outset is rated higher than a blend of all the faces without prior selection. In other words, an average face is attractive but not optimally attractive. Certain dimensions of the face are evidently given more weight in evaluation than others. The analyses then produced a real surprise. When the critical dimensions were identified and exaggerated in artificially modified composites, attractiveness rose still more. Both Caucasian and Japanese female faces had this effect on young British and Japanese subjects of both sexes. The features thought most attractive are relatively high cheek bones, a thin jaw, large eyes relative to the size of the face, and a slightly shorter rather than longer distance between mouth and chin and between nose and chin.

Only a small percentage of young women fall at or close to the average. That is to be expected in a genetically diverse species whose precise combinations of features are created anew within and between families of every generation. What is more puzzling is the divergence of the optimum from the average. Few women—extremely few in fact—approach it. If the perception of facial beauty resulted in the higher survival and reproductive success of the most beautiful conceivable, then the most beautiful should be at or close to the average within the population. Such is the expected result of stabilizing natural selection: Deviations from the optimum dimensions in any direction are disfavored, and the optimum is sustained as the norm through evolutionary time.

The explanation for the rarity of great beauty may be (and I continue to speculate) the behavioral phenomenon known as the supernormal stimulus. Widespread among animal species, it is the preference during communication for signals that exaggerate the norms even if they rarely if ever occur in nature. An instructive example is female attractiveness in the silver-washed fritillary, a silver-dappled orange butterfly found in woodland clearings from western Europe to Japan. During the breeding season males instinctively recognize females of their own species by their unique color and flight movements. They chase them, but they are not what the males really prefer. Researchers found that they could attract male fritillaries with plastic replicas whose wings are flapped mechanically. To their surprise, they also learned that males turn from real females and fly toward the models that have the biggest, brightest, and most rapidly moving wings. No such fritillary superfemale exists in the species' natural environment.

Males of the silver-washed fritillary appear to have evolved to prefer the strongest expression of certain stimuli they encounter, with no upper limit. The phenomenon is widespread in the animal kingdom. While experimenting with anole lizards of the West Indies a few years ago, I found that males display enthusiastically to photographs of other members of the same species, even though the images are the size of a small automobile. Other researchers have learned that herring gulls ignore their own eggs when presented with appropriately painted wooden models so large they cannot even climb on top of them.

In the real world the supernormal response works because the monstrous forms created by experimenters do not exist, and the animals can safely follow an epigenetic rule expressible as follows: "Take the largest (or brightest or most conspicuously moving) individual you find." Female fritillaries cannot be gigantic insects with brilliant whirring wings. Such creatures could not locate enough food to get through the caterpillar stage and survive in the Eurasian woodlands. In parallel manner, women with large eyes and delicate features may have less robust health, especially during the rigors of childbearing, than those closer to the population average. But at the same time—and this could be the adaptive significance—they present physical cues of youth, virginity, and the prospect of a long reproductive period.

The off-center optimum of female attractiveness is no more peculiar than most of the rest of human social behavior. The entire beauty industry can be interpreted as the manufacture of supernormal stimuli. Eyelid shadow and mascara enlarge the eyes, lipstick fills out and brightens the lips, rouge brings a permanent blush to the cheeks, pancake makeup smoothes and reshapes the face toward the innate ideal, fingernail paint adds blood circulation to the hands, and teasing and tinting render the hair full-bodied and youthful. All these touches do more than imitate the natural physiological signs of youth and fecundity. They go beyond the average normal.

The same principle is true for body adornment of all kinds in men and women. Clothing and emblems project vigor and advertise status. Thousands of years before artists painted animals and costumed shamans on the cave walls of Europe, people were fastening beads onto clothing and piercing belts and headbands with carnivore teeth. Such evidence indicates that the original canvas of the visual arts was the human body itself.

Ellen Dissanayake, an American historian of aesthetics, suggests that the primal role of the arts is and always has been to "make special" particular features of humans, animals, and the inanimate environment. Such features, as illustrated by feminine beauty, are the ones toward which human attention is already biologically predisposed. They are among the best places to search for the epigenetic rules of mental development.

THE ARTS, while creating order and meaning from the seeming chaos of daily existence, also nourish our craving for the mystical. We are drawn to the shadowy forms that drift in and out of the subconscious. We dream of the insoluble, of unattainably distant places and times. Why should we so love the unknown? The reason may be the Paleolithic environment in which the brain evolved. In our emotions, I believe, we are still there. As a naturalist, I use an explicit geographic imagery in reveries of this formative world.

At the center of our world is home ground. In the center of the center are shelters backed against a rock wall. From the shelters radiate well-traveled paths where every tree and rock is familiar. Beyond lies opportunity for expansion and riches. Down a river, through a wooded corridor lining the opposite shore, are campsites in grassy places where game and food plants are seasonally abundant. Such opportunities are balanced by risk. We might lose our way on a too-distant foray. A storm can catch us. Neighboring people—poisoners, cannibals, not fully human—will either trade or attack; we can only guess their intentions. In any case they are an impassable barrier. On the other side is the rim of the world, perhaps glimpsed as a mountain front, or a drop toward the sea. Anything could be out there: dragons, demons, gods, paradise, eternal life. Our ancestors came from there. Spirits we know live closer by, and at fall of night are on the move. So much is intangible and strange! We know a little, enough to survive, but all the rest of the world is a mystery.

What is this mystery we find so attractive? It is not a mere puzzle waiting to be solved. It is far more than that, something still too amorphous, too poorly understood to be broken down into puzzles. Our minds travel easily—eagerly!—from the familiar and tangible to the mystic realm. Today the entire planet has become home ground. Global information networks are its radiating trails. But the mystic realm has not vanished; it has just retreated, first from the foreground and then from the distant mountains. Now we look for it in the stars, in the unknowable future, in the still teasing possibility of the supernatural. Both the known and the unknown, the two worlds of our ancestors, nourish the human spirit. Their muses, science and the arts, whisper: *Follow us, explore, find out.*

IN TRYING to comprehend this aura of the ancestral mind, we are not entirely dependent on introspection and fantasy. Anthropologists have carefully studied bands of contemporary hunter-gatherers whose lifeways appear to resemble those of our common Paleolithic forebears. In recording languages, daily activities, and conversations, the researchers have drawn reasonable inferences concerning the thought processes of their subjects.

One such account has been provided by Louis Liebenberg on the San-speaking "Bushman" hunter-gatherers of the central Kalahari, more particularly the Ju/wasi (or !Kung), /Gwi, and !Xo of Botswana and Namibia. He has drawn on his own researches and those of other anthropologists, most prominently Richard B. Lee and George B. Silberbauer, to record the vanishing culture of these remarkable people.

The Kalahari bands, in order to live on the sparse resources of the desert, must plan and act very carefully. Knowledge of the local terrain and of seasonal ecology is particularly important. The bands understand that the distribution of water resources within their territory is most important of all. In Liebenberg's words:

> During the rainy season they live at temporary pools in the midst of nut forests. Only the most palatable and abundant foods that are the least distance from water are collected. As time goes on they have to travel further

> and further to collect food. They usually occupy a camp for a period of weeks or months and eat their way out of it. During the dry season, groups are based at permanent waterholes. They eat out an increasing radius of desirable foods, and as the water-food distances increase the subsistence effort increases.

The Kalahari bands are experts on local geography and the many plants and animals on which their lives depend. Plant gatherers, usually women but also men on their way home from unsuccessful hunts, use knowledge of the botanical communities to pinpoint edible species. They are conservationists by way of necessity. Liebenberg continues:

> They avoid stripping an area of a species, leaving a residue so that regeneration is not imperiled. Locally scarce specimens are not exploited even when these are found while gathering other species.

The hunters are also equally expert on the details of animal life. Their skills at tracking large animals depend on this knowledge.

> When fresh spoor is found, hunters will estimate its age and how fast the animal was moving to decide whether it is worth following up. In thick bush, where there may be no clear footprints, or on hard ground, where only scuff marks may be evident, trackers may not be able to identify the animal. When this happens they will have to follow the trail, looking for signs such as disturbed vegetation and scuff marks, until clear footprints are found. They will reconstruct what the animal was doing and predict where it was going.

In the Kalahari, as throughout all of the hunter-gatherer world for countless millennia, the hunt holds a central place in the social life of the band.

> In storytelling around the campfire at night men give graphic descriptions of hunts of the recent and distant past. To find animals requires all the information on their movements that can be gained from others' observations and the hunter's own interpretation of signs. Hunters will spend many hours discussing the habits and movements of animals.

The life of the Kalahari band, optimally comprising fifty to seventy members, is intensely communal and cooperative. Because the group must move several times a year with all their possessions on their backs, individuals accumulate few material goods not essential to survival.

> Ownership is limited to an individual's clothing, a man's weapons and implements and a woman's household goods. The band's territory and all its assets are not owned individually but communally, by the whole band.

> To hold the group together, decorum and reciprocity are strictly observed.

> While hunting is an important activity in hunter-gatherer subsistence, successful hunters, who may naturally be pleased with themselves, are expected to show humility and gentleness. To the Ju/wasi, for example, announcing a kill is a sign of arrogance and is strongly discouraged. Many good hunters do no hunting for weeks or months at a time. After a run of successful hunts a hunter will stop hunting in order to give other men the chance to reciprocate.

While the Kalahari hunters are close students of animal behavior, they are thoroughly anthropomorphic in their interpretation. They strain to enter the minds of the animals they track. They imagine, they project thoughts directly to the world around them, and they analogize.

> Animal behaviour is perceived as rational and directed by motives based on values (or the negation of those values) that are either held by the hunter-gatherers themselves or by people known to them. The behaviour of animals is seen by the /Gwi as bound by the natural order of *N!adima* (God). Each species is perceived to have characteristic behaviour, which is governed by its *kxodzi* (customs), and each has its particular *kxwisa* (speech, language). Animals are believed to have acquired special capabilities by means of rational thought.

Knowing the belief of preliterate people in the equivalency of the material and immaterial worlds, and of rational and irrational explanation, it is easy to see how they invent narrative forms loaded with myths and totems. The acceptance of mystery is central to their lives.

> The /Gwi believe that some species possess knowledge that transcends that of humans. The bateleur eagle is believed to know when a hunter will be successful and will hover over him, thereby acting as an omen of

> sure success. Some steenbok are thought to have a magical means of protecting themselves from a hunter's arrows, while the duiker is believed to practice sorcery against its animal enemies and even against conspecific rivals. Baboons, because of their legendary love of trickery and teasing, are believed to eavesdrop on hunters and to pass on their plans to the intended prey animals.

The world that preliterate humans factually perceive is only a small fragment of the full natural world. Thus by necessity the primitive mind is continuously tuned to mystery. For the Kalahari and other contemporary hunter-gatherers the experience of daily life grades imperceptibly into their magical surroundings. Spirits dwell in trees and rocks, animals think, and human thought projects outward from the body with a physical force.

We are all still primitives compared to what we might become. Hunter-gatherers and college-educated urbanites alike are aware of fewer than one in a thousand of the kinds of organisms—plants, animals, and microorganisms—that sustain the ecosystems around them. They know very little about the real biological and physical forces that create air, water, and soil. Even the most able naturalist can trace no more than a faint outline of an ecosystem to which he has devoted a lifetime of study.

Yet the great gaps in knowledge are beginning to be filled. That is the strength of cumulative science in a literate world. People learn and forget, they die, and even the strongest institutions they erect deteriorate, but knowledge continues to expand globally while passing from one generation to the next. Any trained person can retrieve and augment any part of it. By this means all the species of organisms in ecosystems such as the Kalahari Desert will eventually come to be known. They will be given scientific names. Their place in the food web will be discovered, their anatomy and physiology penetrated to the level of cell and molecule, the instinctive behavior of the animals reduced to neuron circuitry, then to neurotransmitters and ion exchange. If the history of biology is a guide, all the facts will prove consilient. The explanations can be joined in space from molecule to ecosystem, and in time from microsecond to millennium.

With consilient explanation, the units at different levels of biological organization can be reassembled. Among them will be whole plants and animals as we normally see them—not as collections of molecules in biochemical time, too small and fast-changing to be visible to the unaided eye, not as whole populations living in the slow motion of ecological time, but as individual plants and animals confined to the sliver of organismic time where human consciousness, being organismic itself, is forced to exist.

Returning to that narrow sliver after the science-led grand tour of space-time, we arrive home in the world for which the evolution of the brain prepared us. Now, with science and the arts combined, we have it all.

Poet in my heart, walk with me across the mysterious land. We can still be hunters in the million-year dreamtime. Our minds are filled with calculation and emotion. We are aesthetes tense with anxiety. Once again the bateleur eagle wheels above our heads, trying to tell us something we overlooked, something we forgot. How can we be sure that eagles never speak, that everything can be known about this land? Nearby is spoor of the elusive duiker leading into the scrub: Shall we follow? Magic enters the mind seductively, like a drug in the veins. Accepting its emotive power, we know something important about human nature. And something important intellectually—that in expanded space-time the fiery circle of science and the arts can be closed.

Within the larger scale, the archaic world of myth and passion is perceived as it truly is, across the full range of cause and effect. Every contour of the terrain, every plant and animal living in it, and the human intellect that masters them all, can be understood more completely as a physical entity. Yet in so doing we have not abandoned the instinctual world of our ancestors. By focusing on the peculiarly human niche in the continuum, we can if we wish (and we so desperately wish) inhabit the productions of art with the same sense of beauty and mystery that seized us at the beginning. No barrier stands between the material world of science and the sensibilities of the hunter and the poet.

CHAPTER 11

Ethics and Religion

Centuries of debate on the origin of ethics come down to this: Either ethical precepts, such as justice and human rights, are independent of human experience or else they are human inventions. The distinction is more than an exercise for academic philosophers. The choice between the assumptions makes all the difference in the way we view ourselves as a species. It measures the authority of religion, and it determines the conduct of moral reasoning.

The two assumptions in competition are like islands in a sea of chaos, immovable, as different as life and death, matter and the void. Which is correct cannot be learned by pure logic; for the present only a leap of faith will take you from one to the other. But the true answer will eventually be reached by the accumulation of objective evidence. Moral reasoning, I believe, is at every level intrinsically consilient with the natural sciences.

Every thoughtful person has an opinion on which of the premises is correct. But the split is not, as popularly supposed, between religious believers and secularists. It is between transcendentalists, those who think that moral guidelines exist outside the human mind, and empiricists, who think them contrivances of the mind. The choice between religious or nonreligious conviction and the choice between ethically transcendentalist or empiricist conviction are cross-cutting decisions made in metaphysical thought. An ethical transcendentalist, believing ethics to be independent, can either be an atheist or else assume the existence of a deity. In parallel manner, an ethical empiricist, believing ethics to be a human creation only, can either be an atheist or else believe in a creator deity (though not in a law-giving God in the traditional Judaeo-Christian sense). In simplest terms the option of ethical foundation is as follows:

I believe in the independence of moral values, whether from God or not,

versus

I believe that moral values come from humans alone; God is a separate issue.

Theologians and philosophers have almost always focused on transcendentalism as the means to validate ethics. They seek the grail of natural law, which comprises freestanding principles of moral conduct immune to doubt and compromise. Christian theologians, following St. Thomas Aquinas' reasoning in *Summa Theologiae*, by and large consider natural law to be the expression of God's will. Human beings, in this view, have the obligation to discover the law by diligent reasoning and weave it into the routine of their daily lives. Secular philosophers of transcendental bent may seem to be radically different from theologians, but they are actually quite similar, at least in moral reasoning. They tend to view natural law as a set of principles so powerful as to be self-evident to any rational person, whatever the ultimate origin. In short, transcendentalism is fundamentally the same whether God is invoked or not.

For example, when Thomas Jefferson, following John Locke, derived the doctrine of natural rights from natural law, he was more concerned with the power of transcendental statements than in their divine or secular origin. In the Declaration of Independence he blended the secular and religious presumptions in one transcendentalist sentence, thus deftly covering all bets: "We hold these Truths to be self-evident, that all Men are created equal, that they are endowed by their Creator with certain unalienable Rights, that among these are Life, Liberty, and the Pursuit of Happiness." That assertion became the cardinal premise of America's civil religion, the righteous sword wielded by Lincoln and Martin Luther King, and it endures as the central ethic binding together the diverse peoples of the United States.

So compelling are such fruits of natural law theory, especially when the deity is also invoked, that they may seem to place the transcendentalist assumption beyond question. But to its noble successes must be added appalling failures. It has been perverted many times in the past, used for example to argue passionately for colonial conquest, slavery, and genocide. Nor was any great war ever fought without each side thinking its cause transcendentally sacred in some manner or other. "Oh how we hate one another," observed Cardinal Newman, "for the love of God."

So perhaps we can do better, by taking empiricism more seriously. Ethics, in the empiricist view, is conduct favored consistently enough throughout a society to be expressed as a code of principles. It is driven by hereditary predispositions in mental development—the "moral sentiments" of the Enlightenment philosophers—causing broad convergence across cultures, while reaching precise form in each culture according to historical circumstance. The codes, whether judged by outsiders as good or evil, play an important role in determining which cultures flourish, and which decline.

The importance of the empiricist view is its emphasis on objective knowledge. Because the success of an ethical code depends on how wisely it interprets the moral sentiments, those who frame it should know how the brain works, and how the mind develops. The success of ethics also depends on the accurate prediction of the consequence of particular actions as opposed to others, especially in cases of moral ambiguity. That too takes a great deal of knowledge consilient with the natural and social sciences.

The empiricist argument, then, is that by exploring the biological roots of moral behavior, and explaining their material origins and biases, we should be able to fashion a wiser and more enduring ethical consensus than has gone before. The current expansion of scientific inquiry into the deeper processes of human thought makes this venture feasible.

The choice between transcendentalism and empiricism will be the coming century's version of the struggle for men's souls. Moral reasoning will either remain centered in idioms of theology and philosophy, where it is now, or it will shift toward science-based material analysis. Where it settles will depend on which world view is proved correct, or at least which is more widely *perceived* to be correct.

THE TIME HAS COME to turn the cards face up. Ethicists, scholars who specialize in moral reasoning, are not prone to declare themselves on the foundations of ethics, or to admit fallibility. Rarely do you see an argument that opens with the simple statement: *This is my starting point, and it could be wrong.* Ethicists instead favor a fretful passage from the particular into the ambiguous, or the reverse, vagueness into hard cases. I suspect that almost all are transcendentalists at heart, but they rarely say so in simple declarative sentences. One cannot blame them very much; it is difficult to explain the ineffable, and they evidently do not wish to suffer the indignity of having their personal beliefs clearly understood. So by and large they steer around the foundation issue altogether.

That said, I will of course try to be plain about my own position: I am an empiricist. On religion I lean toward deism but consider its proof largely a problem in astrophysics. The existence of a cosmological God who created the universe (as envisioned by deism) is possible, and may eventually be settled, perhaps by forms of material evidence not yet imagined. Or the matter may be forever beyond human reach. In contrast, and of far greater importance to humanity, the existence of a biological God, one who directs organic evolution and intervenes in human affairs (as envisioned by theism) is increasingly contravened by biology and the brain sciences.

The same evidence, I believe, favors a purely material origin of ethics, and it meets the criterion of consilience: Causal explanations of brain activity and evolution, while imperfect, already cover the most facts known about moral behavior with the greatest accuracy and the smallest number of freestanding assumptions. While this conception is relativistic, in other words dependent on personal viewpoint, it need not be irresponsibly so. If evolved carefully, it can lead more directly and safely to stable moral codes than transcendentalism, which is also, when you think about it, ultimately relativistic.

And yes—lest I forget—I may be wrong.

In order to sharpen the distinction between transcendentalism and empiricism, I have constructed a debate between defenders of the two world views. To add passionate conviction, I have also made the transcendentalist a theist, and the empiricist a skeptic. And to be as fair as possible, I have drawn their arguments from the most closely reasoned sources in theology and philosophy of which I am aware.

The Transcendentalist

"Before taking up ethics, let me affirm the logic of theism, because if the existence of a law-giving God is conceded, the origin of ethics is instantly settled. So please consider carefully the following argument in favor of theism.

"I challenge your rejection of theism on your own empiricist grounds. How can you ever hope to disprove the existence of a personal God? How can you explain away the three thousand years of spiritual testimony from the followers of Judaism, Christianity, and Islam? Hundreds of millions of people, including a large percentage of the educated citizens of industrialized countries, *know* there is an unseen sentient power guiding their lives. The testimony is overwhelming. According to recent polls, nine in ten Americans believe in a personal God who can answer prayers and perform miracles. One in five has experienced His presence and guidance at least once during the year previous to the poll. How can science, the underwriting discipline of ethical empiricism, dismiss such widespread testimony?

"The nucleus of the scientific method, we are constantly reminded, is the rejection of certain propositions in favor of others in strict conformity to fact-based logic. Where are the facts that require the rejection of a personal God? It isn't enough to say that the idea is unnecessary to explain the physical world, at least as scientists understand it. Too much is at stake for theism to be dismissed with that flip of the hand. The burden of proof is on you, not on those who believe in a divine presence.

"Looked at in proper perspective, God subsumes science, science does not subsume God. Scientists collect data on certain subjects and build hypotheses to explain them. In order to extend the reach of objective knowledge as far as they can, they provisionally accept some hypotheses while discarding others. That knowledge, however, can cover only part of reality. Scientific research in particular is not designed to explore all of the wondrous varieties of human mental experience. The idea of God, in contrast, has the capacity to explain *everything*, not just measurable phenomena, but phenomena personally felt and subliminally sensed, including revelations that can be communicated solely through spiritual channels. Why should all mental experience be visible in PET scans? Unlike science, the idea of God is concerned with more than the material world given us to explore. It opens our minds to what lies outside that world. It instructs us to reach out to the mysteries that are comprehensible through faith alone.

"Confine your thoughts to the material world if you wish. Others know that God encompasses the ultimate causes of the Creation. Where do the laws of nature come from if not a power higher than the laws themselves? Science offers no answer to that sovereign question of theology. Put another way, why is there something rather than nothing? The ultimate meaning of existence lies beyond the rational grasp of human beings, and therefore outside the province of science.

"Are you also a pragmatist? There is an urgently practical reason for belief in ethical precepts ordained by a supreme being. To deny such an origin, to assume that moral codes are exclusively man-made, is a dangerous creed. As Dostoyevsky's Grand Inquisitor observed, all things are permitted when there is no ruling hand of God, and freedom turns to misery. In support of that caveat we have nothing less than the authority of the original Enlightenment thinkers themselves. Virtually all believed in a God who created the universe, and many were devout Christians to boot. Almost none was willing to abandon ethics to secular materialism. John Locke said that 'those who deny the existence of the Deity are not to be tolerated at all. Promises, covenants and oaths, which are the bonds of human society, can have no hold upon or sanctity for an atheist; for the taking away of God, even only in thought, dissolves all.' Robert Hooke, a great physicist of the seventeenth century, in composing a brief on the newly created Royal Society, wisely cautioned that the purpose of this quintessential Enlightenment organization should be 'To improve the knowledge of naturall things, and all useful Arts, Manufactures, Mechanick practises, Engynes and Inventions by Experiments—(not meddling with Divinity, Metaphysics, Moralls, Politicks, Grammar, Rhetorick or Logick).'

"These sentiments are just as prevalent among leading thinkers of the modern era, as well as a large minority of working scientists. They are reinforced by queasiness over the idea of organic evolution as espoused by Darwin. This keystone of empiricism presumes to reduce the Creation to the products of random mutations and environmental circumstance. Even George Bernard Shaw, an avowed atheist, responded to Darwinism with despair. He condemned its fatalism and the demoting of beauty, intelligence, honor, and aspiration to an abstract notion of blindly assembled matter. Many writers have suggested, not unfairly in my opinion, that such a sterile view of life, which reduces human beings to little more than intelligent animals, gave intellectual justification to the genocidal horrors of Nazism and communism.

"So surely there is something wrong with the reigning theory of evolution. Even if some form of genetic change occurs within species in the manner proclaimed by the new Darwinism, the full, stupendous complexity of modern organisms could not have been created by blind chance alone. Time and again in the history of science new evidence has overturned prevailing theories. Why are scientists so anxious to stay with autonomous evolution and to discount the possibility of an intelligent design instead? It is all very curious. Design would seem to be a simpler explanation than the random self-assembly of millions of kinds of organisms.

"Finally, theism gains compelling force in the case of the human mind and—I won't shrink from saying it—the immortal soul. Little wonder that a quarter or more of Americans reject totally the idea of any kind of human evolution, even in anatomy and physiology. Science pushed too far is science arrogant. Let it keep its proper place, as the God-given gift to understand His physical dominion."

The Empiricist

"I'll begin by freely acknowledging that religion has an overwhelming attraction for the human mind, and that religious conviction is largely beneficent. Religion rises from the innermost coils of the human spirit. It nourishes love, devotion, and, above all, hope. People hunger for the assurance it offers. I can think of nothing more emotionally compelling than the Christian doctrine that God incarnated himself in testimony of the sacredness of all human life, even of the slave, and that he died and rose again in promise of eternal life for everyone.

"But religious belief has another, destructive side, equaling the worst excesses of materialism. An estimated one hundred thousand belief systems have existed in history, and many have fostered ethnic and tribal wars. Each of the three great Western religions in particular expanded at one time or another in symbiosis with military aggression. Islam, which means "submission," was imposed by force of arms on large portions of the Middle East, Mediterranean perimeter, and southern Asia. Christianity dominated the New World as much by colonial expansion as by spiritual grace. It benefited from a historical accident: Europe, having been blocked to the East by the Muslim Arabs, turned west to occupy the Americas, whereupon the cross accompanied the sword in one campaign of enslavement and genocide after another.

"The Christian rulers had an instructive example to follow in the early history of Judaism. If we are to believe the Old Testament, the Israelites were ordered by God to wipe the promised land clean of heathen. 'Of these peoples which the LORD your God gives you as an inheritance, you shall let nothing that breathes remain alive, but you shall utterly destroy them: the Hittite and the Amorite and the Canaanite and the Perizzite and the Hivite and the Jebusite, as the LORD your God has commanded you,' thus reports Deuteronomy, 20:16–17. Over a hundred cities were consumed by fire and death, beginning with Joshua's campaign against Jericho and ending with David's assault on the ancient Jebusite stronghold of Jerusalem.

"I bring up these historical facts not to cast aspersions on present-day faiths but rather to cast light on their material origins and those of the ethical systems they sponsor. All great civilizations were spread by conquest, and among their chief beneficiaries were the religions validating them. No doubt membership in state-sponsored religions has always been deeply satisfying in many psychological dimensions, and spiritual wisdom has evolved to moderate the more barbaric tenets obeyed in the days of conquest. But every major religion today is a winner in the Darwinian struggle waged among cultures, and none ever flourished by tolerating its rivals. The swiftest road to success has always been sponsorship by a conquering state.

"To be fair, let me now put the matter of cause and effect straight. Religious exclusion and bigotry arise from tribalism, the belief in the innate superiority and special status of the in-group. Tribalism cannot be blamed on religion. The same causal sequence gave rise to totalitarian ideologies. The pagan *corpus mysticum* of Nazism and the class-warfare doctrine of Marxism-Leninism, both essentially dogmas of religions without God, were put to the service of tribalism, not the reverse. Neither would have been so fervently embraced if their devotees had not thought themselves chosen people, virtuous in their mission, surrounded by wicked enemies, and conquerors by right of blood and destiny. Mary Wollstonecraft correctly said, of male domination but extensible to all human behavior, 'No man chooses evil because it is evil; he only mistakes it for happiness, which is the good he seeks.'

"Conquest by a tribe requires that its members make sacrifices to the interests of the group, especially during conflict with competing groups. That is simply the expression of a primal rule of social life throughout the animal kingdom. It arises when loss of personal advantage by submission to the needs of the group is more than offset by gain in personal advantage due to the resulting success of the group. The human corollary is that selfish, prosperous people belonging to losing religions and ideologies are replaced by selfless, poor members of winning religions and ideologies. A better life later on, either an earthly paradise or resurrection in heaven, is the promised reward that cultures invent to justify the subordinating imperative of social existence. Repeated from one generation to the next, submission to the group and its moral codes is solidified in official doctrine and personal belief. But it is not ordained by God or plucked from the air as self-evident truth. It evolves as a necessary device of survival in social organisms.

"The most dangerous of devotions, in my opinion, is the one endemic to Christianity: *I was not born to be of this world.* With a second life waiting, suffering can be endured—especially in other people. The natural environment can be used up. Enemies of the faith can be savaged and suicidal martyrdom praised.

"Is it all an illusion? Well, I hesitate to call it that or, worse, a noble lie, the harsh phrase sometimes used by skeptics, but one has to admit that the objective evidence supporting it is not strong. No statistical proofs exist that prayer reduces illness and mortality, except perhaps through a psychogenic enhancement of the immune system; if it were otherwise the whole world would pray continuously. When two armies blessed by priests clash, one still loses. And when the martyr's righteous forebrain is exploded by the executioner's bullet and his mind disintegrates, what then? Can we safely assume that all those millions of neural circuits will be reconstituted in an immaterial state, so that the conscious mind carries on?

"The smart money in eschatology is on Blaise Pascal's wager: Live well but accept the faith. If there is an afterlife, the seventeenth-century French philosopher reasoned, the believer has a ticket to paradise and the best of both worlds. 'If I lost,' Pascal wrote, 'I would have lost little; if I won I would have gained eternal life.' Now think like an empiricist for a moment. Consider the wisdom of turning the wager around as follows: If fear and hope and reason dictate that you must accept the faith, do so, but treat this world as if there is none other.

"I know true believers will be scandalized by this line of argument. Their wrath falls on outspoken heretics, who are considered at best troublemakers and at worst traitors to the social order. But no evidence has been adduced that nonbelievers are less law-abiding or productive citizens than believers of the same socioeconomic class, or that they face death less bravely. A 1996 survey of American scientists (to take one respectable segment of society) revealed that 46 percent are atheists and 14 percent doubters or agnostics. Only 36 percent expressed a desire for immortality, and most of those only moderately so; 64 percent claimed no desire at all.

"True character arises from a deeper well than religion. It is the internalization of the moral principles of a society, augmented by those tenets personally chosen by the individual, strong enough to endure through trials of solitude and adversity. The principles are fitted together into what we call integrity, literally the integrated self, wherein personal decisions feel good and true. Character is in turn the enduring source of virtue. It stands by itself and excites admiration in others. It is not obedience to authority, and while it is often consistent with and reinforced by religious belief, it is not piety.

"Nor is science the enemy. It is the accumulation of humanity's organized, objective knowledge, the first medium devised able to unite people everywhere in common understanding. It favors no tribe or religion. It is the base of a truly democratic and global culture.

"You say that science cannot explain spiritual phenomena. Why not? The brain sciences are making important advances in the analysis of complex operations of the mind. There is no apparent reason why they cannot in time provide a material account of the emotions and ratiocination that compose spiritual thought.

"You ask where ethical precepts come from if not divine revelation. Consider the alternative empiricist hypothesis, that precepts and religious faith are entirely material products of the mind. For more than a thousand generations they have increased the survival and reproductive success of those who conformed to tribal faiths. There was more than enough time for epigenetic rules—hereditary biases of mental development—to evolve that generate moral and religious sentiments. Indoctrinability became an instinct.

"Ethical codes are precepts reached by consensus under the guidance of the innate rules of mental development. Religion is the ensemble of mythic narratives that explain the origin of a people, their destiny, and why they are obliged to subscribe to particular rituals and moral codes. Ethical and religious beliefs are created from the bottom up, from people to their culture. They do not come from the top down, from God or other nonmaterial source to the people by way of culture.

"Which hypothesis, transcendentalist or empiricist, fits the objective evidence best? The empiricist, by a wide margin. To the extent that this view is accepted, more emphasis in moral reasoning will be placed on social choice, and less on religious and ideological authority.

"Such a shift has in fact been occurring in Western cultures since the Enlightenment, but the pace has been very slow. Part of the reason is a gross insufficiency of knowledge needed to judge the full consequences of our moral decisions, especially for the long term, say a decade or more. We have learned a great deal about ourselves and the world in which we live, but need a great deal more to be fully wise. There is a temptation at every great crisis to yield to transcendental authority, and perhaps that is better for a while. We are still indoctrinable, we still are easily god-struck.

"Resistance to empiricism is also due to a purely emotional shortcoming of the mode of reasoning it promotes: It is bloodless. People need more than reason. They need the poetry of affirmation, they crave an authority greater than themselves at rites of passage and other moments of high seriousness. A majority desperately wish for the immortality the rituals seem to underwrite.

"Great ceremonies summon the history of a people in solemn remembrance. They showcase the sacred symbols. That is the enduring value of ceremony, which in all high civilizations has historically assumed a mostly religious form. Sacred symbols infiltrate the very bones of culture. They will take centuries to replace, if ever.

"So I may surprise you by granting this much: It would be a sorry day if we abandoned our venerated sacral traditions. It would be a tragic misreading of history to expunge *under God* from the American Pledge of Allegiance. Whether atheists or true believers, let oaths be taken with hand on the Bible, and may we continue to hear *So help me God*. Call upon priests and ministers and rabbis to bless civil ceremony with prayer, and by all means let us bow our heads in communal respect. Recognize that when introits and invocations prickle the skin we are in the presence of poetry, and the soul of the tribe, something that will outlive the particularities of sectarian belief, and perhaps belief in God itself.

"But to share reverence is not to surrender the precious self and obscure the true nature of the human race. We should not forget who we are. Our strength is in truth and knowledge and character, under whatever sign. Judaeo-Christians are told by Holy Scripture that pride goeth before destruction. I disagree; it's the reverse: Destruction goeth before pride. Empiricism has turned everything around in the formula. It has destroyed the giddying theory that we are special beings placed by a deity in the center of the universe in order to serve as the summit of Creation for the glory of the gods. We can be proud as a species because, having discovered that we are alone, we owe the gods very little. Humility is better shown to our fellow humans and the rest of life on this planet, on whom all hope really depends. And if any gods are paying attention, surely we have earned their admiration by making that discovery and setting out alone to accomplish the best of which we are capable."

THE ARGUMENT OF the empiricist, to repeat my earlier confession, is my own. It is far from novel, having roots that go back to Aristotle's *Nicomachean Ethics* and, in the beginning of the modern era, to David Hume's *A Treatise of Human Nature* (1739–40). The first clear evolutionary elaboration of it was by Darwin in *The Descent of Man* (1871).

The argument of the religious transcendentalist, on the other hand, is the one I first learned as a child in the Christian faith. I have reflected on it repeatedly since, and am by intellect and temperament bound to respect its ancient traditions.

It is also the case that religious transcendentalism is bolstered by secular transcendentalism, with which it has fundamental similarities. Immanuel Kant, judged by history the greatest of secular philosophers, addressed moral reasoning very much as a theologian. Human beings, he argued, are independent moral agents with a wholly free will capable of obeying or breaking moral law: "There is in man a power of self-determination, independent of any coercion through sensuous impulses." Our minds are subject to a categorical imperative, he said, of what our actions ought to be. The imperative is a good in itself alone, apart from all other considerations, and it can be recognized by this rule: "Act only on that maxim through which you wish also it become a universal law." Most important, and transcendental, *ought* has no place in nature. Nature, Kant said, is a system of cause and effect, while moral choice is a matter of free will, for which there is no cause and effect. In making moral choices, in rising above mere instinct, human beings transcend the realm of nature and enter a realm of freedom that belongs to them exclusively as rational creatures.

Now this formulation has a comforting feel to it, but it makes no sense at all in terms of either material or imaginable entities, which is why Kant, even apart from his tortured prose, is so hard to understand. Sometimes a concept is baffling not because it is profound but because it is wrong. It does not accord, we know now, with the evidence of how the brain works.

In *Principia Ethica* (1903) G. E. Moore, the founder of modern ethical philosophy, essentially agreed with Kant. Moral reasoning in his view cannot dip into psychology and the social sciences in order to locate ethical principles, because they yield only a causal picture and fail to illuminate the basis of moral justification. So to pass from the factual *is* to the normative *ought* commits a basic error of logic, which Moore called the naturalistic fallacy. John Rawls, in *A Theory of Justice* (1971), once again traveled the transcendental road. He offered the very plausible premise that justice be defined as fairness, which is to be accepted as an intrinsic good. It is the imperative we would follow if we had no starting information about our own status in life. But in making such an assumption, Rawls ventured no thought on where the human brain comes from or how it works. He offered no evidence that justice-as-fairness is consistent with human nature, hence practicable as a blanket premise. Probably it is, but how can we know except by blind trial-and-error?

I find it hard to believe that had Kant, Moore, and Rawls known modern biology and experimental psychology they would have reasoned as they did. Yet as this century closes, transcendentalism remains firm in the hearts not just of religious believers but also of countless scholars in the social sciences and humanities who, like Moore and Rawls before them, have chosen to insulate their thinking from the natural sciences.

Many philosophers will respond by saying, But wait! What are you saying? Ethicists don't need that kind of information. You really can't pass from *is* to *ought*. You are not allowed to describe a genetic predisposition and suppose that because it is part of human nature, it is somehow transformed into an ethical precept. We must put moral reasoning in a special category, and use transcendental guidelines as required.

No, we do not have to put moral reasoning in a special category, and use transcendental premises, because the posing of the naturalistic fallacy is itself a fallacy. For if *ought* is not *is*, what is? To translate *is* into *ought* makes sense if we attend to the objective meaning of ethical precepts. They are very unlikely to be ethereal messages outside humanity awaiting revelation, or independent truths vibrating in a nonmaterial dimension of the mind. They are more likely to be physical products of the brain and culture. From the consilient perspective of the natural sciences, they are no more than principles of the social contract hardened into rules and dictates, the behavioral codes that members of a society fervently wish others to follow and are willing to accept themselves for the common good. Precepts are the extreme in a scale of agreements that range from casual assent to public sentiment to law to that part of the canon considered unalterable and sacred. The scale applied to adultery might read as follows:

Let's not go further; it doesn't feel right, and it would lead to trouble. (*We probably ought not.*)

Adultery not only causes feelings of guilt, it is generally disapproved of by society, so these are other reasons to avoid it. (*We ought not.*)

Adultery isn't just disapproved of, it's against the law. (*We almost certainly ought not.*)

God commands that we avoid this mortal sin. (*We absolutely ought not.*)

In transcendental thinking the chain of causation runs downward from the given *ought* in religion or natural law through jurisprudence to education and finally to individual choice. The argument from transcendentalism takes the following general form: *There is a supreme principle, either divine or intrinsic in the order of nature, and we will be wise to learn about it and find the means to conform to it.* Thus John Rawls opens *A Theory of Justice* with a proposition he regards as irrevocable: "In a just society the liberties of equal citizenship are taken as settled; the rights secured by justice are not subject to political bargaining or to the calculus of social interests." As many critiques have made clear, that premise can lead to many unhappy consequences when applied to the real world, including the tightening of social control and decline of personal initiative. A very different premise therefore is suggested by Robert Nozick in *Anarchy, State, and Utopia* (1974): "Individuals have rights, and there are things no person or group may do to them (without violating their rights). So strong and far-reaching are these rights that they raise the question of what, if anything, the state and its officials may do." Rawls would point us toward egalitarianism regulated by the state, Nozick toward libertarianism in a minimalist state.

The empiricist view in contrast, searching for an origin of ethical reasoning that can be objectively studied, reverses the chain of causation. The individual is seen as predisposed biologically to make certain choices. By cultural evolution some of the choices are hardened into precepts, then laws, and if the predisposition or coercion is strong enough, a belief in the command of God or the natural order of the universe. The general empiricist principle takes this form: *Strong innate feeling and historical experience cause certain actions to be preferred; we have experienced them, and weighed their consequences, and agree to conform with codes that express them. Let us take an oath upon the codes, invest our personal honor in them, and suffer punishment for their violation.* The empiricist view concedes that moral codes are devised to conform to some drives of human nature and to suppress others. *Ought* is not the translation of human nature but of the public will, which can be made increasingly wise and stable through the understanding of the needs and pitfalls of human nature. It recognizes that the strength of commitment can wane as a result of new knowledge and experience, with the result that certain rules may be desacralized, old laws rescinded, and behavior that was once prohibited freed. It also recognizes that for the same reason new moral codes may need to be devised, with the potential in time of being made sacred.

IF THE EMPIRICIST WORLD VIEW is correct, *ought* is just shorthand for one kind of factual statement, a word that denotes what society first chose (or was coerced) to do, and then codified. The naturalistic fallacy is thereby reduced to the naturalistic dilemma. The solution of the dilemma is not difficult. It is this: *Ought* is the product of a material process. The solution points the way to an objective grasp of the origin of ethics.

A few investigators are now embarked on just such a foundational inquiry. Most agree that ethical codes have arisen by evolution through the interplay of biology and culture. In a sense they are reviving the idea of moral sentiments developed in the eighteenth century by the British empiricists Francis Hutcheson, David Hume, and Adam Smith.

By moral sentiments is now meant moral instincts as defined by the modern behavioral sciences, subject to judgment according to their consequences. The sentiments are thus derived from epigenetic rules, hereditary biases in mental development, usually conditioned by emotion, that influence concepts and decisions made from them. The primary origin of the moral instincts is the dynamic relation between cooperation and defection. The essential ingredient for the molding of the instincts during genetic evolution in any species is intelligence high enough to judge and manipulate the tension generated by the dynamism. That level of intelligence allows the building of complex mental scenarios well into the future, as I described in the earlier chapter on the mind. It occurs, so far as known, only in human beings and perhaps their closest relatives among the higher apes.

A way of envisioning the hypothetical earliest stages of moral evolution is provided by game theory, particularly the solutions to the famous Prisoner's Dilemma. Consider the following typical scenario of the Dilemma. Two gang members have been arrested for murder and are being questioned separately. The evidence against them is strong but not compelling. The first gang member believes that if he turns state's witness, he will be granted immunity and his partner will be sentenced to life in prison. But he is also aware that his partner has the same option. That is the dilemma. Will the two gang members independently defect so that both take the hard fall? They will not, because they agreed in advance to remain silent if caught. By doing so, both hope to be convicted on a lesser charge or escape punishment altogether. Criminal gangs have turned this principle of calculation into an ethical precept: Never rat on another member; always be a stand-up guy. Honor does exist among thieves. If we view the gang as a society of sorts, the code is the same as that of a captive soldier in wartime obliged to give only name, rank, and serial number.

In one form or another, comparable dilemmas that are solvable by cooperation occur constantly and everywhere in daily life. The payoff is variously money, status, power, sex, access, comfort, and health. Most of these proximate rewards are converted into the universal bottom line of Darwinian genetic fitness: greater longevity and a secure, growing family.

And so it has likely always been. Imagine a Paleolithic hunter band, say composed of five men. One hunter considers breaking away from the others to look for an antelope on his own. If successful he will gain a large quantity of meat and hide, five times greater than if he stays with the band and they are successful. But he knows from experience that his chances of success alone are very low, much less than the chances of a band of five working together. In addition, whether successful alone or not, he will suffer animosity from the others for lessening their own prospects. By custom the band members remain together and share the animals they kill equitably. So the hunter stays. He also observes good manners while doing so, especially if he is the one who makes the kill. Boastful pride is condemned because it rips the delicate web of reciprocity.

Now suppose that human propensities to cooperate or defect are heritable: Some members are innately more cooperative, others less so. In this respect moral aptitude would simply be like almost all other mental traits studied to date. Among traits with documented heritability, those closest to moral aptitude are empathy to the distress of others and certain processes of attachment between infants and their caregivers. To the heritability of moral aptitude add the abundant evidence of history that cooperative individuals generally survive longer and leave more offspring. It is to be expected that in the course of evolutionary history, genes predisposing people toward cooperative behavior would have come to predominate in the human population as a whole.

Such a process repeated through thousands of generations inevitably gave birth to the moral sentiments. With the exception of stone psychopaths (if any truly exist), these instincts are vividly experienced by every person variously as conscience, self-respect, remorse, empathy, shame, humility, and moral outrage. They bias cultural evolution toward the conventions that express the universal moral codes of honor, patriotism, altruism, justice, compassion, mercy, and redemption.

The dark side to the inborn propensity to moral behavior is xenophobia. Because personal familiarity and common interest are vital in social transactions, moral sentiments evolved to be selective. And so it has ever been, and so it will ever be. People give trust to strangers with effort, and true compassion is a commodity in chronically short supply. Tribes cooperate only through carefully defined treaties and other conventions. They are quick to imagine themselves victims of conspiracies by competing groups, and they are prone to dehumanize and murder their rivals during periods of severe conflict. They cement their own group loyalties by means of sacred symbols and ceremonies. Their mythologies are filled with epic victories over menacing enemies.

The complementary instincts of morality and tribalism are easily manipulated. Civilization has made them more so. Only ten thousand years ago, a tick in geological time, when the agricultural revolution began in the Middle East, in China, and in Mesoamerica, populations increased in density tenfold over those of hunter-gatherer societies. Families settled on small plots of land, villages proliferated, and labor was finely divided as a growing minority of the populace specialized as craftsmen, traders, and soldiers. The rising agricultural societies, egalitarian at first, became hierarchical. As chiefdoms and then states thrived on agricultural surpluses, hereditary rulers and priestly castes took power. The old ethical codes were transformed into coercive regulations, always to the advantage of the ruling classes. About this time the idea of law-giving gods originated. Their commands lent the ethical codes overpowering authority, once again—no surprise—to the favor of the rulers.

Because of the technical difficulty of analyzing such phenomena in an objective manner, and because people resist biological explanations of their higher cortical functions in the first place, very little progress has been made in the biological exploration of the moral sentiments. Even so, it is an astonishing circumstance that the study of ethics has advanced so little since the nineteenth century. As a result the most distinguishing and vital qualities of the human species remain a blank space on the scientific map. I think it an error to pivot discussions of ethics upon the free-standing assumptions of contemporary philosophers who have evidently never given thought to the evolutionary origin and material functioning of the human brain. In no other domain of the humanities is a union with the natural sciences more urgently needed.

When the ethical dimension of human nature is at last fully opened to such exploration, the innate epigenetic rules of moral reasoning will probably not prove to be aggregated into simple instincts such as bonding, cooperativeness, or altruism. Instead, the rules most probably will turn out to be an ensemble of many algorithms whose interlocking activities guide the mind across a landscape of nuanced moods and choices.

Such a prestructured mental world may at first seem too complicated to have been created by autonomous genetic evolution alone. But all the evidence of biology suggests that just this process was enough to spawn the millions of species of life surrounding us. Each kind of animal is furthermore guided through its life cycle by unique and often elaborate sets of instinctual algorithms, many of which are beginning to yield to genetic and neurobiological analyses. With all these examples before us, it is not unreasonable to conclude that human behavior originated the same way.

MEANWHILE, the mélanges of moral reasoning employed by modern societies are, to put the matter simply, a mess. They are chimeras, composed of odd parts stuck together. Paleolithic egalitarian and tribalistic instincts are still firmly installed. As part of the genetic foundation of human nature, they cannot be replaced. In some cases, such as quick hostility to strangers and competing groups, they have become generally ill-adapted and persistently dangerous. Above the fundamental instincts rise superstructures of arguments and rules that accommodate the novel institutions created by cultural evolution. These accommodations, which reflect the attempt to maintain order and further tribal interests, have been too volatile to track by genetic evolution; they are not yet in the genes.

Little wonder, then, that ethics is the most publicly contested of all philosophical enterprises. Or that political science, which at foundation is primarily the study of applied ethics, is so frequently problematic. Neither is informed by anything that would be recognizable as authentic theory in the natural sciences. Both ethics and political science lack a foundation of verifiable knowledge of human nature sufficient to produce cause-and-effect predictions and sound judgments based on them. Surely it will be prudent to pay closer attention to the deep springs of ethical behavior. The greatest void in knowledge in such a venture is the biology of the moral sentiments. In time this subject can be understood, I believe, by paying attention to the following topics.

• *The definition of the moral sentiments:* first by precise descriptions from experimental psychology, then by analysis of the underlying neural and endocrine responses.

• *The genetics of the moral sentiments:* most easily approached through measurements of the heritability of the psychological and physiological processes of ethical behavior, and eventually, with difficulty, by identification of the prescribing genes.

• *The development of the moral sentiments as products of the interactions of genes and environment.* The research is most effective when conducted at two levels: the histories of ethical systems as part of the emergence of different cultures, and the cognitive development of individuals living in a variety of cultures. Such investigations are already well along in anthropology and psychology. In the future they will be augmented by contributions from biology.

• *The deep history of the moral sentiments:* why they exist in the first place, presumably by their contributions to survival and reproductive success during the long periods of prehistoric time in which they genetically evolved.

From a convergence of these several approaches, the true origin and meaning of ethical behavior may come into focus. If so, a more certain measure can then be taken of the strengths and flexibility of the epigenetic rules composing the various moral sentiments. From that knowledge, it should be possible to adapt the ancient moral sentiments more wisely to the swiftly changing conditions of modern life into which, willy-nilly and largely in ignorance, we have plunged ourselves.

Then new answers might be found for the truly important questions of moral reasoning. How can the moral instincts be ranked? Which are best subdued and to what degree, which validated by law and symbol? How can precepts be left open to appeal under extraordinary circumstances? In the new understanding can be located the most effective means for reaching consensus. No one can guess the form the agreements will take. The process, however, can be predicted with assurance. It will be democratic, weakening the clash of rival religions and ideologies. History is moving decisively in that direction, and people are by nature too bright and too contentious to abide anything else. And the pace can be confidently predicted: Change will come slowly, across generations, because old beliefs die hard even when demonstrably false.

THE SAME REASONING that aligns ethical philosophy with science can also inform the study of religion. Religions are analogous to superorganisms. They have a life cycle. They are born, they grow, they compete, they reproduce, and, in the fullness of time, most die. In each of these phases religions reflect the human organisms that nourish them. They express a primary rule of human existence, that whatever is necessary to sustain life is also ultimately biological.

Successful religions typically begin as cults, which then increase in power and inclusiveness until they achieve tolerance outside the circle of believers. At the core of each religion is a creation myth, which explains how the world began and how the chosen people—those subscribing to the belief system—arrived at its center. There is often a mystery, a set of secret instructions and formulas available only to hierophants who have worked their way to a higher state of enlightenment. The medieval Jewish cabala, the trigradal system of Freemasonry, and the carvings on Australian Aboriginal spirit sticks are examples of such arcana. Power radiates from the center, gathering converts and binding followers to the group. Sacred places are designated where the gods can be importuned, rites observed, and miracles witnessed.

The devotees of the religion compete as a tribe with those of other religions. They harshly resist the dismissal of their beliefs by rivals. They venerate self-sacrifice in defense of the religion.

The tribalistic roots of religion and those of moral reasoning are similar and may be identical. Religious rites, as evidenced by burial ceremonies, are very old. In the late Paleolithic period of Europe and the Middle East, it appears that bodies were sometimes placed in shallow graves sprinkled with ochre or blossoms, and it is easy to imagine ceremonies performed there that invoked spirits and gods. But, as theoretical deduction and the evidence suggest, the primitive elements of moral behavior are far older than Paleolithic ritual. Religion arose on an ethical foundation, and it has probably always been used in one manner or another to justify moral codes.

The formidable influence of the religious drive is based on far more, however, than just the validation of morals. A great subterranean river of the mind, it gathers strength from a broad spread of tributary emotions. Foremost among them is the survival instinct. "Fear," as the Roman poet Lucretius said, "was the first thing on earth to make gods." Our conscious minds hunger for a permanent existence. If we cannot have everlasting life of the body, then absorption into some immortal whole will serve. *Anything* will serve, as long as it gives the individual meaning and somehow stretches into eternity that swift passage of the mind and spirit lamented by St. Augustine as the short day of time.

The understanding and control of life is another source of religious power. Doctrine draws on the same creative springs as science and the arts, its aim being the extraction of order from the mysteries of the material world. To explain the meaning of life it spins mythic narratives of the tribal history, populating the cosmos with protective spirits and gods. The existence of the supernatural, if accepted, testifies to the existence of that other world so desperately desired.

Religion is also empowered mightily by its principal ally, tribalism. The shamans and priests implore us, in somber cadence, *Trust in the sacred rituals, become part of the immortal force, you are one of us. As your life unfolds, each step has mystic significance that we who love you will mark with a solemn rite of passage, the last to be performed when you enter that second world free of pain and fear.*

If the religious mythos did not exist in a culture, it would be quickly invented, and in fact it has been everywhere, thousands of times through history. Such inevitability is the mark of instinctual behavior in any species. That is, even when learned, it is guided toward certain states by emotion-driven rules of mental development. To call religion instinctive is not to suppose any particular part of its mythos is untrue, only that its sources run deeper than ordinary habit and are in fact hereditary, urged into birth through biases in mental development encoded in the genes.

I have argued in previous chapters that such biases are to be expected as a usual consequence of the brain's genetic evolution. The logic applies to religious behavior, with the added twist of tribalism. There is a hereditary selective advantage to membership in a powerful group united by devout belief and purpose. Even when individuals subordinate themselves and risk death in common cause, their genes are more likely to be transmitted to the next generation than are those of competing groups who lack equivalent resolve.

The mathematical models of population genetics suggest the following rule in the evolutionary origin of such altruism. If the reduction of survival and reproduction of individuals due to genes for altruism is more than offset by the increased probability of survival of the group due to the altruism, the altruism genes will rise in frequency throughout the entire population of competing groups. Put as concisely as possible: The individual pays, his genes and tribe gain, altruism spreads.

LET ME NOW SUGGEST a still deeper significance of the empiricist theory of the origin of ethics and religion. If empiricism is disproved, and transcendentalism is compellingly upheld, the discovery would be quite simply the most consequential in human history. That is the burden laid upon biology as it draws close to the humanities. If the objective evidence accumulated by biology upholds empiricism, consilience succeeds in the most problematic domains of human behavior and is likely to apply everywhere. But if the evidence contradicts empiricism in any part, universal consilience fails and the division between science and the humanities will remain permanent all the way to their foundations.

The matter is still far from resolved. But empiricism, as I have argued, is well supported thus far in the case of ethics. The objective evidence for or against it in religion is weaker, but at least still consistent with biology. For example, the emotions that accompany religious ecstasy clearly have a neurobiological source. At least one form of brain disorder is associated with hyperreligiosity, in which cosmic significance is given to almost everything, including trivial everyday events. Overall it is possible to imagine the biological construction of a mind with religious beliefs, although that alone does not dismiss transcendentalism or prove the beliefs themselves to be untrue.

Equally important, much if not all religious behavior could have arisen from evolution by natural selection. The theory fits—crudely. The behavior includes at least some aspects of belief in gods. Propitiation and sacrifice, which are near-universals of religious practice, are acts of submission to a dominant being. They are one kind of a dominance hierarchy, which is a general trait of organized mammalian societies. Like humans, animals use elaborate signals to advertise and maintain their rank in the hierarchy. The details vary among species but also have consistent similarities across the board, as the following two examples will illustrate.

In packs of wolves the dominant animal walks erect and "proud," stiff-legged, deliberately paced, with head, tail, and ears up, and stares freely and casually at others. In the presence of rivals, the dominant animal bristles its pelt while curling its lips to show teeth, and it takes first choice in food and space. A subordinate uses opposite signals. It turns away from the dominant individual while lowering its head, ears, and tail, and it keeps its fur sleeked and teeth covered. It grovels and slinks, and yields food and space when challenged.

In troops of rhesus monkeys, the alpha male of the troop is remarkably similar in mannerisms to a dominant wolf. He keeps his head and tail up, walks in a deliberate, "regal" manner while casually staring at others. He climbs nearby objects to maintain height above his rivals. When challenged he stares hard at the opponent with mouth open—signaling aggression, not surprise—and sometimes slaps the ground with open palms to signal his readiness to attack. The male or female subordinate affects a furtive walk, holding its head and tail down, turning away from the alpha and other higher-ranked individuals. It keeps its mouth shut except for a fear grimace, and when challenged makes a cringing retreat. It yields space and food and, in the case of males, estrous females.

My point is the following. Behavioral scientists from another planet would notice immediately the semiotic resemblance between animal submissive behavior on the one hand and human obeisance to religious and civil authority on the other. They would point out that the most elaborate rites of obeisance are directed at the gods, the hyperdominant if invisible members of the human group. And they would conclude, correctly, that in baseline social behavior, not just in anatomy, *Homo sapiens* has only recently diverged in evolution from a nonhuman primate stock.

Countless studies of animal species, with instinctive behavior unobscured by cultural elaboration, have shown that membership in dominance orders pays off in survival and lifetime reproductive success. That is true not just for the dominant individuals, but for the subordinates as well. Membership in either class gives animals better protection against enemies and better access to food, shelter, and mates than does solitary existence. Furthermore, subordination in the group is not necessarily permanent. Dominant individuals weaken and die, and as a result some of the underlings advance in rank and appropriate more resources.

It would be surprising to find that modern humans had managed to erase the old mammalian genetic programs and devise other means of distributing power. All the evidence suggests that they have not. True to their primate heritage, people are easily seduced by confident, charismatic leaders, especially males. That predisposition is strongest in religious organizations. Cults form around such leaders. Their power grows if they can persuasively claim special access to the supremely dominant, typically male figure of God. As cults evolve into religions, the image of the supreme being is reinforced by myth and liturgy. In time the authority of the founders and their successors is graven in sacred texts. Unruly subordinates, known as "blasphemers," are squashed.

The symbol-forming human mind, however, never stays satisfied with raw apish feeling in any emotional realm. It strives to build cultures that are maximally rewarding in every dimension. In religion there is ritual and prayer to contact the supreme being directly, consolation from coreligionists to soften otherwise unbearable grief, explanations of the unexplainable, and the oceanic sense of communion with the larger whole that otherwise surpasses understanding.

Communion is the key, and hope rising from it eternal; out of the dark night of the soul there is the prospect of a spiritual journey to the light. For a special few the journey can be taken in this life. The mind reflects in certain ways in order to reach ever higher levels of enlightenment until finally, when no further progress is possible, it enters a mystical union with the whole. Within the great religions, such enlightenment is expressed by the Hindu samadhi, Buddhist Zen satori, Sufi fana, Taoist wu-wei, and Pentecostal Christian rebirth. Something like it is also experienced by hallucinating preliterate shamans. What all these celebrants evidently feel (as I once felt to some degree as a reborn evangelical) is hard to put in words, but Willa Cather came as close as possible in a single sentence. "That is happiness," her fictional narrator says in *My Ántonia*, "to be dissolved into something complete and great."

Of course that is happiness, to find the godhead, or to enter the wholeness of Nature, or otherwise to grasp and hold on to something ineffable, beautiful, and eternal. Millions seek it. They feel otherwise lost, adrift in a life without ultimate meaning. Their predicament is summarized in an insurance advertisement of 1997: *The year is 1999. You are dead. What do you do now?* They enter established religions, succumb to cults, dabble in New Age nostrums. They push *The Celestine Prophecy* and other junk attempts at enlightenment onto the bestseller lists.

Perhaps, as I believe, it can all eventually be explained as brain circuitry and deep, genetic history. But this is not a subject that even the most hardened empiricist should presume to trivialize. The idea of the mystical union is an authentic part of the human spirit. It has occupied humanity for millennia, and it raises questions of utmost seriousness for transcendentalists and scientists alike. What road, we ask, was traveled, what destination reached by the mystics of history?

No one has described the true journey with greater clarity than the great Spanish mystic St. Teresa of Avila, who in her 1563–65 memoir describes the steps she took to attain divine union by means of prayer. At the beginning of the narrative she moves beyond ordinary prayers of devotion and supplication to the second level, the prayer of the quiet. There her mind gathers its faculties inward in order to give "a simple consent to become the prisoner of God." A deep sense of consolation and peace descends upon her when the Lord supplies the "water of grand blessings and graces." Her mind then ceases to care for earthly things.

In the third state of prayer the saint's spirit, "drunk with love," is concerned only with thoughts of God, who controls and animates it.

> *O my King, seeing that I am now, while writing this, still under the power of this heavenly madness . . . grant, I beseech Thee, that all those with whom I may have to converse may become mad through Thy love, or let me converse with none, or order it that I may have nothing to do in the world, or take me away from it.*

In the fourth state of prayer St. Teresa of Avila attains the mystical union:

> *There is no sense of anything, only fruition . . . the senses are all occupied in this function in such a way that not one of them is at liberty. . . . The soul, while thus seeking after God, is conscious, with a joy excessive and sweet, that it is, as it were, utterly fainting away in a trance; breathing, and all the bodily strength fail it. The soul is dissolved into that of God, and with the union at last comes comprehension of the graces bestowed by Him.*

FOR MANY the urge to believe in transcendental existence and immortality is overpowering. Transcendentalism, especially when reinforced by religious faith, is psychically full and rich; it feels somehow *right*. In comparison empiricism seems sterile and inadequate. In the quest for ultimate meaning, the transcendentalist route is much easier to follow. That is why, even as empiricism is winning the mind, transcendentalism continues to win the heart. Science has always defeated religious dogma point by point when the two have conflicted. But to no avail. In the United States there are fifteen million Southern Baptists, the largest denomination favoring literal interpretation of the Christian Bible, but only five thousand members of the American Humanist Association, the leading organization devoted to secular and deistic humanism.

Still, if history and science have taught us anything, it is that passion and desire are not the same as truth. The human mind evolved to believe in the gods. It did not evolve to believe in biology. Acceptance of the supernatural conveyed a great advantage throughout prehistory, when the brain was evolving. Thus it is in sharp contrast to biology, which was developed as a product of the modern age and is not underwritten by genetic algorithms. The uncomfortable truth is that the two beliefs are not factually compatible. As a result those who hunger for both intellectual and religious truth will never acquire both in full measure.

Meanwhile, theology tries to resolve the dilemma by evolving sciencelike toward abstraction. The gods of our ancestors were divine human beings. The Egyptians, as Herodotus noted, represented them as Egyptian (often with body parts of Nilotic animals), and the Greeks represented them as Greeks. The great contribution of the Hebrews was to combine the entire pantheon into a single person, Yahweh—a patriarch appropriate to desert tribes—and to intellectualize His existence. No graven images were allowed. In the process, they rendered the divine presence less tangible. And so in biblical accounts it came to pass that no one, not even Moses approaching Yahweh in the burning bush, could look upon His face. In time the Jews were prohibited even from pronouncing His true full name. Nevertheless, the idea of a theistic God, omniscient, omnipotent, and closely involved in human affairs, has persisted to the present day as the dominant religious image of Western culture.

During the Enlightenment a growing number of liberal Judaeo-Christian theologians, wishing to accommodate theism to a more rationalist view of the material world, moved away from God as a literal person. Baruch Spinoza, the preeminent Jewish philosopher of the seventeenth century, visualized the deity as a transcendent substance present everywhere in the universe. *Deus sive natura*, God or nature, he declared, they are interchangeable. For his philosophical pains he was banished from Amsterdam under a comprehensive anathema, combining all curses in the book. The risk of heresy notwithstanding, the depersonalization of God has continued steadily into the modern era.

For Paul Tillich, one of the most influential Protestant theologians of the twentieth century, the assertion of the existence of God-as-person is not false; it is just meaningless. Among many of the most liberal contemporary thinkers, the denial of a concrete divinity takes the form of process theology. Everything in this most extreme of ontologies is part of a seamless and endlessly complex web of unfolding relationships. God is manifest in everything.

Scientists, the roving scouts of the empiricist movement, are not immune to the idea of God. Those who favor it often lean toward some form of process theology. They ask this question: When the real world of space, time, and matter is well enough known, will that knowledge reveal the Creator's presence? Their hopes are vested in the theoretical physicists who pursue the goal of the final theory, the Theory of Everything, T.O.E., a system of interlocking equations that describe all that can be learned of the forces of the physical universe. T.O.E. is a "beautiful" theory, as Steven Weinberg has called it in his important essay *Dreams of a Final Theory*. Beautiful because it will be elegant, expressing the possibility of unending complexity with minimal laws, and symmetric, because it will hold invariant through all space and time. And inevitable, meaning that once stated no part can be changed without invalidating the whole. All surviving subtheories can be fitted into it permanently, in the manner in which Einstein described his own contribution, the general theory of relativity. "The chief attraction of the theory," Einstein said, "lies in its logical completeness. If a single one of the conclusions drawn from it proves wrong, it must be given up; to modify it without destroying the whole structure seems to be impossible."

The prospect of a final theory by the most mathematical of scientists might seem to signal the approach of a new religious awakening. Stephen Hawking, yielding to the temptation in *A Brief History of Time* (1988), declared that this scientific achievement would be the ultimate triumph of human reason, "for then we would know the mind of God."

Well—perhaps, but I doubt it. Physicists have already laid in place a large part of the final theory. We know the trajectory; we can see roughly where it is headed. But there will be no religious epiphany, at least none recognizable to the authors of Holy Scripture. Science has taken us very far from the personal God who once presided over Western civilization. It has done little to satisfy our instinctual hunger so poignantly expressed by the psalmist:

> *Man liveth his days like a shadow, and he disquieteth himself in vain with prideful delusions; his treasures, he knoweth not who shall gather them. Now, Lord, what is my comfort? My hope is in thee.*

THE ESSENCE OF humanity's spiritual dilemma is that we evolved genetically to accept one truth and discovered another. Is there a way to erase the dilemma, to resolve the contradictions between the transcendentalist and empiricist world views?

No, unfortunately, there is not. Furthermore, a choice between them is unlikely to remain arbitrary forever. The assumptions underlying the two world views are being tested with increasing severity by cumulative verifiable knowledge about how the universe works, from atom to brain to galaxy. In addition, the harsh lessons of history have made it clear that one code of ethics is not as good—at least, not as durable—as another. The same is true of religions. Some cosmologies are factually less correct than others, and some ethical precepts are less workable.

There is a biologically based human nature, and it is relevant to ethics and religion. The evidence shows that because of its influence, people can be readily educated to only a narrow range of ethical precepts. They flourish within certain belief systems, and wither under others. We need to know exactly why.

To that end I will be so presumptuous as to suggest how the conflict between the world views will most likely be settled. The idea of a genetic, evolutionary origin of moral and religious beliefs will be tested by the continuance of biological studies of complex human behavior. To the extent that the sensory and nervous systems appear to have evolved by natural selection or at least some other purely material process, the empiricist interpretation will be supported. It will be further supported by verification of gene-culture coevolution, the essential linking process described in earlier chapters.

Now consider the alternative. To the extent that ethical and religious phenomena do *not* appear to have evolved in a manner congenial to biology, and especially to the extent that such complex behavior cannot be linked to physical events in the sensory and nervous systems, the empiricist position will have to be abandoned and a transcendentalist explanation accepted.

For centuries the writ of empiricism has been spreading into the ancient domain of transcendentalist belief, slowly at the start but quickening in the scientific age. The spirits our ancestors knew intimately first fled the rocks and trees, then the distant mountains. Now they are in the stars, where their final extinction is possible. *But we cannot live without them.* People need a sacred narrative. They must have a sense of larger purpose, in one form or other, however intellectualized. They will refuse to yield to the despair of animal mortality. They will continue to plead in company with the psalmist, *Now, Lord, what is my comfort?* They will find a way to keep the ancestral spirits alive.

If the sacred narrative cannot be in the form of a religious cosmology, it will be taken from the material history of the universe and the human species. That trend is in no way debasing. The true evolutionary epic, retold as poetry, is as intrinsically ennobling as any religious epic. Material reality discovered by science already possesses more content and grandeur than all religious cosmologies combined. The continuity of the human line has been traced through a period of deep history a thousand times older than that conceived by the Western religions. Its study has brought new revelations of great moral importance. It has made us realize that *Homo sapiens* is far more than a congeries of tribes and races. We are a single gene pool from which individuals are drawn in each generation and into which they are dissolved the next generation, forever united as a species by heritage and a common future. Such are the conceptions, based on fact, from which new intimations of immortality can be drawn and a new mythos evolved.

Which world view prevails, religious transcendentalism or scientific empiricism, will make a great difference in the way humanity claims the future. During the time the matter is under advisement, an accommodation can be reached if the following overriding facts are realized. On the one side, ethics and religion are still too complex for present-day science to explain in depth. On the other, they are far more a product of autonomous evolution than hitherto conceded by most theologians. Science faces in ethics and religion its most interesting and possibly humbling challenge, while religion must somehow find the way to incorporate the discoveries of science in order to retain credibility. Religion will possess strength to the extent that it codifies and puts into enduring, poetic form the highest values of humanity consistent with empirical knowledge. That is the only way to provide compelling moral leadership. Blind faith, no matter how passionately expressed, will not suffice. Science for its part will test relentlessly every assumption about the human condition and in time uncover the bedrock of the moral and religious sentiments.

The eventual result of the competition between the two world views, I believe, will be the secularization of the human epic and of religion itself. However the process plays out, it demands open discussion and unwavering intellectual rigor in an atmosphere of mutual respect.

CHAPTER 12

To What End?

It is the custom of scholars when addressing behavior and culture to speak variously of anthropological explanations, psychological explanations, biological explanations, and other explanations appropriate to the perspectives of individual disciplines. I have argued that there is intrinsically only one class of explanation. It traverses the scales of space, time, and complexity to unite the disparate facts of the disciplines by consilience, the perception of a seamless web of cause and effect.

For centuries consilience has been the mother's milk of the natural sciences. Now it is wholly accepted by the brain sciences and evolutionary biology, the disciplines best poised to serve in turn as bridges to the social sciences and humanities. There is abundant evidence to support and none absolutely to refute the proposition that consilient explanations are congenial to the entirety of the great branches of learning.

The central idea of the consilience world view is that all tangible phenomena, from the birth of stars to the workings of social institutions, are based on material processes that are ultimately reducible, however long and tortuous the sequences, to the laws of physics. In support of this idea is the conclusion of biologists that humanity is kin to all other life forms by common descent. We share essentially the same DNA genetic code, which is transcribed into RNA and translated into proteins with the same amino acids. Our anatomy places us among the Old World monkeys and apes. The fossil record shows our immediate ancestor to be either *Homo ergaster* or *Homo erectus*. It suggests that the point of our origin was Africa about two hundred thousand years ago. Our hereditary human nature, which evolved during hundreds of millennia before and afterward, still profoundly affects the evolution of culture.

These considerations do not devalue the determining role of chance in history. Small accidents can have big consequences. The character of individual leaders can mean the difference between war and peace; one technological invention can change an economy. The main thrust of the consilience world view instead is that culture and hence the unique qualities of the human species will make complete sense only when linked in causal explanation to the natural sciences. Biology in particular is the most proximate and hence relevant of the scientific disciplines.

I know that such reductionism is not popular outside the natural sciences. To many scholars in the social sciences and humanities it is a vampire in the sacristy. So let me hasten to dispel the profane image that causes this reaction. As the century closes, the focus of the natural sciences has begun to shift away from the search for new fundamental laws and toward new kinds of synthesis—"holism," if you prefer—in order to understand complex systems. That is the goal, variously, in studies of the origin of the universe, the history of climate, the functioning of cells, the assembly of ecosystems, and the physical basis of mind. The strategy that works best in these enterprises is the construction of coherent cause-and-effect explanations across levels of organization. Thus the cell biologist looks inward and downward to ensembles of molecules, and the cognitive psychologist to patterns of aggregate nerve cell activity. Accidents, when they happen, are rendered understandable.

No compelling reason has ever been offered why the same strategy should not work to unite the natural sciences with the social sciences and humanities. The difference between the two domains is in the magnitude of the problem, not the principles needed for its solution. The human condition is the most important frontier of the natural sciences. Conversely, the material world exposed by the natural sciences is the most important frontier of the social sciences and humanities. The consilience argument can be distilled as follows: The two frontiers are the same.

The map of the material world, including human mental activity, can be thought a sprinkling of charted terrain separated by blank expanses that are of unknown extent yet accessible to coherent interdisciplinary research. Much of what I have offered in earlier chapters has been "gap analysis," a sketch of the position of the blank spaces, and an account of the efforts of scholars to explore them. The gaps of greatest potential include the final unification of physics, the reconstruction of living cells, the assembly of ecosystems, the coevolution of genes and culture, the physical basis of mind, and the deep origins of ethics and religion.

If the consilience world view is correct, the traverse of the gaps will be a Magellanic voyage that eventually encircles the whole of reality. But that view could be wrong: The exploration may be proceeding across an endless sea. The current pace is such that we may find out which of the two images is correct within a few decades. But even if the journey is Magellanic, and even if the boldest excursions of circumscription consequently taper off, so that the broad outline of material existence is well defined, we will still have mastered only an infinitesimal fraction of the internal detail. Exploration will go on in a profusion of scholarly disciplines. There are also the arts, which embrace not only all physically possible worlds but also all conceivable worlds innately interesting and congenial to the nervous system and thus, in the uniquely human sense, true.

Placed in this broader context—of existence coherent enough to be understood in a single system of explanation, yet still largely unexplored—the ambitions of the natural sciences might be viewed in a more favorable light by nonscientists. Nowadays, as polls have repeatedly shown, most people, at least in the United States, respect science but are baffled by it. They don't understand it, they prefer science fiction, they take fantasy and pseudoscience like stimulants to jolt their cerebral pleasure centers. We are still Paleolithic thrill seekers, preferring *Jurassic Park* to the Jurassic Era, and UFOs to astrophysics.

The productions of science, other than medical breakthroughs and the sporadic thrills of space exploration, are thought marginal. What really matters to humanity, a primate species well adapted to Darwinian fundamentals in body and soul, are sex, family, work, security, personal expression, entertainment, and spiritual fulfillment—in no particular order. Most people believe, I am sure erroneously, that science has little to do with any of these preoccupations. They assume that the social sciences and humanities are independent of the natural sciences and more relevant endeavors. Who outside the technically possessed really needs to define a chromosome? Or understand chaos theory?

Science, however, is not marginal. Like art, it is a universal possession of humanity, and scientific knowledge has become a vital part of our species' repertory. It comprises what we know of the material world with reasonable certainty.

If the natural sciences can be successfully united with the social sciences and humanities, the liberal arts in higher education will be revitalized. Even the attempt to accomplish that much is a worthwhile goal. Profession-bent students should be helped to understand that in the twenty-first century the world will not be run by those who possess mere information alone. Thanks to science and technology, access to factual knowledge of all kinds is rising exponentially while dropping in unit cost. It is destined to become global and democratic. Soon it will be available everywhere on television and computer screens. What then? The answer is clear: synthesis. We are drowning in information, while starving for wisdom. The world henceforth will be run by synthesizers, people able to put together the right information at the right time, think critically about it, and make important choices wisely.

And this much about wisdom: In the long haul, civilized nations have come to judge one culture against another by a moral sense of the needs and aspirations of humanity as a whole. In thus globalizing the tribe, they attempt to formulate humankind's noblest and most enduring goals. The most important questions in this endeavor for the liberal arts are the meaning and purpose of all our idiosyncratic frenetic activity: *What are we, Where do we come from, How shall we decide where to go?* Why the toil, yearning, honesty, aesthetics, exaltation, love, hate, deceit, brilliance, hubris, humility, shame, and stupidity that collectively define our species? Theology, which long claimed the subject for itself, has done badly. Still encumbered by precepts based on Iron Age folk knowledge, it is unable to assimilate the great sweep of the real world now open for examination. Western philosophy offers no promising substitute. Its involuted exercises and professional timidity have left modern culture bankrupt of meaning.

The future of the liberal arts lies, therefore, in addressing the fundamental questions of human existence head on, without embarrassment or fear, taking them from the top down in easily understood language, and progressively rearranging them into domains of inquiry that unite the best of science and the humanities at each level of organization in turn. That of course is a very difficult task. But so are cardiac surgery and building space vehicles difficult tasks. Competent people get on with them, because they need to be done. Why should less be expected from the professionals responsible for education? The liberal arts will succeed to the extent that they are both solid in content and as coherent among themselves as the evidence allows. I find it hard to conceive of an adequate core curriculum in colleges and universities that avoids the cause-and-effect connections among the great branches of learning—not metaphor, not the usual second-order lucubrations on why scholars of different disciplines think this or that, but material cause and effect. There lies the high adventure for later generations, often mourned as no longer available. There lies great opportunity.

GRANTED THERE IS also a whiff of brimstone in the consilient world view and a seeming touch of Faust to those committed to its humanistic core. And these too need to be closely examined. What was it that Mephistopheles offered Faust, and how was the ambitious doctor to pay? From Christopher Marlowe's play to Goethe's epic poem the bargain was essentially the same: earthly power and pleasure in exchange for your soul. Then there were the differences. Marlowe's Faust was irrevocably damned when he made the wrong choice; Goethe's Faust was saved because he could not feel the happiness promised him through material gain. Marlowe upheld Protestant piety, Goethe the ideals of humanism.

In our perception of the human condition we have moved beyond Marlowe and Goethe. Today not one but two Mephistophelean bargains can be distinguished. From them, as from the original, hard choices must be made. Both illustrate the value of considering the consilient vision.

The first Faustian choice was actually made centuries ago, when humanity accepted the Ratchet of Progress: The more knowledge people acquire, the more they are able to increase their numbers and to alter the environment, whereupon the more they need new knowledge just to stay alive. In a human-dominated world, the natural environment steadily shrinks, offering correspondingly less and less per capita return in energy and resources. Advanced technology has become the ultimate prosthesis. Take away electric power from a tribe of Australian Aborigines, and little or nothing will happen. Take it away from residents of California, and millions will die. So to understand why humanity has come to relate to the environment in this way is more than a rhetorical question. Greed demands an explanation. The Ratchet should be constantly re-examined, and new choices considered.

The second Mephistophelean promise, generated by the first and strangely echoing the original Enlightenment, is due within a few decades. It says: You may alter the biological nature of the human species in any direction you wish, or you may leave it alone. Either way, genetic evolution is about to become conscious and volitional, and usher in a new epoch in the history of life.

Let us examine the two bargains, the second first for logical coherence, and consider the alternative fates they seem to imply.

It is useful to know, before peering into the future, where we are now. Is genetic change still occurring in the old-fashioned way, or has civilization brought it to a halt? The question can be put more precisely as follows: Is natural selection still operating to drive evolution? Is it forcing our anatomy and behavior to change in some particular direction in response to survival and reproduction?

The answer, like so many responses required in subjects of great complexity, is yes and no. To my knowledge no evidence exists that the human genome is changing in any overall new direction. It may come immediately to your mind that the forces most afflicting humanity, including overpopulation, war, outbreaks of infectious disease, and environmental pollution, must somehow be pushing the species along in a directed manner. But these pressures have existed around the world for millennia, forcing the periodic decline of populations and even the destruction and replacement of entire peoples. Much of the adaptation expected to arise has probably already done so. Contemporary human genes are therefore likely to reflect the necessities these malign forces imposed in the past.

We do not, for example, appear as a species to be acquiring genes for larger or smaller brains, more efficient kidneys, smaller teeth, greater or lesser compassion, or any other important adjustments in body and mind. The one undoubted global change is of lesser consequence. It is the shift occurring worldwide in the frequencies of racial traits such as skin color, hair type, lymphocyte proteins, and immunoglobulins, due to more rapid population growth in developing countries. In 1950, 68 percent of the world's population lived in developing countries. By 2000 the figure will be 78 percent. That amount of change is having an effect on the frequencies of previously existing genes, but none of the traits involved, so far as we know, have world-shaping consequence. None affect intellectual capacity or the fundamentals of human nature.

A few local quirks have been detected as well. There is, for example, brachycephalization. For the past ten thousand years, the heads of people have been growing rounder in populations as far apart as Europe, India, Polynesia, and North America. In rural Poland, between the Carpathian Mountains and the Baltic Sea, anthropologists have documented the trend in skeletons from around 1300 to the early twentieth century, embracing about thirty generations. The change is due principally to the slightly higher survival rate of round-heads, and not to the influx of brachycephalics from outside Poland. The trait has a partial genetic basis, but the reason for its greater Darwinian success, if any, remains unknown.

Many hereditary divergences of local populations have been discovered in blood types, disease resistance, aerobic capacity, and the ability to digest milk and other foodstuffs. Most such differences can at least be tentatively linked to higher survival and reproduction in known conditions of the local environment. The frequency of adults able to digest milk, one of the most thoroughly studied traits, is highest in populations that have relied on dairying for many generations. Another local trend of adaptive nature was reported in 1994 by a group of Russian geneticists. Turkmen-speaking people from the hot deserts of Middle Asia, they discovered, produce more heat shock proteins in their skin fibroblasts (cells that form part of the loose connective tissue) than do people who have lived for many generations in nearby moderate climates. The difference, which is genetically based, confers higher rates of survival following severe heat stress.

None of these regional trends appear to entail properties in anatomy or behavior of major consequence. Even the changes due to differential population growth are likely to prove short-lived if—as in present-day Thailand—birth rates in less developed countries drop to the levels prevailing in North America, Europe, and Japan.

The big story in recent human evolution is not directional change, not natural selection at all, but homogenization through immigration and interbreeding. Populations have been in flux throughout history. Tribes and states have pressed into and around the territories of rivals, often absorbing these neighbors, occasionally extinguishing them altogether. The historical atlases of Europe and Asia, when their pages are flipped chronologically through five millennia, become film clips of changing ethnic boundaries. As we race forward from one decade to the next in the clips, chiefdoms and states spring into existence, expand like hungry two-dimensional amoebae, and vanish as others move in to take their place.

The mixing sharply accelerated when Europeans conquered the New World and transported African slaves to its shores. Homogenization took a smaller leap in the nineteenth century with the European colonization of Australia and Africa. In more recent times it has quickened yet again through the spread of industrialization and democracy, the two signature traits of modernity that render people restless and international borders porous. Most human populations remain differentiated on a geographical basis, and some ethnic enclaves will probably endure for centuries more, but the trend in the opposite direction is unmistakably strong. It is also irreversible.

Homogenization is not dynamic on a global scale. It changes local populations, often swiftly, but cannot by itself consistently drive evolution of the human species as a whole in one direction or another. Its main consequence is the gradual erasure of previous racial differences—those statistical differences in hereditary traits that distinguish whole populations. It also increases the range of individual variation within the populations and across the entire species. Many more combinations of skin color, facial features, talents, and other traits influenced by genes are now arising than ever existed before. Yet the *average* differences between people in different localities around the world, not very great to start with, are narrowing.

Genetic homogenization has similarities to the stirring together of liquid ingredients. The contents change dramatically, and many new kinds of products emerge at the level of gene combinations within individuals. Variance increases, the extremes are extended, new forms of hereditary genius and pathology are more likely to arise. But the most elemental units, the genes, remain unperturbed. They stay about the same in both kind and relative abundance.

Continued over tens or hundreds of generations the present rates of emigration and intermarriage could in theory eliminate all population differences around the world. People residing in Beijing might become statistically the same as those in Amsterdam or Lagos. But this is not the key issue of future genetic trends, because the rules under which evolution can occur are about to change dramatically and fundamentally. Thanks to advances of genetics and molecular biology underway, hereditary change will soon depend less on natural selection than on social choice. Possessing exact knowledge of its own genes, collective humanity in a few decades can, if it wishes, select a new direction in its evolution and move there quickly. Or, if future generations prefer the free market of genetic diversity that existed in the past, they can choose simply to do nothing and live on their million-year-old heritage.

The prospect of this "volitional evolution"—a species deciding what to do about its own heredity—will present the most profound intellectual and ethical choices humanity has ever faced. The dilemma at its core is far from science fantasy. Medical researchers, motivated by the need to understand the genetic basis of disease, have begun in earnest to map the fifty thousand to one hundred thousand human genes. Reproductive biologists have cloned sheep, and presumably could do the same for human beings, if the procedure were allowed. And thanks to the Human Genome Project, geneticists will be able to read off the complete sequence of our DNA letters, 3.6 billion in all, within one or two decades. Scientists are also experimenting with a limited form of molecular engineering, in which genes are altered in a desired direction by substituting snippets of DNA. Still another fast-moving enterprise in the biological sciences is the tracking of individual development from genes to protein synthesis and thence to the final products of anatomy, physiology, and behavior. It is entirely possible that within fifty years we will understand in considerable detail not only our own heredity, but also a great deal about the way our genes interact with the environment to produce a human being. We can then tinker with the products at any level: change them temporarily without altering heredity, or change them permanently by mutating the genes and chromosomes.

If these advances in knowledge are even just partly attained, which seems inevitable unless a great deal of genetic and medical research is halted in its tracks, and if they are made generally available, which is problematic, humanity will be positioned godlike to take control of its own ultimate fate. It can, if it chooses, alter not just the anatomy and intelligence of the species but also the emotions and creative drive that compose the very core of human nature.

The engineering of the genome will be the final of three periods that can be distinguished in the history of human evolution. During almost all of the two-million-year history of the genus *Homo*, culminating in *Homo sapiens*, people were unaware of the ultramicroscopic hereditary codes shaping them. In historical times, over the past ten thousand years, populations still experienced racial differentiation, largely in response to local climatic conditions, just as they had throughout the more distant past.

During this passage through evolutionary time, shared with all other organisms, human populations were also subject to stabilizing selection; gene mutants that caused disease or infertility were weeded out in each generation. These defective alleles were able to persist only when recessive in their expression, which means their effects could be overridden by the activity of dominant genes paired with them. Possession of two recessive genes, however, causes genetic disorders, as exemplified by cystic fibrosis, Tay-Sachs disease, and sickle-cell anemia. Their double-dose carriers die young. Stabilizing selection, in this case through early death, continually sheds the genes from the population, making them mercifully rare.

With the advent of modern medicine, human evolution has entered its second period. More and more of the hereditary defects can be deliberately moderated or averted, even when the genes themselves remain unaltered and present in double dose. Phenylketonuria, for example, until recent time afflicted one out of ten thousand infants with severe mental retardation. Researchers discovered that the cause of phenylketonuria is a single recessive gene, which in double dose prevents normal metabolism of phenylalanine, a common amino acid. Abnormal metabolic products of the substance build up in the blood, causing brain damage. With this elementary fact in their reference books, physicians are now able to prevent the symptoms entirely by restricting phenylketonuric infants to phenylalanine-free diets.

Examples like the circumvention of phenylketonuria are becoming common and will be multiplied many times over in the years immediately ahead. For the first time people are using scientific knowledge to gain conscious control over their heredity, progressing one gene at a time. The evolutionary effect will be to relax stabilizing selection at an increasing rate and thereby increase the genetic variability of humanity as a whole. This second period, the suppression of stabilizing selection, is only beginning. Over many generations, the moderation of the effects of harmful genes could result in a substantial change in human heredity at the population level. The benefits accruing will have to be bought, of course, with a growing dependence on exacting and often expensive medical procedures. The age of gene circumvention is also the age of medical prosthesis.

We should not, however, worry that such destabilizing of selection will go too far. The second period of human evolution is ephemeral. It will not last enough generations to have an important impact on heredity of the species as a whole, because the knowledge that made it possible has brought us swiftly to the brink of the third period, that of volitional evolution. If we understand what changes in the genes cause particular defects, down to the nucleotide letters of the DNA code, then in principle the defect can be permanently repaired. Geneticists are hard at work to make this feat, called gene therapy, a reality. They are hopeful that cystic fibrosis, to cite the most advanced current project, can be at least partly cured by introduction of unimpaired genes into the lung tissues of patients. Another class of defects that seem permanently treatable within a few years includes hemophilia, sickle-cell anemia, and certain other inherited blood diseases.

Progress in gene therapy has admittedly been slow in the early period. But it will accelerate. Too much hope is at stake, and too much venture capital poised, to permit failure. Once established as a practical technology, gene therapy will become a commercial juggernaut. Thousands of genetic defects, many fatal, are already known. More are discovered each year. Each such gene is carried in single or double dose by thousands to millions of people around the world, and each individual person bears on average at least several different kinds of defective genes somewhere on his chromosomes. In most cases the genes are recessive and loaded in single dose; but the carrier, even if he does not suffer the defect, risks having a child with a double dose and full-blown symptoms. It is obvious that when genetic repair becomes safe and affordable, the demand for it will grow swiftly.

Some time in the next century that trend will lead into the full volitional period of evolution. The advance will create a new kind of ethical problem, which will be the Faustian decision of which I spoke: How much should people be allowed to mutate themselves and their descendants? Consider that your descendants, whom you may wish to alter in some beneficent manner, may well be my descendants also through intermarriage in the years ahead. With that in mind, can we ever agree on how much DNA tinkering is moral? In making such choices, there is an important line to be drawn between the remedy of clear-cut genetic defects on one side and the improvement of normal, healthy traits on the other. The scientific imagination will think it but a small step from, say, severe dyslexia (one gene region discovered in 1994 on chromosome number 6) to mild dyslexia, and another short hop to unimpaired learning ability, and, finally one step more to superior learning ability. I suffer from a mild form of dyslexia called visual sequencing disability, habitually reversing numbers (8652 too easily becomes 8562) and struggling to grasp words spelled out to me letter by letter (I apologize and ask to see them in writing). I would certainly prefer not to suffer this minor but inconvenient debility. If it is genetic in origin, I would be pleased to learn instead that it had been fixed when I was an embryo. My parents, had they known and been able, would probably have agreed and taken care of the problem.

Fair enough, but what about altering genes in order to enhance mathematical and verbal ability? To acquire perfect pitch? Athletic talent? Heterosexuality? Adaptability to cyberspace? In a wholly different dimension, citizens of states and then of all humanity might choose to make themselves less variable, in order to increase compatibility. Or the reverse: They might choose to diversify in talent and temperament, aiming for varied personal excellence and thus the creation of communities of specialists able to work together at higher levels of productivity. Above all, they will certainly aim for greater longevity. If such engineering for long life proves even just partly successful, it will create vast social and economic dislocations.

The present trajectory of science ensures that future generations will acquire the technical ability to make such choices. We are not in the volitional period yet, but we are close enough to make the prospect worth thinking about. *Homo sapiens*, the first truly free species, is about to decommission natural selection, the force that made us. There is no genetic destiny outside our free will, no lodestar provided by which we can set course. Evolution, including genetic progress in human nature and human capacity, will be from now on increasingly the domain of science and technology tempered by ethics and political choice. We have reached this point down a long road of travail and self-deception. Soon we must look deep within ourselves and decide what we wish to become. Our childhood having ended, we will hear the true voice of Mephistopheles.

We will also come to understand the true meaning of conservatism. By that overworked and confusing term I do not mean the pietistic and selfish libertarianism into which much of the American conservative movement has lately descended. I mean instead the ethic that cherishes and sustains the resources and proven best institutions of a community. In other words, true conservatism, an idea that can be applied to human nature as well as to social institutions.

I predict that future generations will be genetically conservative. Other than the repair of disabling defects, they will resist hereditary change. They will do so in order to save the emotions and epigenetic rules of mental development, because these elements compose the physical soul of the species. The reasoning is as follows. Alter the emotions and epigenetic rules enough, and people might in some sense be "better," but they would no longer be human. Neutralize the elements of human nature in favor of pure rationality, and the result would be badly constructed, protein-based computers. Why should a species give up the defining core of its existence, built by millions of years of biological trial and error?

What lifts this question above mere futurism is that it reveals so clearly our ignorance of the meaning of human existence in the first place. And illustrates how much more we need to know in order to decide the ultimate question: To what end, or ends, if any in particular, should human genius direct itself?

THE PROBLEM OF collective meaning and purpose is both urgent and immediate because, if for no other reason, it determines the environmental ethic. Few will doubt that humankind has created a planet-sized problem for itself. No one wished it so, but we are the first species to become a geophysical force, altering Earth's climate, a role previously reserved for tectonics, sun flares, and glacial cycles. We are also the greatest destroyer of life since the ten-kilometer-wide meteorite that landed near Yucatán and ended the Age of Reptiles sixty-five million years ago. Through overpopulation we have put ourselves in danger of running out of food and water. So a very Faustian choice is upon us: whether to accept our corrosive and risky behavior as the unavoidable price of population and economic growth, or to take stock of ourselves and search for a new environmental ethic.

That is the dilemma already implicit in current environmental debates. It springs from the clash of two opposing human self-images. The first is the naturalistic self-image, which holds that we are confined to a razor-thin biosphere within which a thousand imaginable hells are possible but only one paradise. What we idealize in nature and seek to re-create is the peculiar physical and biotic environment that cradled the human species. The human body and mind are precisely adapted to this world, notwithstanding its trials and dangers, and that is why we think it beautiful. In this respect *Homo sapiens* conforms to a basic principle of organic evolution, that all species prefer and gravitate to the environment in which their genes were assembled. It is called "habitat selection." There lies survival for humanity, and there lies mental peace, as prescribed by our genes. We are consequently unlikely ever to find any other place or conceive of any other home as beautiful as this blue planet was before we began to change it.

The competing self-image—which also happens to be the guiding theme of Western civilization—is the exemptionalist view. In this conception, our species exists apart from the natural world and holds dominion over it. We are exempt from the iron laws of ecology that bind other species. Few limits on human expansion exist that our special status and ingenuity cannot overcome. We have been set free to modify Earth's surface to create a world better than the one our ancestors knew.

For the committed exemptionalist, *Homo sapiens* has in effect become a new species, which I will now provide with a new name, *Homo proteus*, or "shapechanger man." In the taxonomic classification of Earth's creatures, the diagnosis of hypothetical *Homo proteus* is the following:

Cultural. Indeterminately flexible, with vast potential. Wired and information-driven. Can travel almost anywhere, adapt to any environment. Restless, getting crowded. Thinking about the colonization of space. Regrets the current loss of Nature and all those vanishing species, but it's the price of progress and has little to do with our future anyway.

Now here is the naturalistic, and I believe correct, diagnosis of old *Homo sapiens*, our familiar "wise man":

Cultural. With indeterminate intellectual potential but biologically constrained. Basically a primate species in body and emotional repertory (member of the Order Primates, Infraorder Catarrhini, Family Hominidae). Huge com-

pared to other animals, parvihirsute, bipedal, porous, squishy, composed mostly of water. Runs on millions of coordinated delicate biochemical reactions. Easily shut down by trace toxins and transit of pea-sized projectiles. Short-lived, emotionally fragile. Dependent in body and mind on other earthbound organisms. Colonization of space impossible without massive supply lines. Starting to regret deeply the loss of Nature and all those other species.

The dream of man freed from the natural environment of Earth was tested against reality in the early 1990s with Biosphere 2, a 3.15-acre closed ecosystem built on desert terrain in Oracle, Arizona. Paneled in glass, stocked with soil, air, water, plants, and animals, it was designed to be a miniature working Earth independent of the mother planet. The planners synthesized fragments of rain forest, savanna, thornscrub, desert, pond, marsh, coral reef, and ocean to simulate the natural habitats of home. The only connections to the outside world were electrical power and communication, both reasonable concessions made for a primarily ecological experiment. The design and construction of Biosphere 2 cost $200 million. It incorporated the most advanced scientific knowledge and state-of-the-art engineering. Success of the experiment, if achieved, was expected to prove that human life can be independently sustained in hermetic bubbles anywhere in the solar system not lethally seared by heat or hard radiation.

On September 26, 1991, eight volunteer "Biospherians" walked into the completed enclosure and sealed themselves off. For a while everything went well, but then came a series of nasty surprises. After five months the concentration of oxygen in Biosphere 2 began to drop from its original 21 percent, eventually reaching 14 percent, an amount that normally occurs at 17,500 feet, too low to sustain health. At this point, to keep the experiment going, oxygen was pumped in from the outside. During the same period carbon dioxide levels rose sharply, despite the use of an artificial recycling procedure. Concentrations of nitrous oxide increased to levels dangerous to brain tissue.

Species used to build the ecosystems were drastically affected. Many declined to extinction at an alarmingly high rate. Nineteen of the twenty-five vertebrates and all of the animal pollinators vanished. At the same time, a few species of cockroaches, katydids, and ants multiplied explosively. Morning glory, passionflower, and other vines, planted to serve as a carbon sink, grew so luxuriantly they threatened other plant species, including the crops, and had to be laboriously thinned by hand.

The Biospherians coped heroically with these ordeals, managing to stay inside the enclosure the full two years originally planned. And as an experiment, Biosphere 2 was not at all a failure. It taught us many things, the most important of which is the vulnerability of our species and the living environment on which we depend. Two senior biologists who reviewed the data as part of an independent team, Joel E. Cohen of Rockefeller University and David Tilman of the University of Minnesota, wrote with feeling, "No one yet knows how to engineer systems that provide humans with the life-supporting services that natural ecosystems produce for free," and "despite its mysteries and hazards, Earth remains the only known home that can sustain life."

In its neglect of the fragility of life, exemptionalism fails definitively. To move ahead as though scientific and entrepreneurial genius will solve each crisis arising in turn implies that the decline of the global biosphere can be similarly managed. Perhaps that might be possible in future decades (centuries seem more likely), but the means are not yet in sight. The living world is too complicated to be kept as a garden on a planet that has been converted into an artificial space capsule. No biological homeostat is known that can be worked by humanity. To believe otherwise is to risk reducing Earth to a wasteland, and humanity to a threatened species.

How pressing is the risk? Enough, I think, to change thinking about human self-preservation fundamentally. The current state of the environment can be summarized thus:

The global population is precariously large, and will become much more so before peaking some time after 2050. Humanity overall is improving per capita production, health, and longevity. But it is doing so by eating up the planet's capital, including natural resources and biological diversity millions of years old. Homo sapiens *is approaching the limit of its food and water supply. Unlike any species that lived before, it is also changing the world's atmosphere and climate, lowering and polluting water tables, shrinking forests, and spreading deserts. Most of the stress originates directly or indirectly from a handful of industrialized countries. Their proven formulas for prosperity are being eagerly adopted by the rest of the world. The emulation cannot be sustained, not with the same levels of consumption and waste. Even if the industrialization of developing countries is only partly successful, the environmental aftershock will dwarf the population explosion that preceded it.*

Some will, of course, call this synopsis environmental alarmism. I earnestly wish that accusation were true. Unfortunately, it is the reality-grounded opinion of the overwhelming majority of statured scientists who study the environment. By statured scientists I mean those who collect and analyze the data, build the theoretical models, interpret the results, and publish articles vetted for professional journals by other experts, often including their rivals. I do not mean by statured scientists the many journalists, talk-show hosts, and think-tank polemicists who also address the environment, even though their opinions reach a vastly larger audience. This is not to devalue their professions, which have separate high standards, only to suggest that there are better-qualified sources to consult for factual information about the environment. Seen in this light, the environment is much less a controversial subject than suggested by routine coverage in the media.

Consider, then, the assessment made through the mid-1990s by the statured scientists. Their quantitative estimates differ according to the mathematical assumptions and procedures variously used, but most still fall within limits from which trends can be projected with confidence.

By 1997 the global population had reached 5.8 billion, growing at the rate of 90 million per year. In 1600 there were only about half a billion people on Earth, and in 1940, 2 billion. The amount of increase during the 1990s alone is expected to exceed the entire population alive in 1600. The global growth rate, after reaching a peak during the 1960s, has been dropping ever since. In 1963, for example, each woman bore an average of 4.1 children. In 1996 the number had declined to 2.6. In order to stabilize the world population, the number must be 2.1 children per woman (the extra 0.1 allowing for child mortality). Long-term population size is extremely sensitive to this replacement number, as shown by the following projections. If the number were 2.1, there would be 7.7 billion people on Earth in 2050, leveling off at 8.5 billion in 2150. If 2.0, the population would peak at 7.8 billion, then drop by 2150 to 5.6 billion, the total in the mid-1990s. If 2.2, it would reach 12.5 billion in 2050, 20.8 billion in 2150; and if 2.2 could miraculously be maintained thereafter, the human biomass would eventually equal the weight of the world and then, after a few millennia, expanding outward at the speed of light, it would exceed the mass of the visible universe. Even if the global birth rate were reduced drastically and immediately, say to the Chinese goal of one child per woman, the population would not peak for one or two generations. The overshoot is ensured by the disproportionate number of young people already in existence, who look to long lives ahead.

How many people can the world support for an indefinite period? Experts do not agree, but a majority put the number variously between 4 and 16 billion. The true number will depend on the quality of life that future generations are willing to accept. If everyone agreed to become vegetarian, leaving nothing for livestock, the present 1.4 billion hectares of arable land (3.5 billion acres) would supply about 10 billion people. If humans utilized as food all the energy captured by plant photosynthesis, some 40 trillion watts, Earth could support about 16 billion people. From such a fragile world, almost all other life forms would have to be excluded.

Even if, by *force majeure*, the population levels off at well under 10 billion by mid-century, the relatively extravagant lifestyle now enjoyed by the middle classes of North America, Western Europe, and Japan cannot be attained by most of the rest of the world. The reason is that the impact of each country on the environment is multiplicative. It is dependent, in a complex manner, on the formula called PAT: population size *times* per capita affluence (hence consumption) *times* a measure of the voracity of the technology used in sustaining consumption. The magnitude of PAT can be usefully visualized by the "ecological footprint" of productive land needed to support each member of the society with existing technology. In Europe the footprint is 3.5 hectares (a hectare is 2.5 acres), in Canada 4.3 hectares, and in the United States 5 hectares. In most developing countries it is less than half a hectare. To raise the whole world to the U.S. level with existing technology would require two more planet Earths.

It matters little that North Dakota and Mongolia are mostly empty. It makes no difference that the 5.8 billion people in the world today could be logstacked out of sight in a corner of the Grand Canyon. The datum of interest is the average footprint on productive land, which must somehow be lowered if significantly more people are to achieve a decent standard of living.

To suppose that the living standard of the rest of the world can be raised to that of the most prosperous countries, with existing technology and current levels of consumption and waste, is a dream in pursuit of a mathematical impossibility. Even to level out present-day income inequities would require shrinking the ecological footprints of the prosperous countries. That is problematic in the market-based global economy, where the main players are also militarily the most powerful, and in spite of a great deal of rhetoric largely indifferent to the suffering of others. Few people in industrialized countries are fully aware of how badly off the poor of the world really are. Roughly 1.3 billion people, more than a fifth of the world population, have cash incomes under one U.S. dollar a day. The next tier of 1.6 billion earn $1–3. Somewhat more than 1 billion live in what the United Nations classifies as absolute poverty, uncertain of obtaining food from one day to the next. Each year more than the entire population of Sweden, between 13 and 18 million, mostly children, die of starvation, or the side effects of malnutrition, or other poverty-related causes. In order to gain perspective, imagine the response if Americans and Europeans were told that in the coming year the entire population of Sweden, or Scotland and Wales combined, or New England would die of poverty.

Of course the exemptionalists will say that new technology and the rising tide of the free-market economy can solve the problem. The solution, they explain, is straightforward: Just use more land, fertilizer, and higher-yield crops, and work harder to improve distribution. And, of course, encourage more education, technology transfer, and free trade. Oh, and discourage ethnic strife and political corruption.

All that will certainly help, and should have high priority, but it cannot solve the main problem, which is the finite resources of planet Earth. It is true that only 11 percent of the world's land surface is under cultivation. But that already includes the most arable part. The bulk of the remaining 89 percent has limited use, or none at all. Greenland, Antarctica, most of the vast northern taiga, and the equally vast ultra-dry deserts are not available. The remnant tropical forests and savannas can be cleared and planted, but at the cost of most of the species of plants and animals in the world, with minor agricultural gain. Nearly half their expanse is underlaid by soils of low natural fertility—42 percent of the untapped area of sub-Saharan Africa, for example, and 46 percent of that in Latin America. Meanwhile, cultivated and deforested lands are losing topsoil to erosion at ten times the sustainable level. By 1989, 11 percent of the world's cropland had been classified by soil experts as severely degraded. From 1950 to the mid-1990s the area of cropland per person fell by half, from 0.23 hectare to 0.12 hectare, less than a quarter the size of a soccer field. Widespread starvation was avoided because the Green Revolution during the same forty-year period boosted per hectare yield dramatically with new varieties of rice and other crops, better pesticide application, and increased use of fertilizer and irrigation. But even these technologies have limits. By 1985 the growth in yield slowed; that trend, when combined with the relentless growth of population, initiated a decline in per capita production. The shortfall first became apparent in the developing countries, whose grain self-sufficiency fell from 96 percent in 1969–71, at the height of the Green Revolution, to 88 percent in 1993–95. By 1996 the world grain carryover stocks, humanity's emergency food supply, had declined 50 percent from the all-time peak reached in 1987. At the beginning of the 1990s only a handful of countries—including Canada, the United States, Argentina, the European Union, and Australia—accounted for more than three-fourths of the world's grain resources.

Perhaps all these signs will miraculously disappear. If not, how will the world cope? Perhaps the deserts and nonarable dry grasslands can be irrigated to expand agricultural production. But that remedy also has limitations. Too many people already compete for too little water. The aquifers of the world, on which so much agriculture in drier regions depends, are being drained of their groundwater faster than the reserves can be replaced by natural percolation of rainfall and runoff. The Ogallala aquifer, a principal water source of the central United States, experienced a three-meter drop through a fifth of its area during the 1980s alone. Now it is half depleted beneath a million hectares in Kansas, Texas, and New Mexico. Still worse deficits are building in other countries, and often where they are least affordable. The water table beneath Beijing fell 37 meters between 1965 and 1995. The groundwater reserves of the Arabian peninsula are expected to be exhausted by 2050. In the meantime the oil-rich countries there are making up the deficit in part by desalinizing seawater—trading their precious petroleum for water. On a global scale, humanity is pressing the limit, using a quarter of the accessible water released to the atmosphere by evaporation and plant transpiration, and somewhat more than half that available in rivers and other runoff channels. By 2025, 40 percent of the world's population could be living in countries with chronic water scarcity. New dam construction can add 10 percent to the runoff capture during the next thirty years, but the treadmill opposing it is unceasing: In the same three decades the human population is expected to grow by a third.

As the land gives out, might we turn to Earth's last frontier, the boundless sea? Unfortunately, no. It is not really boundless, having already given most of what it has to offer. All seventeen of the world's oceanic fisheries are being harvested beyond their capacity. Only those in the Indian Ocean have continued to rise in yield, a trend destined to end because the present rate of catch is not sustainable. Several fisheries, including most famously the northwestern Atlantic banks and the Black Sea, have suffered a commercial collapse. The annual world fish catch, after rising fivefold from 1950 to 1990, has leveled off at about 90 million tons.

The history of marine fisheries has been one of increasingly efficient mass capture and on-site processing, which increases yield by cutting ever deeper into existing stocks. By the 1990s proliferating fish farms had taken up part of the slack, adding 20 million tons to the total harvest. But aquaculture, the fin-and-shell revolution, also has limits. Expanding marine farms preempt the mangrove swamps and other coastal wetland habitats that serve as the spawning grounds for many offshore food fishes. Freshwater farms have more growth potential but must compete with conventional agriculture for the shrinking supplies of runoff and aquifer-borne water.

Meanwhile, in accordance with the general principle of life that all large perturbations are bad, Earth's ability to support the voracious human biomass is becoming even dicier through the acceleration of climatic change. During the past 130 years the global average temperature has risen by one degree Celsius. The signs are now strong—some atmospheric scientists say conclusive—that much of the change is due to carbon dioxide pollution. The connection is the greenhouse effect, in which carbon dioxide, along with methane and a few other gases, work like the glass enclosures used by gardeners. They admit sunlight but trap the heat generated by it. For the past 160,000 years, as tests of air bubbles in fossil ice show, the concentration of atmospheric carbon dioxide has been tightly correlated with the global average temperature. Now, boosted by combustion of fossil fuels and the destruction of tropical forests, the carbon dioxide concentration stands at 360 parts per million, the highest measure in the 160,000-year period.

The idea of climatic warming by human activity has been disputed by several scientists, with valid reasons. Atmospheric chemistry and climatic change are both extremely complex subjects. When compounded, they make exact predictions nearly impossible. Nevertheless, trajectories and velocities of the changes can be estimated within broad limits. That has been the goal of the Intergovernmental Panel on Climate Change (IPCC), a group of more than two thousand scientists working worldwide to assess incoming data and build models of future change with the aid of supercomputers. The more difficult variables they must incorporate include the industrial discharge of sulfate aerosols, which counteract the greenhouse effect of carbon dioxide, together with the long-term capture of carbon dioxide by the ocean, which can throw off calculations of atmospheric change, and the tricky idiosyncrasies of local climatic change.

Overall, the IPCC scientists have made the following assessment. There will be an additional rise in the global average temperature of 1.0 to 3.5 degrees Celsius (1.8 to 6.3 degrees Fahrenheit) by the year 2100. Multiple consequences are likely, with few if any likely to be pleasant. Thermal expansion of marine waters and the partial breakup of the Antarctic and Greenland ice shelves will raise the sea level by as much as 30 centimeters (12 inches), causing problems for coastal nations. Kiribati and the Marshall Islands, two small atoll countries in the Western Pacific, risk partial obliteration. Precipitation patterns will change, and most likely as follows: Large increases will be experienced in North Africa, temperate Eurasia and North America, Southeast Asia, and the Pacific coast of South America, and comparable decreases in Australia and most of South America and southern Africa.

Local climates will turn more variable, as heat waves increase in frequency. Even a small rise in average temperature results in many more instances of extremely high temperatures. The reason is a purely statistical effect. A small shift in a normal statistical distribution in one direction lifts the former extreme in that direction from near zero to a proportionately far higher number. (Thus, to take another example, if the average mathematical ability of the human species were raised ten percent, the difference in the mass of people might not be noticeable, but Einsteins would be commonplace.)

Because clouds and storm centers are generated over marine waters heated above 26°C, tropical cyclones will increase in average frequency. The eastern seaboard of the United States, to select one heavily populated region, will thereby suffer both more heat waves in the spring and more hurricanes in the summer. We can expect the hotter climatic zones to expand toward the North and South Poles, with the greatest changes occurring at the highest latitudes. The tundra ecosystems will shrink and may disappear altogether. Agriculture will be affected, in some areas favorably, in others destructively. In general, developing nations can expect to be hit harder than those in the industrialized North. Many natural systems and the species of microorganisms, plants, and animals composing them, unable to adapt to the shift in local conditions or emigrate to newly habitable areas quickly enough, will be extinguished.

To summarize the future of resources and climate, the wall toward which humanity is evidently rushing is a shortage not of minerals and energy, but of food and water. The time of arrival at the wall is being shortened by a physical climate growing less congenial. Humankind is like a household living giddily off vanishing capital. Exemptionalists are risking a lot when they advise us, in effect, that "Life is good and getting better, because look around you, we are still expanding and spending faster. Don't worry about next year. We're such a smart bunch something will turn up. It always has."

They, and most of the rest of us, have yet to learn the arithmetical riddle of the lily pond. A lily pad is placed in a pond. Each day thereafter the pad and then all of its descendants double. On the thirtieth day the pond is covered completely by lily pads, which can grow no more. On which day was the pond half full and half empty? The twenty-ninth day.

Shall we gamble? Suppose the odds are even that humankind will miss the environmental wall. Better, make it two to one: pass on through or collide. To bet on safe passage is a terrible choice, because the stakes on the table are just about everything. You save some time and energy now by making that choice and not taking action, but if you lose the bet down the line, the cost will be ruinous. In ecology, as in medicine, a false positive diagnosis is an inconvenience, but a false negative diagnosis can be catastrophic. That is why ecologists and doctors don't like to gamble at all, and if they must, it is always on the side of caution. It is a mistake to dismiss a worried ecologist or a worried doctor as an alarmist.

At best, an environmental bottleneck is coming in the twenty-first century. It will cause the unfolding of a new kind of history driven by environmental change. Or perhaps an unfolding on a global scale of more of the old kind of history, which saw the collapse of regional civilizations, going back to the earliest in history, in northern Mesopotamia, and subsequently Egypt, then the Mayan and many others scattered across all the inhabited continents except Australia. People died in large numbers, often horribly. Sometimes they were able to emigrate and displace other people, making them die horribly instead.

Archaeologists and historians strive to find the reasons for the collapse of civilizations. They tick off drought, soil exhaustion, overpopulation, and warfare—singly or in some permutation. Their analyses are persuasive. Ecologists add another perspective, with this explanation: The populations reached the local carrying capacity, where further growth could no longer be sustained with the technology available. At that point life was often good, especially for the ruling classes, but fragile. A change such as a drought or depletion of the aquifer or a ravaging war then lowered the carrying capacity. The death rate soared and the birth rate fell (from malnutrition and disease) until lower and more sustainable population levels were reached.

The principle of carrying capacity is illustrated by the recent history of Rwanda, a small and beautiful mountainous land that once rivaled Uganda as the pearl of Central Africa. Until the present century Rwanda supported only a modest population density. For five hundred years a Tutsi dynasty ruled over a Hutu majority. In 1959 the Hutu revolted, causing many of the Tutsi to flee to neighboring countries. In 1994 the conflict escalated, and Rwandan army units massacred over half a million Tutsi and moderate Hutu. Then an army of the Tutsi, the Rwandan Patriotic Front, struck back, capturing the capital town of Kigali. As the Tutsi advanced across the countryside, two million Hutu refugees ran before them, spreading out into Zaire, Tanzania, and Burundi. In 1997 Zaire, newly renamed the Republic of the Congo, forced many of the Hutu refugees back to Rwanda. In the maelstrom, thousands died of starvation and disease.

On the surface it would seem, and was so reported by the media, that the Rwandan catastrophe was ethnic rivalry run amok. That is true only in part. There was a deeper cause, rooted in environment and demography. Between 1950 and 1994 the population of Rwanda, favored by better health care and temporarily improved food supply, more than tripled, from 2.5 million to 8.5 million. In 1992 the country had the highest growth rate in the world, an average of 8 children for every woman. Parturition began early, and generation times were short. But although total food production increased dramatically during this period, it was soon overbalanced by population growth. The average farm size dwindled, as plots were divided from one generation to the next. Per capita grain production fell by half from 1960 to the early 1990s. Water was so overdrawn that hydrologists declared Rwanda one of the world's twenty-seven water-scarce countries. The teenage soldiers of the Hutu and Tutsi then set out to solve the population problem in the most direct possible way.

Rwanda is a microcosm of the world. War and civil strife have many causes, most not related directly to environmental stress. But in general, overpopulation and the consequent dwindling of available resources are tinder that people pile up around themselves. The mounting anxiety and hardship are translated into enmity, and enmity into moral aggression. Scapegoats are identified, sometimes other political or ethnic groups, sometimes neighboring tribes. The tinder continues to grow, awaiting the odd assassination, territorial incursion, atrocity, or other provocative incident to set it off. Rwanda is the most overpopulated country in Africa. Burundi, its war-torn neighbor, is second. Haiti and El Salvador, two of the chronically most troubled nations of the Western Hemisphere, are also among the most densely populated, exceeded only by five tiny island countries of the Caribbean. They are also arguably the most environmentally degraded.

Population growth can justly be called the monster on the land. To the extent that it can be tamed, passage through the bottleneck will be easier. Let us suppose that the last of the old reproductive taboos fade, and family planning becomes universal. Suppose further that governments create population policies with the same earnestness they devote to economic and military policies. And that as a result the global population peaks below ten billion and starts to decline. With NPG (negative population growth) attained, there are grounds for hope. If not attained, humanity's best efforts will fail, and the bottleneck will close to form a solid wall.

Humanity's best efforts will include every technological fix for an overcrowded planet that genius can devise. Endless stand-by schemes are already on the board. Conversion of nitrogenased petroleum to food is one remote possibility. Algal farms in shallow seas is another. The water crisis might be eased by desalinization of seawater with energy from controlled fusion or fuel cell technology. Perhaps as polar ice shelves break up from global warming, more fresh water can be drawn from icebergs herded to dry coasts. With a surplus of energy and fresh water, the agricultural revegetation of arid wasteland is attainable. Pulp production can be increased in such recovered lands with "wood grass," fast-growing, nitrogen-fixing tree species that can be harvested with giant mowers and then sprout new shoots from the severed stocks. Many such schemes will be tried as demand rises, and a few will succeed. They will be driven by venture capital and government subsidy in the global free-market economy. Each advance will reduce the risk of short-term economic calamity.

But be careful! Each advance is also a prosthesis, an artificial device dependent on advanced expertise and intense continuing management. Substituted for part of Earth's natural environment, it adds its own, long-term risk. Human history can be viewed through the lens of ecology as the accumulation of environmental prostheses. As these manmade procedures thicken and interlock, they enlarge the carrying capacity of the planet. Human beings, being typical organisms in reproductive response, expand to fill the added capacity. The spiral continues. The environment, increasingly rigged and strutted to meet the new demands, turns ever more delicate. It requires constant attention from increasingly sophisticated technology.

The Ratchet of Progress seems irreversible. The message then for the primitivists, who dream of nature's balance in Paleolithic serenity: *Too late.* Put away your bow and arrow, forget the harvest of wild berries; the wilderness has become a threatened nature reserve. The message for the environmentalists and exemptionalists: *Get together.* We must plunge ahead and make the best of it, worried but confident of success, our hope well expressed by Hotspur's lines in *Henry IV*: *I tell you, my lord fool, out of this nettle, danger, we pluck this flower, safety.*

The common aim must be to expand resources and improve the quality of life for as many people as heedless population growth forces upon Earth, and do it with minimal prosthetic dependence. That, in essence, is the ethic of sustainable development. It is the dream that acquired general currency at the Earth Summit, the historic United Nations Conference on Environment and Development held in June 1992 in Rio de Janeiro. The representatives of

172 nations, including 106 heads of government, met to establish guidelines by which a sustainable world order might be reached. They signed binding conventions on climate change and the protection of biological diversity. They agreed to the forty nonbinding chapters of Agenda 21, offering procedures by which virtually all of the general problems of the environment can be addressed, if not solved. Most of the initiatives were blunted by political squabbles arising from national self-interest, and global cooperation afterward was principally limited to rhetorical exercise on state occasions. The $600 billion additional expenditure recommended to put Agenda 21 into effect, with $125 billion donated to developing countries by industrialized countries, has not been forthcoming. Still, the principle of sustainable development has been generally accepted, an idea previously little more than the dream of an environmentalist elite. By 1996 no fewer than 117 governments had appointed commissions to develop Agenda 21 strategies.

In the end, the measure of success of the Earth Summit and all other global initiatives will be the diminishment of the total ecological footprint. As the human population soars toward eight billion around 2020, the central question will be the area of productive land required on average to provide each person in the world with an acceptable standard of living. From it, the overriding environmental goal is to shrink the ecological footprint to a level that can be sustained by Earth's fragile environment.

Much of the technology required to reach that goal can be summarized in two concepts. Decarbonization is the shift from the burning of coal, petroleum, and wood to essentially unlimited, environmentally light energy sources such as fuel cells, nuclear fusion, and solar and wind power. Dematerialization, the second concept, is the reduction in bulk of hardware and the energy it consumes. All the microchips in the world, to take the most encouraging contemporary example, can be fitted into the room that housed the Harvard Mark 1 electromagnetic computer at the dawn of the information revolution.

The single greatest intellectual obstacle to environmental realism, as opposed to practical difficulty, is the myopia of most professional economists. In Chapter 9 I described the insular nature of neoclassical economic theory. Its models, while elegant cabinet specimens of applied mathematics, largely ignore human behavior as understood by contemporary psychology and biology. Lacking such a foundation, the conclusions often describe abstract worlds that do not exist. The flaw is especially noticeable in microeconomics, which treats the patterns of choices made by individual consumers.

The weakness of economics is most worrisome, however, in its general failure to incorporate the environment. After the Earth Summit, and after veritable encyclopedias of data compiled by scientists and resource experts have shown clearly the dangerous trends of population size and planetary health, the most influential economists still make recommendations as though there is no environment. Their assessments read like the annual reports of successful brokerage firms. Here, for example, is Frederick Hu, head of the World Economic Forum's competitiveness research team, explaining the conclusions of the Forum's influential *Global Competitiveness Report 1996*:

> Short of military conquest, economic growth is the only viable means for a country to sustain increases in national wealth and living standards . . . An economy is internationally competitive if it performs strongly in three general areas: abundant productive inputs such as capital, labour, infrastructure and technology; optimal economic policies such as low taxes, little interference and free trade and sound market institutions such as the rule of law and the protection of property rights.

This prescription, resonant with the hard-headed pragmatism expected in an economics journal, is true for medium-term growth of individual nations. It is surely the best policy to recommend during the next two decades for Russia (competitiveness index −2.36) and Brazil (−1.73) if they wish to catch up with the United States (+1.34) and Singapore (+2.19). No one can seriously question that a better quality of life for everyone is the unimpeachable universal goal of humanity. Free trade, the rule of law, and sound market practices are the proven means to attain it. But the next two decades will also see the global population leap from six to eight billion, mostly among the poorest nations. That interval will witness water and arable soil running out, forests being stripped, and coastal habitats used up. The planet is already in a precarious state. What will happen as giant China (–0.68) strives to overtake little Taiwan (+0.96) and the other Asian tigers? We tend to forget, and economists are reluctant to stress, that economic miracles are not endogenous. They occur most often when countries consume not only their own material resources, including oil, timber, water, and agricultural produce, but those of other countries as well. And now the globalization of commerce, accelerated by technology and the liquidity of paper assets, has made the mass transfer of material assets far easier. The wood products of Japan are the destroyed forests of tropical Asia, the fuel of Europe the dwindling petroleum reserves of the Middle East.

In national balance sheets economists seldom use full-cost accounting, which includes the loss of natural resources. A country can cut down all its trees, mine out its most profitable minerals, exhaust its fisheries, erode most of its soil, draw down its underground water, and count all the proceeds as income and none of the depletion as cost. It can pollute the environment and promote policies that crowd its populace into urban slums, without charging the result to overhead.

Full-cost accounting is gaining some credibility within the councils of economists and the finance ministers they advise. Ecological economics, a new subdiscipline, has been formed to put a green thumb on the invisible hand of economics. But it is still only marginally influential. Competitive indexes and gross domestic products (GDPs) remain seductive, not to be messed up in conventional economic theory by adding the tricky complexities of environment and social cost. The time has come for economists and business leaders, who so haughtily pride themselves as masters of the real world, to acknowledge the existence of the *real* real world. New indicators of progress are needed to monitor the economy, wherein the natural world and human well-being, not just economic production, are awarded full measure.

TO THE SAME END I count it paramount, and feel obliged to plead, that the new reckoning include a powerful conservation ethic. We hope—surely we must believe—that our species will emerge from the environmental bottleneck in better condition than we entered. But there is another responsibility to meet as we make the passage: preserving the Creation by taking as much of the rest of life with us as possible.

Biological diversity, or biodiversity for short—the full sweep from ecosystems to species within the ecosystems, thence to genes within the species—is in trouble. Mass extinctions are commonplace, especially in tropical regions where most of the biodiversity occurs. Among the more recent are more than half the exclusively freshwater fishes of peninsular Asia, half of the fourteen birds of the Philippine island of Cebu, and more than ninety plant species growing on a single mountain ridge in Ecuador. In the United States an estimated 1 percent of all species have been extinguished; another 32 percent are imperiled.

Conservation experts, responding to what they now perceive as a crisis, have in the past three decades broadened their focus from the panda, tiger, and other charismatic animals to include entire habitats whose destruction endangers the existence of many species.

Familiar "hot spots" of this kind in the United States include the mountain forests of Hawaii, the coastal heath of southern California, and the sandy uplands of central Florida. Arguably the nations with the most hot spots in the world are Ecuador, Madagascar, and the Philippines. Each of these countries has lost two-thirds or more of its biologically rich rain forest, and the remainder is under continuing assault. The logic of conservation experts in addressing the issue is simple: By concentrating conservation efforts on such areas, the largest amount of biodiversity can be saved at the lowest economic cost. If the effort is also made part of the political process during regional planning, the rescue of biodiversity can also gain the widest possible public support.

It is notoriously difficult to estimate the overall rate of extinction, but biologists, by using several indirect methods of analysis, generally agree that on the land at least, species are vanishing at a rate one hundred to a thousand times faster than before the arrival of *Homo sapiens*. Tropical rain forests are the site of most of the known damage. Although they cover only 6 percent of the land surface, they contain more than half the species of plants and animals of the entire world. The rate of clearing and burning of the surviving rain forests averaged about 1 percent a year through the 1980s and into the 1990s, an area about equal to the entire country of Ireland. That magnitude of habitat loss means that each year 0.25 percent or more of the forest species are doomed to immediate or early extinction. How much does the rate translate into absolute numbers? If there are ten million species in the still mostly unexplored forests, which some scientists think possible, the annual loss is in the tens of thousands. Even if there are a "mere" one million species, the loss is still in the thousands.

These projections are based on the known relationships between the area of a given natural habitat and the number of species able to live for indefinite periods within it. Such projections may in fact be on the low side. The outright elimination of habitat, the easiest factor to measure, is the leading cause of extinction. But the introduction of aggressive exotic species and the diseases they carry come close behind in destructiveness, followed in turn by the overharvesting of native species.

All these factors work together in a complex manner. When asked which ones caused the extinction of any particular species, biologists are likely to give the *Murder on the Orient Express* answer: They all did it. A common sequence in tropical countries starts with the building of roads into wilderness, such as those cut across Brazil's Amazonian state of Rondônia during the 1970s and '80s. Land-seeking settlers pour in, clear the forest on both sides of the road, pollute the streams, introduce alien plants and animals, and hunt wildlife for extra food. Many native species become rare, and some disappear entirely. The soil wears out within several years, and the settlers cut and burn their way deeper into the forest.

The ongoing loss of biodiversity is the greatest since the end of the Mesozoic Era sixty-five million years ago. At that time, by current scientific consensus, the impact of one or more giant meteorites darkened the atmosphere, altered much of Earth's climate, and extinguished the dinosaurs. Thus began the next stage of evolution, the Cenozoic Era or Age of Mammals. The extinction spasm we are now inflicting can be moderated if we so choose. Otherwise, the next century will see the closing of the Cenozoic Era and a new one characterized not by new life forms but by biological impoverishment. It might appropriately be called the "Eremozoic Era," the Age of Loneliness.

I have found, during many years of studying biological diversity, that people commonly respond to evidence of species extinction by entering three stages of denial. The first is simply, Why worry? Extinction is natural. Species have been dying out through more than three billion years of life's history without permanent harm to the biosphere. Evolution has always replaced extinct species with new ones.

All these statements are true, but with a terrible twist. Following the Mesozoic spasm, and after each of the four greatest previous spasms spaced over the earlier 350 million years, evolution required about 10 million years to restore the predisaster levels of diversity. Faced with a waiting time that long, and aware that we inflicted so much damage in a single lifetime, our descendants are going to be—how best to say it?—peeved.

Entering the second stage of denial, people commonly ask, Why do we need so many species anyway? Why care, especially since the vast majority are bugs, weeds, and fungi? It is easy to dismiss the creepy-crawlies of the world, forgetting that less than a century ago, before the rise of the modern conservation movement, native birds and mammals around the world were treated with the same callow indifference. Now the value of the little things in the natural world has become compellingly clear. Recent experimental studies on whole ecosystems support what ecologists have long suspected: The more species that live in an ecosystem, the higher its productivity and the greater its ability to withstand drought and other kinds of environmental stress. Since we depend on functioning ecosystems to cleanse our water, enrich our soil, and create the very air we breathe, biodiversity is clearly not something to discard carelessly.

Each species is a masterpiece of evolution, offering a vast source of useful scientific knowledge because it is so thoroughly adapted to the environment in which it lives. Species alive today are thousands to millions of years old. Their genes, having been tested by adversity over so many generations, engineer a staggeringly complex array of biochemical devices to aid the survival and reproduction of the organisms carrying them.

This is why, in addition to creating a habitable environment for humankind, wild species are the source of products that help sustain our lives. Not the least of these amenities are pharmaceuticals. More than 40 percent of all medicinals dispensed by pharmacies in the United States are substances originally extracted from plants, animals, fungi, and microorganisms. Aspirin, for example, the most widely used medicine in the world, was derived from salicylic acid, which in turn was discovered in a species of meadowsweet. Yet only a fraction of the species—probably fewer than 1 percent—have been examined for natural products that might serve as medicines. There is a critical need to press the search for new antibiotics and antimalarial agents. The substances most commonly used today are growing less effective as disease organisms acquire genetic resistance to the drugs. The universal staphylococcus bacterium, for example, has recently re-emerged as a potentially lethal pathogen, and the microorganism that causes pneumonia is growing progressively more dangerous. Medical researchers are locked in an arms race with the rapidly evolving pathogens that is certain to grow more intense. They are obliged to turn to a broader array of wild species in order to acquire new weapons of medicine in the twenty-first century.

Even when all this much is granted, the third stage of denial emerges: Why rush to save all the species right now? Why not keep live specimens in zoos and botanical gardens and return them to the wild later? The grim truth is that all the zoos in the world today can sustain a maximum of only two thousand species of mammals, birds, reptiles, and amphibians out of twenty-four thousand known to exist. The world's botanical gardens would be even more overwhelmed by the quarter-million plant species. These refuges are invaluable in helping to save a few endangered species. So is freezing embryos in liquid nitrogen. But such measures cannot come close to solving the problem as a whole. To add to the difficulty, no one has yet devised a safe harbor for the legion of insects, fungi, and other ecologically vital small organisms.

Even if all that were accomplished, and scientists prepared to return species to independence, the ecosystems in which many lived would no longer exist. Raw land does not suffice. Pandas and tigers, for example, cannot survive in abandoned rice paddies. Can the natural ecosystems be reconstituted by just putting all the species back together again? The feat is at the present time impossible, at least for communities as complex as rain forests. The order of difficulty, as I described it in Chapter 5, is comparable to that of creating a living cell from molecules, or an organism from living cells.

In order to visualize the scope of the problem more concretely, imagine that the last remnant of rain forest in a small tropical country is about to be drowned beneath the rising lake of a hydroelectric project. An unknown number of plant and animal species found nowhere else in the world will disappear beneath the waters. Nothing can be done. The electric power is needed; local political leaders are adamant. People come first! In the final desperate months, a team of biologists scrambles to save the fauna and flora. Their assignment is the following: Collect samples of all the species quickly, before the dam is closed. Maintain the species in zoos, gardens, and laboratory cultures, or else deep-freeze embryos bred from them in liquid nitrogen. Then bring the species back together and resynthesize the community on new ground.

The state of the art is such that biologists cannot accomplish such a task, not if thousands of them came with a billion-dollar budget. They cannot even imagine how to do it. In the forest patch live legions of life forms: perhaps 300 species of birds, 500 butterflies, 200 ants, 50,000 beetles, 1,000 trees, 5,000 fungi, tens of thousands of bacteria and so on down the long roster of major groups. In many of the groups a large minority of the species are new to science, their properties wholly unknown. Each species occupies a precise niche, demanding a certain place, an exact microclimate, particular nutrients, and temperature and humidity cycles by which the sequential phases of the life cycles are timed. Many of the species are locked in symbiosis with other species, and cannot survive unless arrayed with their partners in the correct configurations.

Thus even if the biologists pulled off the taxonomic equivalent of the Manhattan Project, sorting and preserving cultures of all the species, they could not then put the community back together again. Such a task anywhere in the world is like unscrambling an egg with a pair of spoons. Eventually, perhaps in decades, it can be done. But for the present the biology of the microorganisms needed to reanimate the soil is mostly unknown. The pollinators of most of the flowers and the correct timing of their appearance can only be guessed. The "assembly rules," the sequence in which species must be allowed to colonize in order to coexist indefinitely, are still largely in the realm of theory.

In this matter the opinion of biologists and conservationists is virtually unanimous: The only way to save the Creation with existing knowledge is to maintain it in natural ecosystems. Considering how rapidly such habitats are shrinking, even that straightforward solution will be a daunting task. Somehow humanity must find a way to squeeze through the bottleneck without destroying the environments on which the rest of life depends.

THE LEGACY of the Enlightenment is the belief that entirely on our own we can know, and in knowing, understand, and in understanding, choose wisely. That self-confidence has risen with the exponential growth of scientific knowledge, which is being woven into an increasingly full explanatory web of cause and effect. In the course of the enterprise, we have learned a great deal about ourselves as a species. We now better understand where humanity came from, and what it is. *Homo sapiens*, like the rest of life, was self-assembled. So here we are, no one having guided us to this condition, no one looking over our shoulder, our future entirely up to us. Human autonomy having thus been recognized, we should now feel more disposed to reflect on where we wish to go.

In such an endeavor it is not enough to say that history unfolds by processes too complex for reductionistic analysis. That is the white flag of the secular intellectual, the lazy modernist equivalent of The Will of God. On the other hand, it is too early to speak seriously of ultimate goals, such as perfect green-belted cities and robot expeditions to the nearest stars. It is enough to get *Homo sapiens* settled down and happy before we wreck the planet. A great deal of serious thinking is needed to navigate the decades immediately ahead. We are gaining in our ability to identify options in the political economy most likely to be ruinous. We have begun to probe the foundations of human nature, revealing what people intrinsically most need, and why. We are entering a new era of existentialism, not the old absurdist existentialism of Kierkegaard and Sartre, giving complete autonomy to the individual, but the concept that only unified learning, universally shared, makes accurate foresight and wise choice possible.

In the course of all of it we are learning the fundamental principle that ethics is everything. Human social existence, unlike animal sociality, is based on the genetic propensity to form long-term contracts that evolve by culture into moral precepts and law. The rules of contract formation were not given to humanity from above, nor did they emerge randomly in the mechanics of the brain. They evolved over tens or hundreds of millennia because they conferred upon the genes prescribing them survival and the opportunity to be represented in future generations. We are not errant children who occasionally sin by disobeying instructions from outside our species. We are adults who have discovered which covenants are necessary for survival, and we have accepted the necessity of securing them by sacred oath.

The search for consilience might seem at first to imprison creativity. The opposite is true. A united system of knowledge is the surest means of identifying the still unexplored domains of reality. It provides a clear map of what is known, and it frames the most productive questions for future inquiry. Historians of science often observe that asking the right question is more important than producing the right answer. The right answer to a trivial question is also trivial, but the right question, even when insoluble in exact form, is a guide to major discovery. And so it will ever be in the future excursions of science and imaginative flights of the arts.

I believe that in the process of locating new avenues of creative thought, we will also arrive at an existential conservatism. It is worth asking repeatedly: Where are our deepest roots? We are, it seems, Old World, catarrhine primates, brilliant emergent animals, defined genetically by our unique origins, blessed by our newfound biological genius, and secure in our homeland if we wish to make it so. What does it all mean? This is what it all means. To the extent that we depend on prosthetic devices to keep ourselves and the biosphere alive, we will render everything fragile. To the extent that we banish the rest of life, we will impoverish our own species for all time. And if we should surrender our genetic nature to machine-aided ratiocination, and our ethics and art and our very meaning to a habit of careless discursion in the name of progress, imagining ourselves godlike and absolved from our ancient heritage, we will become nothing.

NOTES

CHAPTER 1

The Ionian Enchantment

3 **Autobiographical details** of my introduction through religious experience to scientific synthesis are given in my memoir *Naturalist* (Washington, DC: Island Press/Shearwater Books, 1994).

4 The idea of the **Ionian Enchantment** is introduced and Einstein's expression of it used as an illustration by Gerald Holton in *Einstein, History, and Other Passions* (Woodbury, NY: American Institute of Physics Press, 1995).

7 Arthur Eddington, in order to celebrate boldness and risk-taking as components of major scientific endeavor, narrated the story of **Daedalus and Icarus** in his British Association Address of 1920. The metaphor was then used by Subrahmanyan Chandrasekhar to characterize the research style of his friend in *Eddington: The Most Distinguished Astrophysicist of His Time* (New York: Cambridge University Press, 1983).

CHAPTER 2

The Great Branches of Learning

11–12 The divided and often contentious **nature of the philosophy of science** is graphically revealed in interviews and conversations recorded by Werner Callebaut in *Taking the Naturalistic Turn, or, How Real Philosophy of Science is Done* (Chicago: University of Chicago Press, 1993).

11 **Alexander Rosenberg on science and philosophy**: *The Philosophy of Social Science,* 1st ed. (Oxford: Oxford University Press, 1988), p. 1.

12 Sir Charles Scott Sherrington speaks of the **enchanted loom** as follows: "Swiftly the head-mass becomes an enchanted loom where millions of flashing shuttles weave a dissolving pattern, always a meaningful pattern though never an abiding one; a shifting harmony of subpatterns" (*Man On His Nature,* The Gifford Lectures, Edinburgh, 1937–8; New York: Macmillan, 1941), p. 225.

12 I first presented the concept of **deep history**, a seamless continuity between prehistory and traditional history, in "Deep history," *Chronicles*, 14: 16–18 (1990).

11–13 On **scientific illiteracy** in the United States: Morris H. Shamos, *The Myth of Scientific Literacy* (New Brunswick, NJ: Rutgers University Press, 1995), and David L. Goodstein, "After the big crunch," *The Wilson Quarterly*, 19: 53–60 (1995).

12–13 Data on the history of **general education in the United States** are provided by Stephen H. Balch et al., *The Dissolution of General Education: 1914–1993*, a report prepared by the National Association of Scholars (Princeton, NJ: The Association, 1996).

CHAPTER 3
THE ENLIGHTENMENT

16 Isaiah Berlin praised the **achievements of the Enlightenment** in *The Age of Enlightenment: The Eighteenth-Century Philosophers* (New York: Oxford University Press, 1979).

16–21 My sources on **Condorcet** were: *Sketch for a Historical Picture of the Progress of the Human Mind*, by Jean-Antoine-Nicolas de Caritat, the Marquis de Condorcet 1743–1794, a partial English translation by Edward Goodell (citation below); *The Centenary of Condorcet*, by Henry Ellis (London: William Reeves, 1894); *Condorcet: From Natural Philosophy to Social Mathematics*, by Keith Michael Baker (Chicago: University of Chicago Press, 1975); and *The Noble Philosopher: Condorcet and the Enlightenment*, by Edward Goodell (Buffalo, NY: Prometheus Books, 1994).

23–28 The sketch of the life and work of **Francis Bacon** provided here was drawn from his writings and from many secondary sources, the most important being James Stephens, *Francis Bacon and the Style of Science* (Chicago: University of Chicago Press, 1975); Benjamin Farrington, *Francis Bacon: Philosopher of Industrial Science* (New York: Octagon Books, 1979); Peter Urbach, *Francis Bacon's Philosophy of Science: An Account and a Reappraisal* (La Salle, IL: Open Court, 1987); and Catherine Drinker Bowen, *Francis Bacon: The Temper of a Man* (New York: Fordham University Press, 1993). Urbach, in a highly appreciative analysis, argues that Bacon advocated imaginative hypothesis formation at all stages of research and was not committed to the gathering of raw data at the outset of investigation, thus making him a much more modern thinker than granted by traditional interpretations of his texts.

28 My placement of the Enlightenment founders in **mythic roles** of an epic adventure was inspired by Joseph Campbell's *The Hero with a Thousand Faces* (New York: Pantheon Books, 1949) and its application to popular culture by Christopher Vogler in *The Writer's Journey: Mythic Structures for Screenwriters and Storytellers* (Studio City, CA: Michael Wiese Productions, 1992).

28–29 An excellent recent account of **Descartes' life and achievements** is given by Stephen Gaukroger in *Descartes: An Intellectual Biography* (New York: Oxford University Press, 1995).

30–31 Joseph Needham's interpretation of **Chinese science** was taken from *The Shorter Science and Civilisation in China: An Abridgement of Joseph Needham's Original Text*, volume I, prepared by Colin A. Ronan (New York: Cambridge University Press, 1978).

32 **Einstein's remark to Ernst Straus** is quoted by Gerald Holton in *Thematic Origins of Scientific Thought* (Cambridge, MA: Harvard University Press, 1988).

36 **Goethe on all-seeing Nature** is taken from *Gesammte Werke, Goethe*, volume XXX (Stuttgart: Cotta, 1858), p. 313, as translated by Sir Charles Scott Sherrington in *Goethe On Nature & On Science*, second edition (New York: Cambridge University Press, 1949).

37–38 The translation of **Pico della Mirandola's instruction from God** is one of the more poetically pleasing ones, and is found in *The Renaissance Philosophy of Man*, edited by Ernst Cassirer, Paul O. Kristeller, and John H. Randall, Jr. (Chicago: University of Chicago Press, 1948), p. 225.

38 The **growth of science** since 1700 is documented and discussed by David L. Goodstein in "After the big crunch," *The Wilson Quarterly*, 19: 53–60 (1995).

39–40 On **modernism**: Carl E. Schorske in *Fin-de-Siècle Vienna: Politics and Culture* (New York: Knopf, 1980). Howard Gardner examines modernism from a psychologist's perspective in *Creating Minds: An Anatomy of Creativity Seen Through the Lives of Freud, Einstein, Picasso, Stravinsky, Eliot, Graham, and Gandhi* (New York: BasicBooks, 1993), p. 397.

40 **C. P. Snow** deplored the separation between the literary and scientific cultures in his celebrated tract *The Two Cultures and the Scientific Revolution* (New York: Cambridge University Press, 1959), based on his 1959 Rede Lecture.

41 The works by **Jacques Derrida** on which I have based my admittedly less-than-enthusiastic impressions are *Of Grammatology*, translated by Gayatri Chakravorty Spivak (Baltimore: Johns Hopkins University Press, 1976); *Writing and Difference*, translated by Alan Bass (Chicago: University of Chicago Press, 1978); and *Dissemination*, translated by Barbara Johnson (Chicago: University of Chicago Press, 1981). Given Derrida's deliberately surreal style, much is owed the exegeses given by the translators in their introductions.

42 On **root metaphors in psychology**: Kenneth J. Gergen, "Correspondence versus autonomy in the language of understanding human action," in Donald W. Fiske and Richard A. Shweder, eds., *Metatheory in Social Science: Pluralisms and Subjectivities* (Chicago: University of Chicago Press, 1986), pp. 145–146.

43 George Scialabba wrote about **Michel Foucault** in "The tormented quest of Michel Foucault," a review of *The Passion of Michel Foucault*, by James Miller (New York: Simon & Schuster, 1993), in *The Boston Sunday Globe*, 3 January 1993, p. A12. An earlier and fuller account of Foucault's scholarship, including his "archeology of knowledge," is provided by Alan Sheridan in *Michel Foucault: The Will to Truth* (London: Tavistock, 1980).

CHAPTER 4
THE NATURAL SCIENCES

45–47 Among the many textbooks and other introductory accounts of **animal senses** available, one of the best and most widely used is John Alcock's *Animal Behavior: An Evolutionary Approach*, fifth edition (Sunderland, MA: Sinauer Associates, 1993).

48–49 Eugene P. Wigner's description of **mathematics as the natural language of physics** is in "The unreasonable effectiveness of mathematics in the natural sciences," *Communications on Pure and Applied Mathematics*, 13: 1–14 (1960).

49–50 The account of **quantum electrodynamics** (Q.E.D.) and measurement of properties of the electron is taken from David J. Gross, "Physics and mathematics at the frontier," *Proceedings of the National Academy of Sciences, USA*, 85: 8371–5 (1988), and John R. Gribbin's *Schrödinger's Kittens and the Search for Reality: Solving the Quantum Mysteries* (Boston: Little, Brown, 1995). To Gribbin I owe the imagery of the flight of a needle across the United States to illustrate the accuracy of Q.E.D.

50–51 The prospects of **nanotechnology**, along with scanning-tunneling and atomic force microscopy, are described by the multiple authors of *Nanotechnology: Molecular Speculations on Global Abundance*, edited by B. C. Crandall (Cambridge, MA: MIT Press, 1996). The manufacture of **high-density ROMs** is described in *Science News*, 148: 58 (1995). The exact timing of **chemical reactions** is described by Robert F. Service in "Getting a reaction in close-up," *Science*, 268: 1846 (1995); and membranelike **self-assembled monolayers** of molecules by George M. Whitesides in "Self-assembling materials," *Scientific American*, 273: 146–57 (1995).

57 **Einstein's tribute to Planck** has been often quoted. I do not know the original attribution, but the words can be found, for example, in Walter Kaufmann's *The Future of the Humanities* (New York: Reader's Digest Press, distributed by Thomas Y. Crowell, 1977).

57–59 The **individuality of the scientist**, his frailties, and his pursuit of research as an art form are searchingly probed by Freeman Dyson in "The scientist as rebel," *The New York Review of Books*, 25 May 1995, pp. 31–3. His views on the subject, independently evolved as a physicist, are in many respects closely similar to my own.

59–60 The original report on **conserved DNA duplication** was published by Matthew S. Meselson and Franklin W. Stahl in *Proceedings of the National Academy of Sciences, USA*, 44: 671–82 (1958). I am grateful to Meselson for a personal discussion of the experiment.

61–64 My synopsis of the history and content of **logical positivism** and the quest for objective truth is based on many texts and informal discussions with scientists and others, but has been most influenced in recent years by Gerald Holton's *Science and Anti-Science* (Cambridge, MA: Harvard University Press, 1993), and Alexander Rosenberg's *Economics: Mathematical Politics or Science of Diminishing Returns?* (Chicago: University of Chicago Press, 1992).

64 Herbert A. Simon has written on the psychology of **creative thought** in "Discovery, invention, and development: human creative thinking," *Proceedings of the National Academy of Sciences, USA* (Physical Sciences), 80: 4569–71 (1983).

CHAPTER 5
Ariadne's Thread

66–67 The **Cretan labyrinth and Ariadne's thread** have been given diverse metaphorical interpretations over the years. The closest to my own, yet different in key respects, is Mary E. Clark's *Ariadne's Thread: The Search for New Modes of Think-*

ing (New York: St. Martin's Press, 1989). Clark perceives the labyrinth as humanity's complex environmental and social problems and the thread as the objective truths and realistic thinking needed to solve them.

68–71 The details of **ant communication** can be found in *The Ants* (1990) and *Journey to the Ants: A Story of Scientific Exploration* (1994), by Bert Hölldobler and Edward O. Wilson (Cambridge, MA: Belknap Press of Harvard University Press).

72–74 Ancestor-summoning by the **Jívaro** is described by Michael J. Harner in *The Jívaro: People of the Sacred Waterfalls* (Garden City, NY: Doubleday/Natural History Press, 1972). The dreams and art of **Pablo Amaringo** are presented in *Ayahuasca Visions: The Religious Iconography of a Peruvian Shaman*, by Luis Eduardo Luna and Pablo Amaringo (Berkeley, CA: North Atlantic Books, 1991).

74–78 Current understanding of the **biology of dreaming** is explained by J. Allan Hobson in *The Chemistry of Conscious States: How the Brain Changes Its Mind* (Boston: Little, Brown, 1994) and *Sleep* (New York: Scientific American Library, 1995). Many of the technical details of current studies of the structure and physiology of dreaming are reviewed in "Dream consciousness: a neurocognitive approach," a special issue of *Consciousness and Cognition*, 3: 1–128 (1994). Recent research on the adaptive function of sleep is reported by Avi Karni et al. in "Dependence on REM sleep of overnight improvement of a perceptual skill," *Science*, 265: 679–82 (1994).

78–81 The relation between **live snakes and dream serpents** in the origin of dreams and myth given here is based largely on Balaji Mundkur's important monograph *The Cult of the Serpent: An Interdisciplinary Survey of Its Manifestations and Origins* (Albany, NY: State University of New York Press, 1983) plus, with little modification, the extensions I made in *Biophilia* (Cambridge, MA: Harvard University Press, 1984).

81–83 I first used the imagery of changing **space-time scales** as magical cinematography in *Biophilia* (Cambridge, MA: Harvard University Press, 1984).

83–84 In characterizing the difficulty of **predicting protein structure** from the interaction of its constituent atoms, I benefited greatly from an unpublished paper presented by S. J. Singer at the American Academy of Arts and Sciences in December 1993; he has also kindly reviewed my account.

84–85 **Higher-order interactions** in rain forests is described in my book *The Diversity of Life* (Cambridge, MA: Belknap Press of Harvard University Press, 1992), and in ecosystems generally in a special section edited by Peter Kareiva in the journal *Ecology*, 75: 1527–59 (1994).

87–91 An excellent introduction to the meaning and goals of **complexity theory** is given by Harold Morowitz in the main journal of the discipline, of which he is editor, *Complexity*, 1: 4–5 (1995); and by Murray Gell-Mann in the same issue, pp. 16–19. Among the many full-scale expositions of the subject that have appeared in the 1990s, the best include *The Origins of Order: Self-Organization and Selection in Evolution*, by Stuart A. Kauffman (New York: Oxford University Press, 1993); and *The Collapse of Chaos: Discovering Simplicity in a Complex World*, by Jack Cohen and Ian Stewart (New York: Viking, 1994).

92 The **cell as a system of genetic networks** is described by William F. Loomis and Paul W. Sternberg in "Genetic networks," *Science*, 269: 649 (1995). Their account

is based on the longer, more technical report by Harley H. McAdams and Lucy Shapiro in the same issue (pp. 650–6).

93 The exponential rise in **computer performance** is described by Ivars Peterson in "Petacrunchers: setting a course toward ultrafast supercomputing," *Science News*, 147: 232–5 (1995); and by David A. Patterson in "Microprocessors in 2020," *Scientific American*, 273: 62–7 (1995). Peta- refers to the order of magnitude 10^{15}, or a thousand trillion.

93–94 The **opinions of cell biologists** on the most important problems of **cell and organismic development** are reported by Marcia Barinaga in "Looking to development's future," *Science*, 266: 561–4 (1994).

CHAPTER 6
THE MIND

96–124 Many of the leading brain scientists have written recent accounts of their subject for the broader public. Fortunately, those of most recent vintage contain among them the full range of views held by members of the research community. The best such works on the structure of the brain and the neural and biochemical correlates of behavior include *The Engine of Reason, the Seat of the Soul: A Philosophical Journey into the Brain*, by Paul M. Churchland (Cambridge, MA: MIT Press, 1995); *The Astonishing Hypothesis: The Scientific Search for the Soul*, by Francis Crick (New York, Scribner, 1994); *Descartes' Error: Emotion, Reason, and the Human Brain*, by Antonio R. Damasio (New York: G. P. Putnam, 1994); *Bright Air, Brilliant Fire: On the Matter of the Mind*, by Gerald M. Edelman (New York: BasicBooks, 1992); *The Chemistry of Conscious States: How the Brain Changes Its Mind*, by J. Allan Hobson (Boston: Little, Brown, 1994); *Image and Brain: The Resolution of the Imagery Debate*, by Stephen M. Kosslyn (Cambridge, MA: MIT Press, 1994); *Wet Mind: The New Cognitive Neuroscience*, by Stephen M. Kosslyn and Olivier Koenig (New York: Free Press, 1992); *How the Mind Works*, by Steven Pinker (New York: W. W. Norton, 1997); and *Images of Mind*, by Michael I. Posner and Marcus E. Raichle (New York: Scientific American Library, 1994). A thoroughgoing review of contemporary research on emotion is provided by multiple authors in *The Nature of Emotion: Fundamental Questions*, edited by Paul Ekman and Richard J. Davidson (New York: Oxford University Press, 1994). The poetic allusion to the divisions of the brain as heartbeat, heartstrings, and heartless was made by Robert E. Pool in *Eve's Rib: The Biological Roots of Sex Differences* (New York: Crown, 1994).

The **contemporary view of conscious experience** is explored to varying degrees of penetration by the above works. The many ramifications in philosophy opened by neurobiological research are a principal focus in the following notable works: *Neurophilosophy: Toward a Unified Science of the Mind-Brain*, by Patricia S. Churchland (Cambridge, MA: MIT Press, 1986); *Consciousness Explained*, by Daniel C. Dennett (Boston: Little, Brown, 1991); *Darwin's Dangerous Idea: Evolution and the Meanings of Life*, by Daniel C. Dennett (New York: Simon & Schuster, 1995); and *The Rediscovery of the Mind*, by John R. Searle (Cambridge, MA: MIT Press, 1992).

Roger Penrose, in *Shadows of the Mind: A Search for the Missing Science of Consciousness* (New York: Oxford University Press, 1994), argues that neither conventional science nor artificial computation will solve the problem of mind. He visualizes a radical new approach, arising from quantum physics and a new look at cellular physiology; few brain scientists, however, feel any urgency to depart from the present course of investigation, which has progressed so dramatically to the present time.

Other **special aspects of modern research on consciousness** are explored in *The Creative Mind: Myths & Mechanisms*, by Margaret A. Boden (New York: BasicBooks, 1991); *Emotional Intelligence*, by Daniel Goleman (New York: Bantam Books, 1995); *The Emotional Computer*, by José A. Jáuregui (Cambridge, MA: Blackwell, 1995); *The Sexual Brain*, by Simon LeVay (Cambridge, MA: MIT Press, 1993); and *The Language Instinct: The New Science of Language and Mind*, by Steven Pinker (New York: W. Morrow, 1994).

In constructing **my own brief account of the physical basis of mind**, I have drawn to varying degrees on each of the foregoing works and on consultation with some of the authors as well as other researchers in the brain sciences. I have also used the outstanding reviews and peer commentaries published in the journal *Behavioral and Brain Sciences*.

97 The number of **genes engaged in human brain development** is reported in *Nature* magazine's *The Genome Directory*, 28 September 1995, p. 8, table 8.

100–118 References to certain specific examples cited in the chapter are the following. On the **Phineas Gage** case and the role of the prefrontal lobe: Hanna Damasio et al., "The return of Phineas Gage: clues about the brain from the skull of a famous patient," *Science*, 264: 1102–5 (1994); and Antonio Damasio in *Descartes' Error*; and on **Karen Ann Quinlan** and the role of the thalamus, Kathy A. Fackelmann in "The conscious mind," *Science News*, 146: 10–11 (1994). On the **exploration of brain neurons**: Santiago Ramón y Cajal, *Recollections of My Life* (Memoirs of the American Philosophical Society, v. 8) (Philadelphia: American Philosophical Society, 1937), p. 363. On the **brain's categorical processing** of animals as opposed to tools: Alex Martin, Cheri L. Wiggs, Leslie G. Ungerleider, and James V. Haxby, "Neural correlates of category-specific knowledge," *Nature*, 379: 649–52 (1996). The imaginary example of **interaction of body and brain** is adapted from one given by Antonio Damasio in *Descartes' Error*. The **"hard problem"** of the brain science is explained by David J. Chalmers in "The puzzle of conscious experience," *Scientific American*, 273: 80–6 (December 1995). Daniel C. Dennett has thoroughly explored and independently solved it in *Consciousness Explained* (Boston: Little, Brown, 1991). Simon Leys' interpretation of **Chinese calligraphy** is presented in his review of *The Chinese Art of Writing*, by Jean François Billeter (New York: Skira/Rizzoli, 1990), in *The New York Review of Books*, 43: 28–31 (1996).

120–124 The definition of **artificial intelligence** (AI) used is from an essay by Gordon S. Novak, Jr., in the *Academic Press Dictionary of Science and Technology*, edited by Christopher Morris (San Diego: Academic Press, 1992), p. 160. An excellent account of the use of AI in playing chess and other deterministic games (checkers, go, and bridge) is provided by Fred Guterl in "Silicon Gambit," *Discover*, 17: 48–56 (June 1996).

CHAPTER 7
FROM GENES TO CULTURE

126–128 The full conception of **gene-culture coevolution** (and the term) was introduced by Charles J. Lumsden and myself in *Genes, Mind, and Culture: The Coevolutionary Process* (Cambridge, MA: Harvard University Press, 1981) and *Promethean Fire* (Cambridge, MA: Harvard University Press, 1983). Key models of the interaction of heredity and culture leading to this formulation were constructed by Robert Boyd and Peter J. Richerson in 1976, Mark W. Feldman and L. Luca Cavalli-Sforza in 1976, William H. Durham in 1978, and myself in 1978. Recent reviews of gene-culture coevolution as advanced to date include those by William H. Durham, *Coevolution: Genes, Culture, and Human Diversity* (Stanford, CA: Stanford University Press, 1991); "The mathematical modelling of human culture and its implications for psychology and the human sciences," by Kevin N. Laland, *British Journal of Psychology*, 84: 145–69 (1993); and "Sociobiology and sociology," by François Nielsen, *Annual Review of Sociology*, 20: 267–303 (1994). These authors have all made important original contributions. Each places different emphases and interpretations on the different sections of the coevolutionary cycle, and would no doubt question some details in the brief interpretation presented here; but I believe the core of my argument closely approaches the consensus.

128 **Jacques Monod's** book *Chance and Necessity: An Essay on the Natural Philosophy of Modern Biology* (New York: Knopf, 1971) contains as epigraph this statement by Democritus: "Everything existing in the Universe is the fruit of chance and necessity."

130 On the **definition of culture**, see Alfred L. Kroeber, *Anthropology*, with supplements 1923–33 (New York: Harcourt, Brace and World, 1933); Alfred L. Kroeber and Clyde K. M. Kluckhohn, "Culture: a critical review of concepts and definitions" (Papers of the Peabody Museum of American Archaeology and Ethnology, Harvard University, v. 47, no. 12, pp. 643–4, 656) (Cambridge, MA: The Peabody Museum, 1952); and Walter Goldschmidt, *The Human Career: The Self in the Symbolic World* (Cambridge, MA: B. Blackwell, 1990). For an account of the corruption of the term "culture" in recent popular literature, consult "Welcome to post-culturalism," by Christopher Clausen in *The American Scholar*, 65: 379–88 (1996).

131–133 The nature of **intelligence in bonobos and other great apes**, as well as culture (or absence of it), is the subject of a large recent literature. The topics I have covered here are presented in greater detail and in various parts by E. Sue Savage-Rumbaugh and Roger Lewin in *Kanzi: The Ape at the Brink of the Human Mind* (New York: Wiley, 1994); *Chimpanzee Cultures*, edited by Richard W. Wrangham, W. C. McGrew, Frans de Waal, and Paul G. Heltne (Cambridge, MA: Harvard University Press, 1994); two general reviews by Frans de Waal from Harvard University Press, *Peacemaking among Primates* (1989) and *Good Natured: The Origins of Right and Wrong in Humans and Other Animals* (1996); and "New clues surface about the making of the mind," by Joshua Fischman in *Science*, 262: 1517 (1993). The silence of chimpanzees in contrast to the compulsive volubility of humans is described by John L. Locke in "Phases in the child's development of language," *American Scientist*, 82: 436–45 (1994). The evaluation of speech and bonding is examined by Anne Fernald in "Human maternal vocalizations to infants as biologically relevant signals: an evolutionary perspective," in Jerome H. Barkow, Leda Cosmides, and John Tooby, eds., *The Adapted Mind: Evolutionary Psychology and the Generation of Culture* (New York: Oxford University Press, 1992), pp. 391–428.

133 The **precocity of infant imitation** is described by Andrew N. Meltzoff and M. Keith Moore in "Imitation of facial and manual gestures by human neonates," *Science*, 19: 75–8 (1977); and "Newborn infants imitate adult facial gestures," *Child Development*, 54: 702–9 (1983).

133–134 The early stages of **human culture** as revealed by recent archaeological discoveries are reported in "Old dates for modern behavior," by Ann Gibbons, *Science*, 268: 495–6 (1995); "Did *Homo erectus* tame fire first?," by Michael Balter, in *Science*, 268: 1570 (1995); and "Did Kenya tools root birth of modern thought in Africa?," by Elizabeth Culotta, in *Science*, 270: 1116–7 (1995). The modern proliferation of material culture is described by Henry Petroski in "The evolution of artifacts," *American Scientist*, 80: 416–20 (1992).

134–135 The distinction between the **two basic classes of memory** was made by Endel Tulving in E. Tulving and Wayne Donaldson, eds., *Organization of Memory* (New York: Academic Press, 1972), pp. 382–403.

136 The definition of memes, the **units of culture**, as nodes in semantic memory was proposed by Charles J. Lumsden and Edward O. Wilson in "The relation between biological and cultural evolution," *Journal of Social and Biological Structures*, 8: 343–59 (1985).

137–142 An introduction to the measures of **norm of reaction and heritability** is now standard in introductory textbooks on genetics, as well as in many on general biology. More detailed accounts and applications are provided, among numerous references available, in *Introduction to Quantitative Genetics*, fourth edition, by Douglas S. Falconer and Trudy F. C. Mackay (Essex, England: Longman, 1996); *Human Heredity: Principles and Issues*, fourth edition, by Michael R. Cummings (New York: West Publishing Company, 1997); and *Behavioral Genetics*, third edition, by Robert Plomin et al. (New York: W. H. Freeman, 1997). A summary of some important recent research on the heritability of human behavioral traits is given by Thomas J. Bouchard, Jr., et al. in "Sources of human psychological differences: the Minnesota study of twins reared apart," *Science*, 250: 223–8 (1990).

143–145 Recent research on the **biological basis of schizophrenia** is summarized by Leena Peltonen in "All out for chromosome six," *Nature*, 378: 665–6 (1995); by B. Brower in "Schizophrenia: fetal roots for GABA loss," *Science News*, 147: 247 (1995); and, on brain activity during psychotic episodes, by D. A. Silbersweig et al., "A functional neuroanatomy of hallucinations in schizophrenia," *Nature*, 378: 176–9 (1995), and R. J. Dolan et al., "Dopaminergic modulation of impaired cognitive activation in the anterior cingulate cortex in schizophrenia," *Nature*, 378: 180–2 (1995).

146 The estimated number of polygenes determining **human skin color** is discussed by Curt Stern in *Principles of Human Genetics*, third edition (San Francisco: W. H. Freeman, 1973).

147 The **universals of culture** were identified by George P. Murdock in "The common denominator of cultures," in Ralph Linton, ed., *The Science of Man in the World Crisis* (New York: Columbia University Press, 1945). An excellent update

and evaluation with the aid of anthropological and sociobiological principles is provided by Donald E. Brown in *Human Universals* (Philadelphia: Temple University Press, 1991).

148 My imaginary exercise on **termite civilization**, presented to emphasize the uniqueness of human nature, is taken from "Comparative social theory," *The Tanner Lectures on Human Values*, v. I (Salt Lake City: University of Utah Press, 1980), pp. 49–73.

149 The **convergence of institutions** in advanced societies of the Old and New Worlds was characterized by Alfred V. Kidder in "Looking backward," *Proceedings of the American Philosophical Society*, 83: 527–37 (1940).

150 The **principle of prepared learning** was formulated by Martin E. P. Seligman and others in *Biological Boundaries of Learning*, compiled by Seligman and Joanne L. Hager (New York: Appleton-Century-Crofts, 1972).

150–154 The **epigenetic rules** of human social behavior were enumerated and classified by Charles J. Lumsden and Edward O. Wilson in *Genes, Mind, and Culture* in 1981 (Cambridge, MA: Harvard University Press). Among the best comprehensive treatments of the rules in recent years have been *Human Ethology*, by Irenäus Eibl-Eibesfeldt (Hawthorne, NY: Aldine de Gruyter, 1989); *Coevolution: Genes, Culture, and Human Diversity*, by William H. Durham (Stanford, CA: Stanford University Press, 1991); and the authors of *The Adapted Mind*, edited by Jerome H. Barkow, Leda Cosmides, and John Tooby (New York: Oxford University Press, 1992), and especially the essay by Tooby and Cosmides, "The psychological foundations of culture," pp. 19–136.

151–152 The transition from **Moro's reflex** of newborns to the life-long startle reflex is drawn from Luther Emmett Holt and John Howland, *Holt's Diseases of Infancy and Childhood*, eleventh edition, revised by L. E. Holt, Jr., and Rustin McIntosh (New York: D. Appleton-Century, 1940). The universal audiovisual bias in vocabularies of the senses is based on research by C. J. Lumsden and E. O. Wilson, presented in *Genes, Mind, and Culture* (Cambridge, MA: Harvard University Press, 1981), pp. 38–40. The swift fixation by newborns on the mother's face was first established in experiments by Carolyn G. Jirari, reported in a Ph.D. thesis cited by Daniel G. Freedman in *Human Infancy: An Evolutionary Perspective* (Hillsdale, NJ: L. Erlbaum Associates, 1974). The results were confirmed and extended in *Biology and Cognitive Development: The Case of Face Recognition*, by Mark Henry Johnson and John Morton (Cambridge, MA: B. Blackwell, 1991).

152–153 The cross-cultural pattern of **smiling** is from the account by Melvin J. Konner in "Aspects of the developmental ethology of a foraging people," in Nicholas G. Blurton Jones, ed., *Ethological Studies of Child Behavior* (New York: Cambridge University Press, 1972), p. 77; two contributions by Irenäus Eibl-Eibesfeldt, "Human ethology: concepts and implications for the sciences of man," *Behavioral and Brain Sciences*, 2: 1–57 (1979), and *Human Ethology* (Hawthorne, NY: Aldine de Gruyter, 1989). The combined account given here is taken with little change from C. J. Lumsden and E. O. Wilson, *Genes, Mind, and Culture* (Cambridge, MA: Harvard University Press, 1981)., pp. 77–8.

153–154 The account of **reification and the dyadic principle** is based on C. J. Lumsden and E. O. Wilson, ibid., pp. 93–5, with the example of the Dusun of Borneo taken from Thomas Rhys Williams' *Introduction to Socialization: Human Culture Transmitted* (St. Louis, MO: C. V. Mosby, 1972).

155 **The heredity of dyslexia** is discussed by Chris Frith and Uta Frith in "A biological marker for dyslexia," *Nature*, 382: 19–20 (1996). The current status of behavioral genetics of both animals and humans is authoritatively evaluated in a series of articles published under the heading "Behavioral genetics in transition" in *Science*, 264: 1686–739 (1994).

155 The Dutch "**aggression gene**" is analyzed by H. G. Brunner et al. in "X-linked borderline mental retardation with prominent behavioral disturbance: phenotype, genetic localization, and evidence for disturbed monoamine metabolism," *American Journal of Human Genetics*, 52: 1032–9 (1993). The gene associated with novelty seeking is reported by Richard P. Ebstein et al. in "Dopamine D4 receptor (*D4DR*) exon III polymorphism associated with the human personality trait of Novelty Seeking," *Nature Genetics*, 12: 78–80 (1996).

158–159 The account of **paralanguage** is based on a comprehensive study by Irenäus Eibl Eibesfeldt, *Human Ethology* (Hawthorne, NY: Aldine de Gruyter, 1989), pp. 424–92.

159–163 The account given here on the origin of **color vocabularies** has been assembled from many sources, but mostly from the recently published and important series of articles by Denis Baylor, John Gage, John Lyons, and John Mollon in *Colour: Art & Science*, edited by Trevor Lamb and Janine Bourriau (New York: Cambridge University Press, 1995). The description of the cross-cultural studies of color vocabulary has been modified from C. J. Lumsden and E. O. Wilson, *Promethean Fire* (Cambridge, MA: Harvard University Press, 1983). I have also weighed (and recommend) an informative critique of the mainstream psychophysiological explanation provided by multiple authors, and stoutly defended by others, forming the majority, in the peer-commentary review journal *Behavioral and Brain Sciences*, 20 (2): 167–228 (1997). I am grateful to William H. Bossert and George F. Oster for calculating the theoretical maximum and the actual, constrained maximum number of color vocabularies that can be created from eleven basic colors.

CHAPTER 8

The Fitness of Human Nature

164–168 Many of the ideas concerning **human nature and the role of epigenetic rules** presented here were first developed by Charles J. Lumsden and Edward O. Wilson in *Genes, Mind, and Culture* (Cambridge, MA: Harvard University Press, 1981) and *Promethean Fire* (Cambridge, MA: Harvard University Press, 1983). Epigenetic rules are also a focus of *The Adapted Mind*, edited by Jerome H. Barkow, Leda Cosmides, and John Tooby (New York: Oxford University Press, 1992).

168–173 The "**classical" approach of sociobiology** to the evolution of culture is the subject of an excellent collection of articles and critiques in *Human Nature: A Critical Reader*, edited by Laura L. Betzig (New York: Oxford University Press, 1997). Much of the research published and synthesized in the 1980s and 1990s has appeared in the journals *Ethology and Sociobiology*, *Behavioral and Brain Sciences*, and *Human Nature*. The intellectual history of sociobiology and other evolutionary approaches to human behavior is ably analyzed by Carl N. Degler, *In Search of Human Nature: The Decline & Revival of Darwinism in American Social Thought* (New York: Oxford University Press, 1991).

168–169 The **origins of kin selection theory** and theory of the family, due chiefly to William D. Hamilton and Robert L. Trivers, are reviewed in Edward O. Wilson, *Sociobiology: The New Synthesis* (Cambridge, MA: Belknap Press of Harvard University Press, 1975), and in many later textbooks and reviews, including, most recently, Laura L. Betzig, ed., *Human Nature: A Critical Reader* (New York: Oxford University Press, 1997).

169–170 Well-documented accounts of **gender differences and mating strategies** in particular are the subjects of *Despotism and Differential Reproduction: A Darwinian View of History*, by Laura L. Betzig (New York: Aldine, 1986); *The Evolution of Desire: Strategies of Human Mating*, by David M. Buss (New York: BasicBooks, 1994); and *Eve's Rib*, by Robert E. Pool (New York: Crown Publishers, 1994).

170–171 The conception of **territorial aggression** arising as a density-dependent factor of population regulation was introduced by Edward O. Wilson in "Competitive and aggressive behavior," in *Man and Beast: Comparative Social Behavior*, John F. Eisenberg and Wilton S. Dillon, eds. (Washington, DC: Smithsonian Institution Press, 1971), pp. 183–217. The deep roots of tribal strife and war are effectively illustrated in preliterate societies by Laurence H. Keeley in *War Before Civilization* (New York: Oxford University Press, 1996) and in more recent history by R. Paul Shaw and Yuwa Wong in *Genetic Seeds of Warfare: Evolution, Nationalism, and Patriotism* (Boston: Unwin Hyman, 1989); Daniel Patrick Moynihan in *Pandaemonium: Ethnicity in International Politics* (New York: Oxford University Press, 1993); and Donald Kagan in *On the Origins of War and the Preservation of Peace* (New York: Doubleday, 1995).

171–172 The evidence for specialized **cheater recognition** in human mental development is presented in "Cognitive adaptations for social exchange," by Leda Cosmides and John Tooby, in Jerome H. Barkow et al., eds., *The Adapted Mind* (New York: Oxford University Press, 1992), pp. 163–228.

173–180 **Human incest avoidance,** as well as that of nonhuman primates, is authoritatively reviewed by Arthur P. Wolf in *Sexual Attraction and Childhood Association: A Chinese Brief for Edward Westermarck* (Stanford, CA: Stanford University Press, 1995). The evidence for direct recognition of inbreeding depression by traditional societies, which serves as an enhancement of the Westermarck effect in the formation of incest taboos, is given by William H. Durham in *Coevolution: Genes, Culture, and Human Diversity* (Stanford, CA: Stanford University Press, 1991).

CHAPTER 9
THE SOCIAL SCIENCES

185–186 The ambivalence of the **American Anthropological Association** toward the sources of human diversity was expressed by James Peacock, AAA president, in "Challenges Facing the Discipline" (*Anthropology Newsletter*, v. 35, no. 9, pp. 1, 3), as follows: "The May 1994 retreat included heads of all Sections and representatives from the Long-Range Planning and Finance committee. The assembly subcommittees . . . both separately and as a body addressed two questions: whither the discipline and whither the AAA. The participants affirmed the strength of abiding commitments to biological and cultural variation and to the refusal to biologize or otherwise essentialize diversity. At the same time, the group expressed a goal of reaching out and strengthening the discipline's relevance."

185–186 For a sample of **histories and critiques of anthropology** from widely differing viewpoints, see Herbert Applebaum, ed., *Perspectives in Cultural Anthropology* (Albany, NY: State University of New York, 1987); Donald E. Brown, *Human Universals* (Philadelphia: Temple University Press, 1991); Carl N. Degler, *In Search of Human Nature: The Decline & Revival of Darwinism in American Social Thought* (New York: Oxford University Press, 1991); Robin Fox, *The Search for Society: Quest for a Biosocial Science and Morality* (New Brunswick, NJ: Rutgers University Press, 1989); Clifford Geertz, *The Interpretation of Cultures: Selected Essays* (New York: BasicBooks, 1973); Walter R. Goldschmidt, *The Human Career: The Self in the Symbolic World* (Cambridge, MA: B. Blackwell, 1990); Marvin Harris, *The Rise of Anthropological Theory: A History of Theories of Culture* (New York: Thomas Y. Crowell, 1968); Jonathan Marks, *Human Biodiversity: Genes, Race, and History* (Hawthorne, NY: Aldine de Gruyter, 1995); and Alexander Rosenberg, *Philosophy of Social Science*, second edition (Boulder, CO: Westview Press, 1995).

186–188 Within academic sociology, the **heresy of foundational biology and psychology** has been promoted by, among a few others, Joseph Lopreato in *Human Nature & Biocultural Evolution* (Boston: Allen & Unwin, 1984); Pierre L. van den Berghe in *The Ethnic Phenomenon* (New York: Elsevier, 1981); and Walter L. Wallace, *Principles of Scientific Sociology* (Hawthorne, NY: Aldine de Gruyter, 1983). A thoroughgoing history of the discipline in its classical period is Robert W. Friedrichs' *A Sociology of Sociology* (New York: Free Press, 1970). The later, model-building period, in which a partial attempt is being made to connect individual behavior to social pattern in the manner of economic theory, is epitomized by James S. Coleman's *Foundations of Social Theory* (Cambridge, MA: Belknap Press of Harvard University Press, 1990).

187 Robert Nisbet explores the **roots of the sociological imagination** in *Sociology as an Art Form* (New York: Oxford University Press, 1976).

188 The felicitous expression **Standard Social Science Model (SSSM)** was introduced by John Tooby and Leda Cosmides in "The Psychological Foundations of Culture," in J. A. Barkow et al., eds., *The Adapted Mind* (New York: Oxford University Press, 1992), pp. 19–136. That it still flourishes within the social sciences is well illustrated by the strongly constructivist tone of *Open the Social Sciences: Report of the Gulbenkian Commission on the Restructuring of the Social Sciences* (Stanford, CA: Stanford University Press, 1996). The central conception within it has been well characterized by many earlier writers, including Donald E. Brown—see *Human Universals* (Philadelphia: Temple University Press, 1991)—and the multiple contributors to *Metatheory in Social Science: Pluralisms and Subjectivities*, edited by Donald W. Fiske and Richard A. Shweder (Chicago: University of Chicago Press, 1986). Tooby and Cosmides, whose assessment is by far the most thorough and persuasive, also introduce the Integrated Causal Model (ICM) to denote the new causal linkage of psychology and evolutionary biology to the study of cultures.

189 The conception of **hermeneutics** as a thick description crafted from differing perspectives is well represented in Fiske and Shweder (ibid.), especially in the articles

"Three scientific world views and the covering law model" by Roy D'Andrade, pp. 19–41, and "Science's social system of validity-enhancing collective belief change and the problems of the social sciences," pp. 108–35.

190 **Richard Rorty's interpretation of hermeneutics** is given in *Philosophy and the Mirror of Nature* (Princeton, NJ: Princeton University Press, 1979).

190–191 The **personalized characterizations of disciplines** in the natural and social sciences is based loosely on my earlier account in "Comparative social theory," *The Tanner Lectures on Human Values*, v. I (Salt Lake City: University of Utah Press, 1980), pp. 49–73.

194–195 Stephen T. Emlen's synthesis of **parent-offspring relations** in birds and mammals is given in "An evolutionary theory of the family," *Proceedings of the National Academy of Sciences, USA*, 92: 8092–9 (1995).

202–204 I have based my interpretation of **Gary S. Becker's research** on his major work *A Treatise on the Family*, enlarged edition, and collection of essays, *Accounting for Tastes* (both from Cambridge, MA: Harvard University Press, 1991 and 1996). I have also benefited from Alexander Rosenberg's insightful *Economics: Mathematical Politics or Science of Diminishing Returns?* (Chicago: University of Chicago Press, 1992). We have substantially different assessments, however, of the prospects for linking the models of economics to psychology and biology, with Rosenberg being the more pessimistic, for reasons described in the text.

206 **Rational choice theory** is often called by other names in the social sciences, including public choice, social choice, and formal theory. Its weaknesses, especially its excessive reliance on abstract and data-free models, have recently been explored by Donald P. Green and Ian Shapiro in *Pathologies of Rational Choice Theory: A Critique of Applications in Political Science* (New Haven: Yale University Press, 1994).

206–208 The examples of **heuristics** ("rules of thumb") used by people during intuitive quantitative reasoning are taken from "Judgment under uncertainty: heuristics and biases," by Amos Tversky and Daniel Kahneman, in *Science*, 185: 1124–31 (1974). An updated explanation of the concept, with other case studies, is provided by the same authors in "On the reality of cognitive illusions," *Psychological Review*, 103: 582–91 (1996).

208 On **reasoning in preliterate people**: Christopher Robert Hallpike in *The Foundations of Primitive Thought* (New York: Oxford University Press, 1979).

208–209 For bleak views by **leading philosophers** of the reductionist approach to human social behavior, and hence the entire program of uniting biology and the social sciences, see Philip Kitcher in *Vaulting Ambition: Sociobiology and the Quest for Human Nature* (Cambridge, MA: MIT Press, 1985) and Alexander Rosenberg in his trilogy: *Philosophy of Social Science* (Boulder, CO: Westview Press, 1988), *Economics: Mathematical Politics or Science of Diminishing Returns?* (Chicago: University of Chicago Press, 1992), and *Instrumental Biology, or the Disunity of Science* (Chicago: University of Chicago Press, 1994). Generally more favorable stances are taken, for example, by the philosophers who contributed to *Sociobiology and Epistemology*, edited by James H. Fetzer (Boston: D. Reidel, 1985), and by Michael Ruse in *Taking Darwin Seriously: A Naturalistic Approach to Philosophy* (Cambridge, MA: B. Blackwell, 1986).

CHAPTER 10

THE ARTS AND THEIR INTERPRETATION

210 The 1979–80 **Report of the Commission on the Humanities** was published as a book: Richard W. Lyman et al., *The Humanities in American Life* (Berkeley: University of California Press, 1980).

211 **George Steiner on the arts** is quoted from his commencement address at Kenyon College, published in *The Chronicle of Higher Education*, 21 June 1996, p. B6.

213 Brain development in the **musically gifted** is reported by G. Schlaug and coworkers in "Increased corpus callosum size in musicians," *Neuropsychologia*, 33: 1047–55 (1995) and "In vivo evidence of structural brain asymmetry in musicians," *Science*, 267: 699–701 (1995).

214 **Harold Bloom on postmodernism** is cited from *The Western Canon: The Books and School of the Ages* (Orlando, FL: Harcourt Brace, 1994).

215–216 **The mood swings of literary history** are described by Edmund Wilson in "Mo ern literature: between the whirlpool and the rock," *New Republic* (November 1926), reprinted in *From the Uncollected Edmund Wilson*, selected and introduced by Janet Groth and David Castronovo (Athens, OH: Ohio University Press, 1995).

215 **Frederick Turner diagnoses literary postmodernism** in "The birth of natural classicism," *Wilson Quarterly*, pp. 26–32 (Winter 1996). The impact of postmodernism on literary theory is lucidly described in historical context by M. H. Abrams in "The transformation of English studies," *Daedalus*, 126: 105–31 (1997).

216–218 Among the principal works contributing to the **biological theory of arts interpretation and history** are, in chronological order, Charles J. Lumsden and Edward O. Wilson, *Genes, Mind, and Culture* (Cambridge, MA: Harvard University Press, 1981); E. O. Wilson, *Biophilia* (Cambridge, MA: Harvard University Press, 1984); Frederick Turner, *Natural Classicism: Essays on Literature and Science* (New York: Paragon House Publishers, 1985), *Beauty: The Value of Values* (Charlottesville: University Press of Virginia, 1991), and *The Culture of Hope: A New Birth of the Classical Spirit* (New York: Free Press, 1995); Ellen Dissanayake, *What Is Art For?* (Seattle, WA: University of Washington Press, 1988) and *Homo Aestheticus: Where Art Comes From and Why* (New York: Free Press, 1992); Irenäus Eibl-Eibesfeldt, *Human Ethology* (New York: Aldine de Gruyter, 1989); Margaret A. Boden, *The Creative Mind: Myths & Mechanisms* (New York: BasicBooks, 1991); Alexander J. Argyros, *A Blessed Rage for Order: Deconstruction, Evolution, and Chaos* (Ann Arbor: University of Michigan Press, 1991); Kathryn Coe, "Art: the replicable unit—an inquiry into the possible origin of art as a social behavior," *Journal of Social and Evolutionary Systems*, 15: 217–34 (1992); Walter A. Koch, *The Roots of Literature*, and W. A. Koch, ed., *The Biology of Literature* (Bochum: N. Brockmeyer, 1993); Robin Fox, *The Challenge of Anthropology: Old Encounters and New Excursions* (New Brunswick, NJ: Transaction, 1994); Joseph Carroll, *Evolution and Literary Theory* (Columbia, MO: University of Missouri Press, 1995); Robert Storey, *Mimesis and the Human Animal: On the Biogenetic Foundations of Literary Representation* (Evanston, IL: Northwestern University Press, 1996); Brett Cooke, "Utopia and the art of the visceral response," in Gary Westfahl, George Slusser, and Eric S. Rabin, eds., *Foods of the Gods: Eating and the Eaten in Fantasy and Science Fiction* (Athens, GA: University of Georgia

Press, 1996), pp. 188–99; Brett Cooke and Frederick Turner, eds., *Biopoetics: Evolutionary Explorations in the Arts* (New York: Paragon Press, in press).

219 The **metaphors of art and literary history** are taken from an article by John Hollander, "The poetry of architecture," *Bulletin of the American Academy of Arts and Sciences,* 49: 17–35 (1996).

219 **Edward Rothstein's comparison of music and mathematics** is from his *Emblems of Mind: The Inner Life of Music and Mathematics* (New York: Times Books, 1995).

219–220 **Hideki Yukawa described creativity in physics** in *Creativity and Intuition: A Physicist Looks East and West,* translated by John Bester (Tokyo: Kodansha International, distributed in U.S. by Harper & Row, New York, 1973).

220 **Picasso on the origin of art** was quoted by Brassaï (originally Gyula Halasz) in *Picasso & Co.* (London: Thames and Hudson, 1967).

220 The idea of **metapatterns** was originated by Gregory Bateson in *Mind and Nature: A Necessary Unity* (New York: Dutton, 1979) and expanded into biology and art by Tyler Volk in *Metapatterns across Space, Time, and Mind* (New York: Columbia University Press, 1995).

220 **Vincent Joseph Scully's conception of the evolution of architecture** is outlined in *Architecture: The Natural and the Man-made* (New York: St. Martin's Press, 1991).

221 Excellent accounts of **the evolution of Mondrian's art**, among many available, include John Milner's *Mondrian* (New York: Abbeville Press, 1992) and Carel Blotkamp's *Mondrian: The Art of Destruction* (New York: H. N. Abrams, 1995). The neurobiological interpretation I have given it is my own.

221–222 The history of **Chinese and Japanese script** is detailed by Yujiro Nakata in *The Art of Japanese Calligraphy* (New York: Weatherhill/Heibonsha, 1973).

222 **The metaphor of eternity** by Elizabeth Spires is given in her *Annonciade* (New York: Viking Penguin, 1989), and is quoted by permission of the publisher.

223–224 The listing of **archetypes** is largely my own contrivance, with its elements gleaned from many sources, including especially Joseph Campbell's *The Hero with a Thousand Faces* (New York: Pantheon Books, 1949) and *The Masks of God: Primitive Mythology* (New York: Viking Press, 1959); Anthony Stevens' *Archetypes: A Natural History of the Self* (New York: William Morrow, 1982); Christopher Vogler's *The Writer's Journey: Mythic Structure for Storytellers & Screenwriters* (Studio City, CA: Michael Wise Productions, 1992); and Robin Fox's *The Challenge of Anthropology: Old Encounters and New Excursions* (New Brunswick, NJ: Transaction, 1994).

224–229 Of the many descriptions of **European cavern art** and other Paleolithic art, and its interpretation, may be cited *Homo Aestheticus: Where Art Comes From and Why,* by Ellen Dissanayake (New York: Free Press, 1992); *Dawn of Art: The Chauvet Cave, the Oldest Known Paintings in the World,* by Jean-Marie Chauvet, Eliette Brunel Deschamps, and Christian Hillaire (New York: H. N. Abrams, 1996); "Images of the Ice Age," by Alexander Marshack, *Archaeology,* July/August 1995, pp. 29–39; and "The miracle at Chauvet," by E. H. J. Gombrich, *New York Review of Books,* 14 November 1996, pp. 8–12.

229–230 **Gerda Smets' neurobiological study of visual arousal** is described in *Aesthetic Judgment and Arousal: An Experimental Contribution to Psycho-aesthetics* (Leuven, Belgium: Leuven University Press, 1973).

230–231 The experimental studies of **optimum female facial beauty** are reported in "Facial shape and judgements of female attractiveness," by D. I. Perrett, K. A. May, and S. Yoshikawa, *Nature*, 368: 239–42 (1994). Other studies on ideal physical characteristics are described by David M. Buss in *The Evolution of Desire* (New York: BasicBooks, 1994).

233–236 The account of the **Kalahari hunter-gatherers** used here is given by Louis Liebenberg in *The Art of Tracking* (Claremont, South Africa: D. Philip, 1990). A comparable description of Australian Pleistocene and modern-day Aborigines is provided by Josephine Flood in *Archaeology of the Dreamtime: The Story of Prehistoric Australia and Its People*, revised edition (New York: Angus & Robetson, 1995).

237 Some of the themes of the chapter on arts and criticism, particularly the significance of **mythic archetypes and the relation of science to the arts**, are brilliantly anticipated in Northrop Frye's *Anatomy of Criticism: Four Essays* (Princeton, NJ: Princeton University Press, 1957). Frye could not, however, relate his subject to the brain sciences and sociobiology, which did not exist in their present form in the 1950s.

CHAPTER 11

ETHICS AND RELIGION

238–265 Among key references to the **foundations of moral reasoning**, and particularly to the role of the natural sciences in defining the empiricist world view, are, alphabetically by author: Richard D. Alexander, *The Biology of Moral Systems* (Hawthorne, NY: Aldine de Gruyter, 1987); Larry Arnhart, "The new Darwinian naturalism in political theory," *American Political Science Review*, 89: 389–400 (1995); Daniel Callahan and H. Tristram Engelhardt, Jr., eds., *The Roots of Ethics: Science, Religion, and Values* (New York: Plenum Press, 1976); Abraham Edel, *In Search of the Ethical: Moral Theory in Twentieth Century America* (New Brunswick, NJ: Transaction, 1993); Paul L. Farber, *The Temptations of Evolutionary Ethics* (Berkeley: University of California Press, 1994); Matthew H. Nitecki and Doris V. Nitecki, eds., *Evolutionary Ethics* (Albany: State University of New York Press, 1993); James G. Paradis and George C. Williams, *Evolution & Ethics: T. H. Huxley's* Evolution and Ethics *with New Essays on Its Victorian and Sociobiological Context* (Princeton, NJ: Princeton University Press, 1989); Van Rensselaer Potter, *Bioethics: Bridge to the Future* (Englewood Cliffs, NJ: Prentice-Hall, 1971); Matt Ridley, *The Origins of Virtue: Human Instincts and the Evolution of Cooperation* (New York: Viking, 1997); Edward O. Wilson, *Sociobiology: The New Synthesis* (Cambridge, MA: Belknap Press of Harvard University Press, 1975), *On Human Nature* (Cambridge, MA: Harvard University Press, 1978), and *Biophilia* (Cambridge, MA: Harvard University Press, 1984); Robert Wright, *The Moral Animal: Evolutionary Psychology and Everyday Life* (New York: Pantheon Books, 1994).

The scholarly sources on the **relation of science to religion** from which I have drawn ideas and information include Walter Burkert, *Creation of the Sacred: Tracks of Biology in Early Religion* (Cambridge, MA: Harvard University

Press, 1996); James M. Gustafson, *Ethics from a Theocentric Perspective; Volume One, Theology and Ethics* (Chicago: University of Chicago Press, 1981); John F. Haught, *Science and Religion: From Conflict to Conversation* (New York: Paulist Press, 1995); Hans J. Mol, *Identity and the Sacred: A Sketch for a New Social-Scientific Theory of Religion* (Oxford: Blackwell, 1976); Arthur R. Peacocke, *Intimations of Reality: Critical Realism in Science and Religion* (Notre Dame, IN: University of Notre Dame Press, 1984); Vernon Reynolds and Ralph E. S. Tanner, *The Biology of Religion* (Burnt Mill, Harlow, Essex, England: Longman, 1983); Conrad H. Waddington, *The Ethical Animal* (New York: Atheneum, 1961); Edward O. Wilson, *On Human Nature* (Cambridge, MA: Harvard University Press, 1978).

241–243 I have based the **argument of the religious transcendentalist** on my own early experience in the Southern Baptist tradition, and upon many other sources, including excellent expositions by Karen Armstrong in *A History of God: The 4,000-Year Quest of Judaism, Christianity, and Islam* (New York: Alfred A. Knopf/Random House, 1993); Paul Johnson in *The Quest for God: A Personal Pilgrimage* (New York: HarperCollins, 1996); Jack Miles in *God: A Biography* (New York: Alfred A. Knopf, 1995); and Richard Swinburne in *Is There a God?* (New York: Oxford University Press, 1996).

242–243 John Locke's **condemnation of atheists** is in *A Letter on Toleration*, Latin text edited by Raymond Klibansky and translated by J. W. Gough (Oxford: Clarendon Press, 1968).

243 **Robert Hooke on the limits of science** is quoted by Charles Richard Weld in *A History of The Royal Society, with Memoirs of the Presidents*, compiled from documents, in two volumes (London: John Parker, West Strand, 1848), vol. 1, p. 146.

244 The estimate cited of the **number of religions** throughout human history (100,000) was made by Anthony F. C. Wallace in *Religion: An Anthropological View* (New York: Random House, 1966).

245 **Mary Wollstonecraft on evil:** *A Vindication of the Rights of Woman* (London: J. Johnson, 1792).

246 The survey of the **religious belief of scientists** was conducted by Edward J. Larson and Larry Witham and is reported in *The Chronicle of Higher Education*, 11 April 1997, p. A16.

250–251 The model of **the evolution of moral behavior** follows similar reasoning in my first work on the subject, *On Human Nature* (Cambridge, MA: Harvard University Press, 1978), and is consistent with the theory of gene-culture coevolution detailed in Chapters 7 and 8 of the present work.

252–254 The fundamentals of the **evolution of cooperation**, including the use of the Prisoner's Dilemma, is given by Robert M. Axelrod in *The Evolution of Cooperation* (New York: BasicBooks, 1984) and Martin A. Nowack, Robert M. May, and Karl Sigmund in "The arithmetics of mutual help," *Scientific American*, June 1995, pp. 76–81. Proto-ethical behavior in chimpanzees, including cooperation and retribution toward those failing to cooperate, is described by Frans de Waal in *Peacemaking Among Primates* (Cambridge, MA: Harvard University Press, 1989), and *Good Natured: The Origins of Right and Wrong in Humans and Other Animals* (Cambridge, MA: Harvard University Press, 1996).

253 Evidence for **inherited differences among people in empathy and infant-caregiver bonding** is cited by Robert Plomin et al. in *Behavioral Genetics*, third edition (New York: W. H. Freeman, 1997).

259–260 **Dominance communication in mammals** is described widely in the literature on animal behavior, for example in some detail in my *Sociobiology: The New Synthesis* (Cambridge, MA: Belknap Press of Harvard University Press, 1975).

261 The account by St. Teresa of Avila (1515–1583) of her **mystical experience of prayer** is provided in *The Life of St. Teresa of Jesus of the Order of Our Lady of Carmel, Written by Herself*, translated from the Spanish by David Lewis; it is compared with the original autograph text and re-edited with additional notes and introduction by Benedict Zimmerman, fifth edition (Westminster, MD: The Newman Press, 1948).

264–265 The closing statement on the **relation of science and religion** is drawn from the 1991–92 Dudleian Lecture I gave at the Harvard Divinity School, which was published as "The return to natural philosophy," *Harvard Divinity Bulletin*, 21: 12–15 (1992).

CHAPTER 12

To What End?

266 The **genetic kinship by common descent** of all organisms on Earth is detailed at the molecular level by J. Peter Gogarten in "The early evolution of cellular life," *Trends in Ecology and Evolution*, 10: 147–51 (1995).

267 The **descent of modern humanity** from earlier species of *Homo* is authoritatively reviewed by multiple authors in *The First Humans: Human Origins and History to 10,000 BC*, Göran Burenhult ed. (New York: HarperCollins, 1993).

267–268 **Gap analysis** is a term borrowed from the study of biological diversity and conservation; it refers to the method of mapping the distribution of plant and animal species, overlaying them with maps of biological reserves, and using the information to select the best sites for future reserves. See "Gap analysis for biodiversity survey and maintenance," by J. Michael Scott and Blair Csuti in Marjorie L. Reaka-Kudla, Don E. Wilson, and Edward O. Wilson, eds., *Biodiversity II: Understanding and Protecting Our Biological Resources* (Washington, DC: Joseph Henry Press, 1997), pp. 321–40.

270–277 The section on **present and future human genetic evolution** has been modified from my article "Quo Vadis, Homo Sapiens?," *Geo Extra*, no. 1, pp. 176–9 (1995). The evolution in head shape during the past millennium is documented by T. Bielicki and Z. Welon in "The operation of natural selection in human head form in an East European population," in Carl J. Bajema, ed., *Natural Selection in Human Populations: The Measurement of Ongoing Genetic Evolution in Contemporary Societies* (New York: Wiley, 1970). The evidence for recent evolution in heat-shock proteins is given by V. N. Lyashko et al. in "Comparison of the heat shock response in ethnically and ecologically different human populations," *Proceedings of the National Academy of Sciences, USA*, 91:12492–5 (1994).

279–280 The results of the **Biosphere 2 experiment** are discussed by Joel E. Cohen and David Tilman in "Biosphere 2 and Biodiversity: The Lessons So Far," *Science* 274:

1150–1 (1996). A first-hand account of the two-year adventure has been published by two of the Biospherians, Abigail Alling and Mark Nelson, in *Life Under Glass: The Inside Story of Biosphere 2* (Oracle, AZ: Biosphere Press, 1993).

281–282 The most thorough and authoritative recent account of **human population growth** written for a broad audience is Joel E. Cohen's *How Many People Can the Earth Support?* (New York: W. W. Norton, 1995). It is very difficult to estimate the total number of humans who can exist sustainably on Earth, due, as Cohen argues, to factors as spongy as the ultimate levels of food production technology and average acceptable quality of life. Yet an absolute limit exists and it is not much greater than ten billion. The estimated limit of sixteen billion people based on total energy capture by photosynthesis converted solely to human use is taken from John M. Gowdy and Carl N. McDaniel in "One world, one experiment: addressing the biodiversity-economics conflict," *Ecological Economics,* 15: 181–92 (1995).

282 The **PAT formula** for estimating impact of population on the environment was developed originally by Paul R. Ehrlich and John P. Holdren in "Impact of population growth," *Science,* 171: 1212–17 (1971), and has been discussed in many aspects since. "It is a rough approximation, since the three multiplicative factors are not independent . . . It is especially useful in assessing global impacts, where we normally must fall back on using per-capita energy use in place of AT": Paul Ehrlich, "The scale of the human enterprise," in Denis A. Saunders et al., *Nature Conservation 3: Reconstruction of Fragmented Ecosystems* (Chipping Norton, NSW, Australia: Surrey Beatty & Sons, 1993), pp. 3–8.

282 The concept of **ecological footprints** as a measure of environment impact was introduced by William E. Rees and Mathis Wackernagel in "Ecological footprints and appropriated carrying capacity: measuring the natural capital requirements of the human economy," in AnnMari Jansson et al., eds., *Investing in Natural Capital: The Ecological Economics Approach to Sustainability* (Washington, DC: Island Press, 1994), pp. 362–90.

282–283 An important general statement on **population and environment**, coauthored by eleven leading scientists whose expertise covers virtually all the relevant disciplines, is "Economic growth, carrying capacity, and the environment," by Kenneth Arrow et al., *Science,* 268: 520–1 (1995).

283–286 The most comprehensive, up-to-date, and accessible summaries of the immense **databases on the global environment** are provided by the reports of the Worldwatch Institute, headquartered in Washington, D.C. They include the two annual series *State of the World* and *Vital Signs: The Trends That Are Shaping Our Future,* published by W. W. Norton (New York), as well as occasional specialized Worldwatch Papers, published by the Institute. An independent assessment of available data by environmental scientists, confirming the same trends I have described here, are reported in "Land resources: On the edge of the Malthusian precipice?," proceedings of a conference organized by D. J. Greenland et al., *Philosophical Transactions of the Royal Society of London, Series B,* 352: 859–1033 (1997).

287–288 These recent works on environmental factors in the **rise and fall of civilizations** are among those that can be recommended out of a large literature: "The genesis and collapse of third millennium North Mesopotamian civilization," by H. Weiss et al., *Science,* 261: 995–1004 (1993); "Climate and the collapse of civilization," by

Tom Abate in *BioScience*, 44: 516–19 (1994); and the exceptionally broad and biologically insightful *Guns, Germs, and Steel: The Fates of Human Societies*, by Jared Diamond (New York: W. W. Norton, 1997).

289–290 An excellent account of the 1992 **United Nations Conference on Environment and Development** (UNCED), including a history of the meeting and the substance of the binding conventions and of Agenda 21, is Adam Rogers' *The Earth Summit: A Planetary Reckoning* (Los Angeles: Global View Press, 1993).

290 On the **accommodation of technology and economic growth to the natural environment**, see the U.S. National Research Council's special report, *Linking Science and Technology to Society's Environmental Goals*, John F. Ahearne and H. Guyford Stever, co-chairs (Washington, DC: National Academy Press, 1996). Incisive descriptions of particular technological solutions are given by Jesse H. Ausubel in "Can technology spare the earth?," *American Scientist*, 84: 166–78 (1996), and the multiple authors of the Summer 1996 issue of *Daedalus* (Journal of the American Academy of Arts and Sciences) entitled "Liberation of the Environment."

290–292 The relations between **economics and the environment** is the subject of a rapidly expanding library of journals and books. Excellent introductions to the subject are provided by James Eggert, *Meadowlark Economics: Work & Leisure in the Ecosystem* (Armonk, NY: M. E. Sharpe, 1992); R. Kerry Turner, David Pearce, and Ian Bateman, *Environmental Economics: An Elementary Introduction* (Baltimore, MD: Johns Hopkins University Press, 1993); Paul Hawken, *The Ecology of Commerce: A Declaration of Sustainability* (New York: HarperCollins, 1993); and Thomas Michael Power, *Lost Landscapes and Failed Economies: The Search for a Value of Place* (Washington, DC: Island Press, 1996).

291 **Frederick Hu** on the economic growth of nations: "What is competition?," *World Link*, July/August 1996, pp. 14–17.

292–297 The account of **biodiversity and extinction** is modified from portions of two of my own articles, "Is humanity suicidal?," *The New York Times Magazine*, 30 May 1993, pp. 24–9; and "Wildlife: legions of the doomed," *Time* (*International*), 30 October 1995, pp. 57–9.

294–295 On the moral argument for the **preservation of biodiversity**: see my earlier presentations in *Biophilia* (Cambridge, MA: Harvard University Press, 1984) and *The Diversity of Life* (Cambridge, MA: Belknap Press of Harvard University Press, 1992); and Stephen R. Kellert, *The Value of Life: Biological Diversity and Human Society* (Washington, DC: Island Press/Shearwater Books, 1996) and *Kinship to Mastery: Biophilia in Human Evolution and Development* (Washington, DC: Island Press, 1997).

297 On the ultimately **moral foundations of society**: *Democracy and Disagreement*, by Amy Gutmann and Dennis Thompson (Cambridge, MA: Belknap Press of Harvard University Press, 1996).

ACKNOWLEDGMENTS

FOR FORTY-ONE YEARS, ending with my retirement in 1997, I taught large classes in elementary and intermediate biology at Harvard University. In the second half of that period the presentations were part of the core curriculum, commissioned by the Faculty of Arts and Sciences to teach the basic substance and "ways of thinking" of each of the great branches of learning. The subject for which I had particular responsibility, evolutionary biology, is an intellectual caravanserai located near the boundary of the natural and social sciences. It is a logical meeting place for scholars of diverse interests who wish to trade back and forth. Given that my primary research interests also include the evolution of social behavior, I felt comfortable discussing the key issues of consilience with experts across a large part of the academy.

It would be almost impossible to list all those I consulted during the three years it took to write *Consilience*. They range in interests from a scholar in Slavic literature to the speaker of the U.S. House of Representatives, from Nobel Laureates in the physical sciences and economics to the chief executive officer of an international insurance company. Instead, I will take space here to acknowledge only those who read portions of the manuscript. In expressing my gratitude for their invaluable help, I also exonerate them from errors and misconceptions that might remain as the book goes to press (September 1997).

Gary S. Becker
(economics)

Rodney A. Brooks
(artificial intelligence)

Terence C. Burnham (economics)
Joseph Carroll (literary theory)
I. Bernard Cohen (history of science)
Joel E. Cohen (ecology)
Brett Cooke (literary theory)
William R. Crout (religion)
Antonio R. Damasio (neurobiology)
Daniel C. Dennett (philosophy of science, brain sciences)
Ellen Dissanayake (arts theory)
George B. Field (physical sciences)
Newt Gingrich (general)
Paul R. Gross (general)
J. Allan Hobson (psychology)
Joshua Lederberg (general)
Barbara K. Lewalski (literary criticism)
Charles J. Lumsden (general)
Myra A. Mayman (the arts)
Michael B. McElroy (atmospheric physics)
Peter J. McIntyre (evolution)
Matthew S. Meselson (molecular biology)
Harold J. Morowitz (complexity theory)
William R. Page (general)
Robert Plomin (psychology)
William E. Rees (ecology)
Angelica Z. Rudenstine (arts history)
Loyal Rue (general)
Michael Ruse (general)
Sue Savage-Rumbaugh (primatology)
S. J. Singer (molecular biology)
James M. Stone (general)
Frank J. Sulloway (general)
Martin L. Weitzman (economics)
Irene K. Wilson (poetry, theology)
Arthur P. Wolf (anthropology)

Finally, as I have for all my books and articles back to 1966, I acknowledge with pleasure the meticulous and invaluable work of Kathleen M. Horton in bibliographic research and preparation of the manuscript. I am also grateful to John Taylor Williams, agent and adviser, whose wise counsel helped make the project a reality, and to my editor at Knopf, Carol Brown Janeway, for her important moral support and help in steering past at least some of the more dangerous reefs unavoidable in such a synthesis.

Index

A

Abate, Tom, 318
adaptation, *see* evolution
Adler, Mortimer, 121
aesthetics, *see* arts; facial beauty; optimum complexity
Agassiz, Louis, 37
aggression, 155, 170–1, 309–10
agriculture, 283–4
Ahearne, John F., 319
Alcock, John, 301
alcoholism, 142
Alexander, Richard D., 315
Alling, Abigail, 318
altruism, 150, 171–2, 253, 316–17
Amaringo, Pablo, 72–3, 81–2, 303
American Anthropological Association, 185–6, 310–11
American Humanist Association, 262
American Philosophical Society, 39
anaconda, 72
anthropic principle, 33
anthropology, 184–6, 190, 310–11
ants, 68–71, 303
Apollo, 135, 212
Applebaum, Herbert, 311
aquaculture, 284–5
aquifers, 284
Aquinas, St. Thomas, 239
archetypes, in creative arts, 223–4, 314–15
architecture, 219–21, 314
Argyros, Alexander J., 313
Aristotle, 248
Armstrong, Karen, 316
Armstrong, Louis, 222
Arnhardt, Larry, 315
Arrow, Kenneth J., 318
artificial emotion (AE), 123
artificial intelligence (AI), 120–4, 305
arts, the, 12, 72–3, 116–17, 210–37, 268, 313–15; adaptive advantage, 224–6, 314; relation to sociology, 187
astrology, 54, 228
astronomy, 30, 53–4
Augustine of Hippo, St., 257
Australian aboriginals, 315
Ausubel, Jesse H., 319
autism, 142
autonomic nervous system, 111–12
Axelrod, Robert M., 316
ayahuasca, 72–4
Aztec gods, 80–1

B

Bacon, Francis, 9, 23–8, 39, 300
Bajema, Carl J., 317
Balch, Steven H., 300
Bali, culture and fauna, 189
Balter, Michael, 307
baptism (Christian practice), 227–8
Baptists (Christian denomination), 6, 262
Barinaga, Marcia, 304
Barkow, Jerome H., 307–11
bat echolocation, 47
Bateson, Gregory, 220, 314
Baylor, Denis, 309
beauty, facial, in women, 230–2, 315
Becker, Gary S., 202–4, 312
behavioral genetics, 142–7, 154–7, 160–6, 172–3
Benedict, Ruth, 184

Berlin, Brent, 161–3
Berlin, Isaiah, 16
Betzig, Laura L., 309–10
Bible, 6, 73, 95, 248, 263
Bielicki, T., 317
Billeter, Jean F., 305
biochemistry, 83–4, 91–5
biodiversity (biological diversity), 292–7, 319
biology, general, 78, 86, 90–1; *see also* biochemistry; biomedicine; cell biology; ecology; evolutionary biology; genetics; molecular biology; neurobiology
biomedicine, 53–4, 181–2, 274–6
Biosphere 2, 279–80, 317–18
birth order, effects of, 138
Blackburn, Simon, 183
Bloom, Harold, 214, 313
Blotkamp, Carel, 314
Blurton Jones, Nicholas G., 308
Boas, Franz, 184
Boden, Margaret, 217, 305, 313
body adornment, 232
Boltzmann, Ludwig, 85
bonding, mother-infant, 152, 306–7
bonobos (pygmy chimpanzees), 131–3, 306
borderland disciplines, great branches of learning, 192, 208–9
Bose-Einstein condensate, 5
Bossert, William H., 69–70, 309
bottleneck, environmental, 287
Bouchard, Thomas J., Jr., 307
Bowen, Catherine Drinker, 300
Boyd, Robert, 306
brain, 81–2, 97–108, 160, 165; *see also* color vision; mind
brain imaging, 108, 117–18, 144, 156, 242
brain sciences (cognitive neuroscience), 99–100, 216, 246, 266, 304; *see also* brain; mind
Brassaï, 220, 314
Breuil, Abbé, 227
bridging disciplines, *see* borderland disciplines
Brooks, Rodney A., 122–3
Brower, B., 307
Brown, Donald E., 308, 311
Burenhult, Göran, 317
burial ceremonies, 256–7
Burkert, Walter, 315
Buss, David M., 310, 315
butterflies, 46, 231

C

cabala, 256
Callahan, Daniel, 315
Callebaut, Werner, 299
Campbell, Joseph, 300, 314
Carnap, Rudolf, 62–3
Carroll, Joseph, 217, 313
carrying capacity, 287–8
Cassirer, Ernst, 301
Castañeda, Carlos, 73
Castronovo, David, 313
Cather, Willa, 260
cause and effect, 66–95
Cavalli-Sforza, L. Luca, 306
cavern art, Europe, 226–9, 314
cell biology, 51, 55, 68, 91–5, 303–4
ceremony, 153, 247, 256
Ceres, 212
Chalmers, David J., 115–16, 305
chance, in history, 267
Chandrasekhar, S., 7, 299
chaos theory, *see* complexity theory
character, personal, 246, 248
Chauvet, Jean-Marie, 314
Chauvet cavern, art of, 226, 314
cheater detection, 171–2, 310
chemistry, 51, 55, 69–71, 83–4, 86
Chermock, Ralph L., 4
chess, 122
chimpanzees, 79, 131–2, 174, 306
Chinese calligraphy, 117–18, 221–2, 230, 314
Chinese science, 30–1
Christianity, 33–4, 243–5
Christian Science, 54
Churchill, Winston, 34
Churchland, Patricia S., 304
Churchland, Paul M., 304
civilization, origin of, 148–9, 253–4, 308, 318–19
Clark, Mary E., 302–3
Clausen, Christopher, 306
climate change, 285–6
Coe, Kathryn, 313
cognitive neuroscience, *see* brain sciences
Cohen, Jack, 303
Cohen, Joel E., 280, 317–18
Coleman, James S., 186–7, 311
color, skin, 196
color vision, 46, 107, 116–17, 151, 159–63, 309
color vocabulary, 161–5, 309
Commission on the Humanities, 210
communication, animal, 131–3; arts and science, 116–17; chemical, 68–71; electroreceptive, 47; olfactory, 158; touch, 158; *see also* color vision; facial expressions; hearing; language; paralanguage
communion, religious, 260
complexity theory, 87–95, 303

computers, 93, 290, 304; *see also* complexity theory
Comte, Auguste, 30
Condorcet, Marquis de, 14–21, 300
Conrad, Joseph, 215
consciousness, *see* mind
conservation, biological, 292–7
consilience (interlocking of causal explanation across disciplines), 8–13, 55, 125, 136–7, 154, 182, 191–2, 195, 204–5, 216–26, 236–7, 266–70, 297–8
constructivism, 40
contract formation, 171–2, 310, 316
Cooke, Brett, 217, 313–14
Cosmides, Leda, 171–2, 307–11
cosmologies, 264–5
Crandall, B. C., 302
creation science (Creationism), 54, 129–30, 198
creative arts, *see* arts, the
creativity, 58–9, 64–5, 190, 213, 223–4, 267–8, 302
Cretan labyrinth, 66–7, 86, 302–3
Crick, Francis, 304
crime, 203
Csuti, Blair, 317
Culotta, Elizabeth, 307
cultural relativism, 184–6
culture, 12, 130–6, 166, 306; equivalency, 184–6; origins, 149–50, 223; universals, 147–9, 307–8; *see also* gene-culture coevolution
Cummings, Michael R., 307

D

Daedalus, 7, 299
Damasio, Antonio R., 113–15, 304–5
Damasio, Hanna, 305
Dani (New Guinea people), 161–2
Daphne, 135, 212
Darwin, Charles, 37, 75, 248
Davidson, Richard J., 304
Decatur, Stephen, 170
decision-making, neurobiology of, 112–15
deconstruction (literary theory), 41, 214–15, 301
Deep Blue (computer), 122
deep history, 12, 300
Degler, Carl N., 310–11
deism, 32–4
de Man, Paul, 214
Democritus, 50, 128, 306
Dennett, Daniel C., 110, 304–5
depression, clinical, 146
Derrida, Jacques, 41, 214, 301
Descartes, René, 28–9, 39, 96, 98, 300
determinism: genetic, 137–47, 166–7, 188, 276–7; mental, 118–20
Diamond, Jared, 319
Dillon, Wilton S., 310
Dis, 212
Dissanayake, Ellen, 217, 232, 313–14
DNA (deoxyribonucleic acid), 50, 60, 87, 91, 128–9, 160, 266, 273–4, 302
Dolan, R. J., 307
dominance behavior, 259–60, 317
Dostoyevsky, Fyodor, 242
dreams, 74–8, 303
drugs, 72–4; *see also* neurotransmitters
dualism: mind-body, 98–9; in concept formation, 153
Durham, William H., 179–80, 306, 308, 310
Durkheim, Émile, 184, 187
Dusun (Borneo people), 153, 308–9
Dutch aggression gene, 155, 309
dyadic instinct, 153, 308–9
dyslexia, 155, 309
Dyson, Freeman J., 302

E

Earth Summit, Rio de Janeiro, 289–90, 319
Ebstein, Richard P., 309
ecological footprint, 282, 318
ecology, 84–5, 170–1, 205, 277–97, 317–19
economic growth, 291
economics, 195–205, 290–2, 312, 319
ecosystems, 84–5
ecstasy, religious, 258, 260–1, 317
Eddington, Arthur S., 7, 299
Edel, Abraham, 315
Edelman, Gerald M., 304
Eden, 211–12
edge of chaos, theory, 89–90
education, 12–13
Eggert, James, 319
Egypt, empire, 287; incest, 177–8
Ehrlich, Paul R., 318
Eibl-Eibesfeldt, Irenäus, 159, 217, 308–9, 313
Einstein, Albert, 5, 31–3, 57, 97, 263, 299, 301–2
Eisenberg, John F., 310
Ekman, Paul, 152, 304
electroencephalograms, 156, 221–2, 229–30
electromagnetic spectrum, 46–7
electroreception, fish, 47, 116
elephants, 166
Ellis, Henry, 300
Ellis, Lee, 186
Emerson, Ralph Waldo, 36

Emlen, Stephen T., 194–5, 312
emotion, 112–15
empathy, 253, 316–17
empiricism, ethical, 238–51
Engelhardt, H. Tristram, Jr., 315
Enlightenment, 8, 14–44, 61, 215, 247, 262, 300–1
environment: current status, 277–97, 317–19; interaction with genes, 137–43, 142–3, 188, 277–97; technological maintenance, 289–90
enzymes, 83–4, 91–2, 303
epic, religious vs. evolutionary, 264–5
epigenetic rules (hereditary regularities in development, including mental), 150–83, 164–80, 193, 308, 309–10; in the arts, 213, 229–32; defined, 150; in ethics and religion, 246–7, 254, 257–8
episodic memory, 134–5
epistasis, genetic, 156
epistemology, 190, 258, 267–8
ethics, 33, 38, 238–56, 315–17
ethnicity, 183–4, 287–8
eugenics, 184, 276–7
evolution: human, 97–8, 133–4, 167–8, 224–9, 266–7, 270–7, 307, 317; natural selection, 52, 77–8, 96–7, 102, 122–3, 126–30, 165–8, 200–1, 204–5, 224–6, 243, 254, 257–8
evolutionary biology, 48, 266
evolutionary epic, 264–5
evolutionary psychology, *see* sociobiology
exemptionalism (environmental ethic), 278
exogamy, 174
explanation, scientific, 66–95
exploratory instinct, 232–3
extinction, species, 292–4, 319
extrasensory perception (ESP), 118, 228
eyebrow flashing, 259

F

facial expressions, 151–2, 158–9
facial beauty, 230–1, 315
Fackelmann, Kathy A., 305
Falconer, Douglas S., 307
fallacies: affirming the consequent, 87; naturalistic, 249
family theory, 194–5, 312
Farber, Paul L., 315
Farrington, Benjamin, 300
Faust, 270
Federal Reserve Board, 197
Feldman, Mark W., 306
feminism, 215
Fernald, Anne, 306–7
Fetzer, James H., 312
Fischman, Joshua, 306
fisheries, global, 284–5
Fiske, Donald W., 301, 311–12
Flood, Josephine, 315
folk psychology (untested by science), 183, 202–3
footprint, ecological, 282, 318
forest management, 10
Foucault, Michel, 43, 301
Fox, Robin, 217, 311, 313–14
Frazer, James G., 178
Freedman, Daniel G., 308
Freemasonry, 256
free will, 118–20
Freud, Sigmund, 40, 74–5, 78, 178–9, 184
Friedrich, Robert W., 311
Frith, Chris and Uta, 309
Frye, Northrop, 315
fundamentalism, Muslim, 184

G

GABA (neurotransmitter), 144
Gage, John, 309
Gage, Phineas P., 100–1, 305
Galileo Galilei, 33
gap analysis, 267–8, 317
Gardner, Howard, 301
Gaukroger, Stephen, 300
Geertz, Clifford, 311
Gell-Mann, Murray, 303
gender differences, genetic, 156–7, 169–70, 214–15, 310
gene-culture coevolution (linkage of genetic and cultural evolution), 126–8, 157–68, 217–22, 254–5, 306
genes, *see* determinism: genetic; genetics
gene therapy, 275–6
genetic determinism, *see* determinism: genetic
genetic disease, 144–5, 173–4, 273–5
genetic leash, 157–8
genetics, 89–91, 143–7, 154–7, 258, 270–7; population genetics, 198–201; *see also* DNA; gene-culture coevolution; heritability; interaction
genius, 213
genotype-environment correlation, 140–1
Gergen, Kenneth J., 42, 301
Gibbons, Ann, 307
God, 32–4, 198, 211–12, 238–48, 259–63
gods, 80–1, 212, 221, 262
Goethe, Johann Wolfgang von, 36, 270, 301
Gogarten, J. Peter, 317
Goldschmidt, Walter, 306, 311

Goleman, Daniel, 305
Gombrich, Ernst H., 314
Goodell, Edward, 300
Goodstein, David L., 300–1
Gowdy, John M., 318
Grand Inquisitor, Dostoyevsky's, 242
Greece, ancient, myth, 66–7, 212; philosophy, 34, 61, 183; religion, 262
Green, Donald P., 312
Greenland, D. J., 318
Gribben, John, 302
Gross, David J., 302
Grossman, Marcel, 5
Groth, Janet, 313
Gustafson, James M., 316
Guterl, Fred, 305
Gutmann, Amy, 319

H

habit, biological basis of, 107
habitat selection, 278
Hallpike, Christopher Robert, 208, 312
hallucination, 72–4
Hamilton, William D., 310
Hanunóo (Philippine language), 163
Hardy-Weinberg principle, 199–200
Harner, Michael J., 303
Harris, Marvin, 311
Haught, John F., 316
Hawken, Paul, 319
Hawking, Stephen, 263
head shape, evolution of, 271
hearing, 151
hereditarianism, 142–3
heredity, *see* genetics; gene-culture coevolution
heritability, 139–43, 307
hermeneutics, 189–90, 311–12
Herodotus, 262
Herrnstein, Richard J., 140
heuristics, 206–8, 312
Hilbert, David, 44
Hirshleifer, Jack, 204
historical materialism, 38
history, 11, 138, 166, 256, 267, 287, 297, 314
Hobson, J. Allan, 75, 303–4
Holdren, John P., 318
holism, *see* synthesis
Hollander, John, 314
Hölldobler, Bert, 303
Holt, Luther E., 308
Holton, Gerald, 4, 299, 301–2
Hooke, Robert, 243, 316
Hu, Frederick, 291, 319
human epic, 264–5
Human Genome Project, 273
humanism, movement, 34, 262
humanities, 12
human nature, 164–80, 216, 218, 309–10; defined, 164
Human Relations Area Files, 147
Hume, David, 248, 251
hunter-gatherers, 148–9, 167–8, 208, 233–6, 315
Hutcheson, Francis, 251
hyperreligiosity, 258, 261
hypothesis formation, 59–60

I

Icarus, 7, 299
imitation, infant, 133, 307
incest avoidance and taboos, 173–80, 194, 310
incomplete penetrance, in heredity, 146–7
induction, Francis Bacon on, 26, 300
information, in arts and science, compared, 116–17; *see also* communication
inspiration, *see* creativity
insulin, 83–4
intelligence, optimum for science, 58
interaction, of genes and environment, 137–43
Ionian Enchantment (preoccupation with unification of knowledge), 4–7, 299
Islam, early conquests of, 244
Israeli kibbutzim, 175–6

J

Jackson, Frank, 116
James, William, 59
Jansson, AnnMari, 318
Japanese calligraphy, 222, 230, 314
Jáuregui, José A., 305
jazz, 222
Jefferson, Thomas, 239
Jesus, 73, 119
Jirari, Carolyn G., 308
Jívaro, 72, 303
Johnson, Mark H., 308
Johnson, Paul, 316
Judaism, origins, 244, 256, 262
Jung, Carl, 78

K

Kac, Mark, 59
Kagan, Donald, 310
Kahneman, Daniel, 206–8, 312

Kalahari hunter-gatherers, 153, 233–6, 315
Kant, Immanuel, 20, 96, 248–9
Kanzi (bonobo), 131–2
Kareiva, Peter M., 303
Karni, Avi, 303
Kasparov, Gary, 122
Kauffman, Stuart A., 88–90, 303
Kaufmann, Walter, 302
Kay, Paul, 161–3
Keats, John, 121
Keeley, Laurence H., 310
Kekule von Stradonitz, Friedrich August, 80
Kellert, Stephen R., 319
kibbutzim, Israeli, 175–6
Kidder, Alfred V., 149, 308
kin selection, 168–9, 310
King, Martin Luther, Jr., 239
Kitcher, Philip, 312
Kluckhohn, Clyde, 130, 306
knowledge, nature of, 60–5
Koch, Walter, 217, 313
Koenig, Olivier, 304
Konner, Melvin J., 308
Kosslyn, Stephen M., 304
Kroeber, Alfred, 130, 306
!Kung (Ju/Wasi), Kalahari Desert, 153, 233–6, 315

L

labyrinth, of knowledge, 66–8
Laland, Kevin N., 306
Lamartine, Bruce, 51
Langton, Christopher, 88
language, 132–3, 152, 161–3, 306–7
Larson, Edward J., 316
Lavoisier, Antoine, 51
Leary, Timothy, 73
Lee, Richard B., 233
Leeuwenhoek, Anton van, 50
Leibniz, Gottfried, 20
Lespinasse, Julie de, 17
Leucippus, 50
LeVay, Simon, 305
Lévi-Strauss, Claude, 154
Lewis, David, 317
Leys, Simon, 118, 305
liberal arts, 12–13, 269–70
Liebenberg, Louis, 233–6, 315
light, visible, 46
Lincoln, Abraham, 239
Linnaeus, Carolus, 3–4
Locke, John, 239, 242–3, 316
Locke, John L., 306
logical positivism, 61–5, 302
Loomis, William F., 92
Lopreato, Joseph, 186, 311
love: biological origin, 169–70; role in religion, 243
Lucretius, 257
Lumsden, Charles J., 136, 217, 306–9
Luna, Luis Eduardo, 303
Lyashko, V. N., 317
Lyman, Richard W., 313
Lyons, John, 163, 309

M

Mackay, Trudy F. C., 307
magic, 227–9
Malthus, Thomas, 195
Marks, Jonathan, 311
Marlowe, Christopher, 270
Marshack, Alexander, 314
Marshall, Alfred, 196
Martin, Alex, 305
Marx, Karl, 184, 187
Marxism-Leninism, 245
mathematics: basic nature, 62–3, 95, 301–2, 314; comparison with music, 219; role in science, 48, 196
mating strategy, 169–70
Maxwell, James Clerk, 85
May, Robert M., 316
Mayas, 287
Mayr, Ernst, 4
McDaniel, Carl N., 318
Mead, Margaret, 184
meaning, neurobiology of, 115, 134–5
Meltzoff, Andrew N., 307
memory, 110–11, 134–7, 307; units, 136–7
meme, 136, 307
Meselson, Matthew S., 60, 302
Mesopotamia, 287
metapatterns, in art, 220, 314
metaphor, 163, 219–20, 301, 314
microscopy, history of, 50
Miles, Jack, 316
Mill, John Stuart, 44
Milner, John, 314
Milton, John, 211
mimicry, in human infants, 133
mind, nature of, 61–5, 96–124, 215–16, 304; Francis Bacon on, 26–7; dreaming, 74–8; drug effects, 72–4
mind script, 117–18
“minor marriages,” Taiwan, 175
Minsky, Marvin L., 123
modernism, in the arts, 38–40, 301
Mol, Hans J., 316

molecular biology, 55, 60, 68, 83–4
Mollon, John, 309
Mondrian, Piet, 221, 230, 314
monkeys, 78–9, 174, 259
Monod, Jacques, 128, 306
mood, 76, 115, 146
Moore, G. E., 249
Moore, M. Keith, 307
Moro reflex, 151, 308
Morowitz, Harold, 303
Morris, Christopher, 305
Morton, John, 308
Moses, 262
mother-infant bonding, 152, 306–7
Moynihan, Daniel Patrick, 310
multiculturalism, 41, 184–6
multiple competing hypotheses, 59–60
Mundkur, Balaji, 78, 303
Murdock, George P., 147, 307–8
Murray, Charles, 140
music, in ceremony, 227
music, qualities of, 219, 314
musical ability, 140–1, 213, 313
mutations, 144–7, 173–4
mystical experience, 232–3, 260–1
myth, 66–7, 212, 257, 300

N

Nabokov, Vladimir, 222
Nakata, Yujiro, 314
nanotechnology, 51, 302
narrative, sacred, 264–5
National Academy of Sciences, U.S.A., 39
natural history, 189–90
naturalism (environmental ethic), 278
naturalistic fallacy, 249
natural resources, global, 283–5
natural sciences, 26–7, 45–65, 191–2, 218–19, 242, 266–70
natural scientists, qualities of, 38–9, 53, 57–8, 61–5, 126, 195, 209, 246, 263, 302
Nazism, 245
Needham, Joseph, 31, 300
Nelson, Mark, 318
neurobiology, 77, 103–5, 315; *see also* brain; mind
neurotransmitters, 73, 75–8, 144, 155, 307
New Age philosophy, 43, 260
New Critics, in literature, 215
New Guinea, culture, 152, 161–2
Newman, John Henry, Cardinal, 239
Newton, Isaac, 29–30, 36
Niehans, Jürg, 195–7
Nielsen, François, 306
Nisbet, Robert, 187, 311
Nitecki, Matthew and Doris V., 315
NK model, evolution, 89–90
Nobel Prize, 195, 220, 229
norm of reaction, in genetics, 137–9, 307
Novak, Gordon S., Jr., 305
novelty-seeking gene, 155, 309
Nowack, Martin A., 316
Nozick, Robert, 250
nurturism, 142–3, 188

O

obesity, 138
Occam's razor, 53
odor, human, 158
OGOD (one gene, one disease) principle, 145–7
ophidiophobia (fear of snakes), 79
optimum complexity, in the arts, 229–30, 315
Oster, George F., 309
Ouroboros, 80
Ovid, 135

P

painting, 72–3
Paradis, James G., 315
paralanguage (nonverbal communication), 158–9, 309
parental investment, 169
Pascal, Blaise, 246
Pascal's wager, 246
PAT (environmental formula), 282, 318
Patterson, David A., 304
Peacock, James, 310–11
Peacocke, Arthur R., 316
Peltonen, Leena, 307
Penfield, Wilder, 102
Penrose, Roger, 304–5
Perrett, D. I., 315
personality, heredity of, 155; *see also* heritability; mood
PET (positron emission tomography) imaging, 108, 144
Peterson, Ivars, 304
Petroski, Henry, 307
phenylketonuria, 156, 274–5
pheromones, human, 158
philosophy, general qualities of, 11–12, 96, 208–9, 269, 299, 312
phobias, 79
physical sciences, 66–8
physics, 49–50, 55, 66–8, 85–6, 219–20, 314; unification in, 5, 263
Picasso, Pablo, 220, 314

Pico della Mirandola, Giovanni, 38, 301
Pinker, Steven, 304–5
Planck, Max, 31, 57, 302
Plomin, Robert, 307, 317
poetry, spiritual effect of, 247–8
polygenes, 156
Pool, Robert E., 304, 310
Pope, Alexander, 30, 214
population genetics, 198–201, 258, 271–7, 318
population growth, 272, 281–3, 287–90
positivism, 61–5
Posner, Michael I., 304
postmodernism, 40–4, 214–15, 301, 313
Potter, Van Rensselaer, 315
Power, Thomas Michael, 319
pragmatism, 38, 61, 242
prediction, scientific, 68
prepared learning, 79; *see also* epigenetic rules
primitive thought, 208, 233–6, 312, 315
Prisoner's Dilemma, 252–3
progress, as concept, 98
Promethean knowledge, 34
propitiation, 258–9
Proserpine, 212
proteins, 83–4, 91–4, 272, 303
psychoanalysis, 216; *see also* dreams; Freud, Sigmund
psychology, 53–4, 77; Bacon on, 26–7; in the arts, 216; in economics, 202–8

Q

quantum electrodynamics (Q.E.D.), 49–50, 52, 302
Quetelet, Adolphe, 30
Quetzalcoatl, 80
Quinlan, Karen Ann, 101, 305

R

race, 146
racism, 34, 184
Raichle, Marcus E., 304
rain forest, 84–5, 293
Ramón y Cajal, Santiago, 104, 305
Ratchet of Progress, 270, 289
rational choice theory, 206–8, 312
Rawls, John, 249–50
Reaka-Kudla, Marjorie L., 317
reductionism, 30–1, 54–5, 67–8, 83–5, 186–7, 211, 267
Rees, William E., 318
reflexes, 112
Regnier, Fred, 69
reification (turning concepts into imagined objects), 153, 308–9
Reign of Terror, France, 15
relativism, cultural, 184–6
religion, 238–48, 256–65, 315–17; author's, 5–6, 248, 316; of scientists, 57
religious ecstasy, 258, 260–1, 317
revelation, divine, 241, 246–7
Reynolds, Vernon, 316
Ricardo, David, 195
Richerson, Peter J., 306
Ridley, Matt, 315
Rio Conference, on environment, 289–90, 319
ritual, 153–4, 227, 256–7
ritualization, 159
r–K continuum, in evolutionary biology, 205
Robespierre, Maximilien de, 15
Rogers, Adam, 319
Romanticism, 35–7, 44
ROMs (read-only memories), 51, 302
Ronan, Colin A., 300
Rorty, Richard, 190, 312
Rosenberg, Alexander, 11, 299, 302, 311–12
Rothstein, Edward, 219, 314
Rousseau, Jean-Jacques, 15, 35–6, 41
Roux, Wilhelm, 93
Ruse, Michael, 312
Russell, Bertrand, 216
Rwanda, 287–8
Ryle, Gilbert, 61

S

sacred narrative, 264–5
sacrifice, 227–8, 245, 258–9
Samuelson, Paul, 52, 197
Santa Fe Institute, 88
Satan, 211–12, 270, 277
satisficing, in rational choice, 206
Saunders, Denis A., 318
Savage-Rumbaugh, E. Sue, 131, 306
scaling, space-time, 4, 81–3, 204–5, 236–7, 303
Schelling, Friedrich, 36
Schelling, Thomas, 204
schizophrenia, 142–5, 307
Schlaug, G., 313
Schlick, Moritz, 61–2
Schorske, Carl E., 40, 301
Scialabba, George, 43, 301
science, *see* natural sciences; social sciences
science fiction, 268
scientists, *see* natural scientists

Scott, J. Michael, 317
Scully, Vincent, 220–1, 314
Searle, John R., 304
self, neurobiology of, 119–120
self-assembled monolayers (SAMs), 51
Seligman, Martin E. P., 308
semantic memory, 134–5
semiotics, 183
Sen, Amartya K., 204
serpents, *see* snakes
Service, Robert F., 302
Shamos, Morris H., 300
Shapiro, Ian, 312
Shaw, George Bernard, 243
Shaw, R. Paul, 310
Shepher, Joseph, 175–6
Sheridan, Alan, 301
Sherrington, Charles, 12, 299, 301
Shweder, Richard A., 301, 311–12
Sigmund, Karl, 316
Silberbauer, George B., 233
Silbersweig, D. A., 307
Simmel, Georg, 187
Simon, Herbert A., 64, 206, 302
Singer, S. J., 110, 303
sleep, 75–6
Smets, Gerda, 229–30, 315
smile, 112, 152–3, 308
Smith, Adam, 195–6, 251
snakes, 71–2, 78–81, 127–8, 218, 303
Snow, C. P., 40, 125–6, 301
Social Darwinism, 184
socialism, 34
social sciences, 12, 37, 181–209, 310–12
sociobiology, 150–80, 309–10; *see also* gene-culture coevolution
sociology, 186–8, 190, 311
sorcery, 227–9
Soviet empire, 183
space-time scale, *see* scaling, space-time
Spinoza, Baruch, 262
Spires, Elizabeth, 222–3, 314
spirit sticks, Australian, 256
Stahl, Franklin W., 60, 302
Standard Social Sciences Model (SSSM), 188, 204, 311
status, 170
Steiner, George, 211, 313
Stephens, James, 300
Stern, Curt, 307
Sternberg, Paul W., 303
Stevens, Anthony, 314
Stever, H. Guyford, 319
Stewart, Ian, 303
Stigler, George J., 204
Storey, Robert, 217, 313
Straus, Ernst, 32–3
strong inference, 59
structuralism, 154
Stutz, Roger, 51
subjective experience, neurobiology of, 115–18
Sulloway, Frank J., 138
supernormal stimulus, 231–2
Swedenborg, Emanuel, 73–4
Swinburne, Richard, 316
symbolism, 134–5; *see also* language; mind
Symbolists, in literature, 215
synthesis, 54–5, 67–8, 83–5, 267, 269

T

taboos, incest, 177
Tanner, Ralph E. S., 316
taste, neurobiological aspects of, 151–2
technology, and environment, 289–90, 319
Teotihuacán, 220
Teresa of Avila, St., 261, 317
termites, 148, 166, 308
territorial instinct, 170–1, 244–5, 310
Thailand, 272
theism, 33–4, 241–8, 262
theology, 119, 269
theory, fundamental nature of, 52–3
Theory of Everything (T.O.E.), in physics, 263
Thompson, Dennis, 319
Thoreau, Henry David, 36–7
Tiger, Lionel, 217
Tillich, Paul, 263
Tilman, G. David, 280, 317–18
Tlaloc (Aztec god), 80–1, 227
Tocqueville, Alexis de, 187
Toennies, Ferdinand J., 187
Tooby, John, 171–2, 307–11
tool-making: animal, 132–3; human, 133–4
totems, 228–9
touch, communication by, 158
transcendentalism: ethical, 238–51; New England, 36–7
tribalism, 244–5, 253–4, 256–7, 272, 287–8
trigradal system, in Freemasonry, 256
Trivers, Robert L., 310
truth, criteria of, 60–5
Tulving, Endel, 134–5, 307
Turgot, Anne Robert Jacques, 20
Turing, Alan, 121
Turing test, 121, 124
Turkmen protein, evolution of, 272
Turner, Frederick, 215, 217, 313–14
Turner, R. Kerry, 319
Tversky, Amos, 206–8, 312

Two Cultures (literary, scientific), 40, 125–6, 211, 301

U

ufology, 54
universals, cultural, 147–9
Urbach, Peter, 300

V

van den Berghe, Pierre L., 186, 311
variable expressivity, heredity, 147
Veblen, Thorstein, 196
Vienna Circle, 61–4
vision, color, 46
Vogler, Christopher, 300, 314
volitional evolution, 273–7
Volk, Tyler, 220, 314

W

Waal, Frans de, 132, 306, 316
Wackernagel, Mathis, 318
Waddington, Conrad H., 316
Wallace, Anthony F. C., 316
Wallace, Walter L., 186, 311
war, 62, 170–1, 183–4, 272, 287–8, 310
water supplies, global, 284
Weber, Max, 187
Weinberg, Steven, 263
Weiss, H., 318
Weld, Charles Richard, 316
Welon, Z., 317
Westermarck, Edward A., 174–80
Westermarck effect, 174–80, 218, 310
Whewell, William, 8
Whitehead, Alfred North, 56, 216
Whitesides, George M., 51, 302
Wightman, Mark, 51
Wigner, Eugene P., 48–9, 301–2
Williams, George C., 315
Williams, Thomas Rhys, 309
Wilson, Edmund (literary critic), 215–16, 313
Wilson, Edward O., publication references, 303, 306–8, 310, 312–13, 315–17, 319
Witham, Larry, 316
Wolf, Arthur P., 175–9, 310
Wollstonecraft, Mary, 245, 316
wolves, social behavior, 259
Wong, Yuwa, 310
Wordsworth, William, 35
World Economic Forum, 291
World War II, 62
Worldwatch Institute, 318
Wrangham, Richard W., 306
Wright, Robert, 315

X

xenophobia, 253–4

Y

Yukawa, Hideki, 219–20, 314

Grateful acknowledgment is made to the following for permission to reprint previously published material:

GEO Extra Nr. 1: "Quo Vadis, Homo Sapiens" by Edward O. Wilson, reprinted courtesy of *GEO Extra Nr. 1*, Hamburg, Germany.

Harvard University Press: Excerpts from *Biophilia*; *The Diversity of Life*; *Genes, Mind, and Culture*; and *Promethean Fire* by Edward O. Wilson, reprinted courtesy of Harvard University Press.

The New York Times Company: Excerpt from "Is Humanity Suicidal?" by Edward O. Wilson (*The New York Times Magazine*, May 20, 1993), copyright © 1993 by The New York Times Company, reprinted courtesy of The New York Times Company.

David Philip Publishers (Pty) Ltd.: Excerpts from *The Art of Tracking: The Origin of Science* by Louis Liebenberg. Reprinted by permission of David Philip Publishers (Pty) Ltd., Claremont, South Africa.

Time International: Excerpts from "Legions of the Doomed" by Edward O. Wilson (*Time International*, October 30, 1995). Adapted by permission of *Time International*.

Viking Penguin and *Elizabeth Spires*: Excerpt from "Falling Away" from *Annonciade* by Elizabeth Spires, copyright © 1985, 1986, 1987, 1988, 1989 by Elizabeth Spires. Reprinted by permission of Viking Penguin, a division of Penguin Books USA Inc., and of Elizabeth Spires.

A Note About the Author

Edward O. Wilson was born in Birmingham, Alabama, in 1929. He received his B.S. and M.S. in biology from the University of Alabama and, in 1955, his Ph.D. in biology from Harvard, where he has since taught, and where he has received both of its college-wide teaching awards. He is currently Pellegrino University Research Professor and Honorary Curator in Entomology of the Museum of Comparative Zoology at Harvard. The author/coauthor of two Pulitzer Prize–winning books, *On Human Nature* (1978) and *The Ants* (1990, with Bert Hölldobler), he is also the recipient of many fellowships, honors, and awards, including the 1977 National Medal of Science, the Crafoord Prize from the Royal Swedish Academy of Sciences (1990), the International Prize for Biology from Japan (1993), and, for his conservation efforts, the Gold Medal of the Worldwide Fund for Nature (1990) and the Audubon Medal of the National Audubon Society (1995). He is on the Board of Directors of The Nature Conservancy, Conservation International, and the American Museum of Natural History, and has lectured frequently throughout the world. He lives in Lexington, Massachusetts, with his wife, Irene.

A Note on the Type

This book was set in Electra, a typeface designed by William Addison Dwiggins (1880–1956) for the Mergenthaler Linotype Company and first made available in 1935. Electra cannot be classified as either "modern" or "old style." It is not based on any historical model, and hence does not echo any particular period or style of type design. It avoids the extreme contrast between thick and thin elements that marks most modern faces, and it is without eccentricities that catch the eye and interfere with reading. In general, Electra is a simple, readable typeface that attempts to give a feeling of fluidity, power, and speed.

W. A. Dwiggins was born in Martinsville, Ohio, and studied art in Chicago. In the late 1920s he moved to Hingham, Massachusetts, where he built a solid reputation as a designer of advertisements and as a calligrapher. He began an association with the Mergenthaler Linotype Company in 1929 and over the next twenty-seven years designed a number of book types of which Metro, Electra, and Caledonia have been used widely.

Composed by Creative Graphics, Inc.
Allentown, Pennsylvania

Printed and bound by Quebecor Printing,
Martinsburg, West Virginia

Designed by Cassandra J. Pappas